Current Trends in Symmetric Polynomials with their Applications

Current Trends in Symmetric Polynomials with their Applications

Special Issue Editor

Taekyun Kim

MDPI • Basel • Beijing • Wuhan • Barcelona • Belgrade

Special Issue Editor
Taekyun Kim
Kwangwoon University
Republic of Korea

Editorial Office
MDPI
St. Alban-Anlage 66
4052 Basel, Switzerland

This is a reprint of articles from the Special Issue published online in the open access journal *Symmetry* (ISSN 2073-8994) from 2018 to 2019 (available at: https://www.mdpi.com/journal/symmetry/special_issues/Current_Trends_Symmetric_Polynomials_Their_Applications)

For citation purposes, cite each article independently as indicated on the article page online and as indicated below:

LastName, A.A.; LastName, B.B.; LastName, C.C. Article Title. *Journal Name* **Year**, *Article Number*, Page Range.

ISBN 978-3-03921-620-8 (Pbk)
ISBN 978-3-03921-621-5 (PDF)

© 2019 by the authors. Articles in this book are Open Access and distributed under the Creative Commons Attribution (CC BY) license, which allows users to download, copy and build upon published articles, as long as the author and publisher are properly credited, which ensures maximum dissemination and a wider impact of our publications.

The book as a whole is distributed by MDPI under the terms and conditions of the Creative Commons license CC BY-NC-ND.

Contents

About the Special Issue Editor vii

Preface to "Current Trends in Symmetric Polynomials with their Applications" ix

Lee Jinwoo
Fluctuation Theorem of Information Exchange between Subsystems that Co-Evolve in Time
Reprinted from: *Symmetry* **2019**, *11*, 433, doi:10.3390/sym11030433 1

Jin Zhang and Zhuoyu Chen
A Note on the Sequence Related to Catalan Numbers
Reprinted from: *Symmetry* **2019**, *11*, 371, doi:10.3390/sym11030371 10

Ran Duan and Shimeng Shen
Bernoulli Polynomials and Their Some New Congruence Properties
Reprinted from: *Symmetry* **2019**, *11*, 365, doi:10.3390/sym11030365 15

Taekyun Kim, Kyung-Won Hwang, Dae San Kim and Dmitry V. Dolgy
Connection Problem for Sums of Finite Products of Legendre and Laguerre Polynomials
Reprinted from: *Symmetry* **2019**, *11*, 317, doi:10.3390/sym11030317 21

Taekyun Kim, Dae San Kim and Gwan-Woo Jang
On Central Complete and Incomplete Bell Polynomials I
Reprinted from: *Symmetry* **2019**, *11*, 288, doi:10.3390/sym11020288 35

Can Kızılateş, Bayram Çekim, Naim Tuğlu and Taekyun Kim
New Families of Three-Variable Polynomials Coupled with Well-Known
Polynomials and Numbers
Reprinted from: *Symmetry* **2019**, *11*, 264, doi:10.3390/sym11020264 47

Dojin Kim
A Modified PML Acoustic Wave Equation
Reprinted from: *Symmetry* **2019**, *11*, 177, doi:10.3390/sym11020177 60

Wenpeng Zhang and Li Chen
On the Catalan Numbers and Some of Their Identities
Reprinted from: *Symmetry* **2019**, *11*, 62, doi:10.3390/sym11010062 75

Taekyun Kim, Dae San Kim, Lee-Chae Jang and Dmitry V. Dolgy
Representation by Chebyshev Polynomials for Sums of Finite Products of
Chebyshev Polynomials
Reprinted from: *Symmetry* **2018**, *10*, 742, doi:10.3390/sym10120742 84

Joohee Jeong, Dong-Jin Kang and Seog-Hoon Rim
Symmetry Identities of Changhee Polynomials of Type Two
Reprinted from: *Symmetry* **2018**, *10*, 740, doi:10.3390/sym10120740 98

Serkan Araci, Waseem Ahmad Khan and Kottakkaran Sooppy Nisar
Symmetric Identities of Hermite-Bernoulli Polynomials and Hermite-BernoulliNumbers
Attached to a Dirichlet Character χ
Reprinted from: *Symmetry* **2018**, *10*, 675, doi:10.3390/sym10120675 107

Serkan Araci, Mumtaz Riyasat, Shahid Ahmad Wani and Subuhi Khan
A New Class of Hermite-Apostol Type Frobenius-Euler Polynomials and Its Applications
Reprinted from: *Symmetry* **2018**, *10*, 652, doi:10.3390/sym10110652 117

Yunjae Kim, Byung Moon Kim and Jin-Woo Park
Symmetric Properties of Carlitz's Type q-Changhee Polynomials
Reprinted from: *Symmetry* **2018**, *10*, 634, doi:10.3390/sym10110634 133

Chen Li
On Classical Gauss Sums and Some of Their Properties
Reprinted from: *Symmetry* **2018**, *10*, 625, doi:10.3390/sym10110625 142

Dmitry Victorovich Dolgy
Connection Problem for Sums of Finite Products of Chebyshev Polynomials of the Third and Fourth Kinds
Reprinted from: *Symmetry* **2018**, *10*, 617, doi:10.3390/sym10110617 148

Yunyun Qu, Jiwen Zeng and Yongfeng Cao
Fibonacci and Lucas Numbers of the Form $2^a + 3^b + 5^c + 7^d$
Reprinted from: *Symmetry* **2018**, *10*, 509, doi:10.3390/sym10100509 162

YunJae Kim, Byung Moon Kim, Lee-Chae Jang and Jongkyum Kwon
A Note on Modified Degenerate Gamma and Laplace Transformation
Reprinted from: *Symmetry* **2018**, *10*, 471, doi:10.3390/sym10100471 169

Dae San Kim, Taekyun Kim, Cheon Seoung Ryoo and Yonghong Yao
On p-adic Integral Representation of q-Bernoulli Numbers Arising from Two Variable q-Bernstein Polynomials
Reprinted from: *Symmetry* **2018**, *10*, 451, doi:10.3390/sym10100451 177

Wenpeng Zhang and Xin Lin
A New Sequence and Its Some Congruence Properties
Reprinted from: *Symmetry* **2018**, *10*, 359, doi:10.3390/sym10090359 188

Lee-Chae Jang, Taekyun Kim, Dae San Kim and D.V. Dolgy
On p-Adic Fermionic Integrals of q-Bernstein Polynomials Associated with q-Euler Numbers and Polynomials
Reprinted from: *Symmetry* **2018**, *10*, 311, doi:10.3390/sym10080311 194

Zhao Jianhong and Chen Zhuoyu
Some Symmetric Identities Involving Fubini Polynomials and Euler Numbers
Reprinted from: *Symmetry* **2018**, *10*, 303, doi:10.3390/sym10080303 203

Taekyun Kim, Dae San Kim, Dmitry V. Dolgy and Cheon Seoung Ryoo
Representing Sums of Finite Products ofChebyshev Polynomials of Third and FourthKinds by Chebyshev Polynomials
Reprinted from: *Symmetry* **2018**, *10*, 258, doi:10.3390/sym10070258 209

Taekyun Kim, Dae San Kim, Gwan-Woo Jang and Jongkyum Kwon
Symmetric Identities for Fubini Polynomials
Reprinted from: *Symmetry* **2018**, *10*, 219, doi:10.3390/sym10060219 219

About the Special Issue Editor

Taekyun Kim completed his PhD at the Department of Mathematics in Kyushu University, Japan (1994). He was Lecturer at Kyungpook National University in 1994–1996, Research Professor at the Institute of Science Education, Kongju National University, in 2001–2006, Professor (BK) at the Department of Electrical and Computer Engineering, Kyungpook National University, in 2006–2008, and Chair Professor at Tianjin Polytechnic University in 2015–2019. He has been Professor at the Department of Mathematics in Kwangwoon University since his appointment in 2008.

Preface to "Current Trends in Symmetric Polynomials with their Applications"

Special numbers and polynomials play an extremely important role in various applications within such diverse areas as mathematics, probability and statistics, mathematical physics, and engineering. Due to their powerful expressions, the combinations of special numbers and polynomials can be almost ubiquitously seen as the solutions for differential equations in the diverse fields of orthogonality condition, generating functions, recurrence relations, and bosonic and fermionic p-adic integrals, to name but a few. Furthermore, their importance can be also seen in the developments of classical analysis, number theory, mathematical analysis, mathematical physics, symmetric functions, combinatorics, and other sections of the natural sciences. A great amount of effort has been exerted by a multitude of researchers over the years in attempting to find new representations of families of special functions and polynomials along with associated practical applications. This Special Issue will cover the modern trends in the fields of special functions and orthogonal polynomials (or q-special functions and orthogonal polynomials).

Taekyun Kim
Special Issue Editor

Article

Fluctuation Theorem of Information Exchange between Subsystems that Co-Evolve in Time

Lee Jinwoo

Department of Mathematics, Kwangwoon University, Seoul 01897, Korea; jinwoolee@kw.ac.kr

Received: 27 February 2019; Accepted: 22 March 2019; Published: 22 March 2019

Abstract: Sagawa and Ueda established a fluctuation theorem of information exchange by revealing the role of correlations in stochastic thermodynamics and unified the non-equilibrium thermodynamics of measurement and feedback control. They considered a process where a non-equilibrium system exchanges information with other degrees of freedom such as an observer or a feedback controller. They proved the fluctuation theorem of information exchange under the assumption that the state of the other degrees of freedom that exchange information with the system does not change over time while the states of the system evolve in time. Here we relax this constraint and prove that the same form of the fluctuation theorem holds even if both subsystems co-evolve during information exchange processes. This result may extend the applicability of the fluctuation theorem of information exchange to a broader class of non-equilibrium processes, such as a dynamic coupling in biological systems, where subsystems that exchange information interact with each other.

Keywords: fluctuation theorem; thermodynamics of information; stochastic thermodynamics; mutual information; non-equilibrium free energy; entropy production

1. Introduction

Biological systems possess information processing mechanisms for their survival and heredity [1–3]. They, for example, sense external ligand concentrations [4,5], transmit information through signaling networks [6–8], and coordinate gene expressions [9] by secreting and sensing signaling molecules [10]. Cells even implement time integration by copying states of environment into molecular states inside the cells to reduce their sensing errors [11,12]. Therefore it is crucial to reveal the role of information in thermodynamics to properly understand complex biological information processes.

Historically, information has entered into the realm of thermodynamics by the name of Maxwell's demon. The demon observes the speed of molecules in a box that is divided into two portions by a partition in which there is a small hole, and lets the fast particles pass from the lower-half of the box to the upper-half, and only the slow particles pass from the upper-half to the lower-half by opening/closing the hole without expenditure of work (see Figure 1a). This results in raising the temperature of the upper-half of the box and lower that of the lower-half, indicating that the second law of thermodynamics, which implies heat flows spontaneously from hotter to colder places, might hypothetically be violated [13]. This paradox shows that information can affect thermodynamics of a physical system, or information is a physical element [14].

Szilard has devised a much simpler model that carries the essential role of information in Maxwell's thought experiment. The Szilard engine consists of a single particle in a box which is surrounded by a heat reservoir of constant temperature. A cycle of the engine begins with inserting a partition in the middle of the box. Depending on whether the particle is in the left-half or in the right-half of the box, one controls a lever such that a weight can be lifted during the wall moves quasi-statically in the direction that the particle pushes (see Figure 1b). If the partition reaches an end of the box, the partition is removed and a new cycle begins again with inserting a partition at the center. Since the energy required for lifting the weight comes from the heat reservoir, this engine corresponds to a perpetual-motion machine of the second kind, where the single heat reservoir is spontaneously cooled and the corresponding thermal energy is converted into mechanical work cyclically, which is prohibited by the second-law of thermodynamics [15].

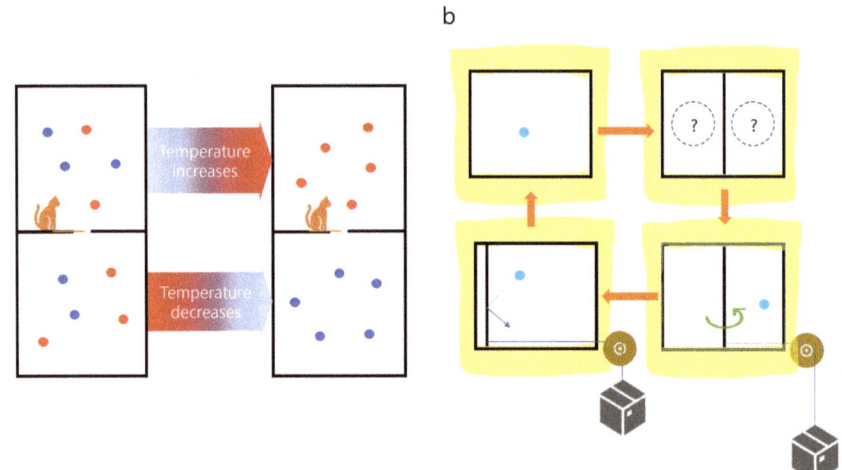

Figure 1. Paradox in thermodynamics of information (**a**) Maxwell's demon (orange cat) uses information on the speed of the particles in the box: He opens/closes the small hole (orange line) without expenditure of energy such that fast particles (red filled circles) are gathered in the upper-half of the box and slow particles (blue filled circles) are gathered in the lower-half of the box. Since temperature is the average velocity of the particles, the demon's action results in spontaneous flow of heat from colder places to hotter places, which violates the second-law of thermodynamics. (**b**) A cycle of Szilard's engine is represented. A lever (green curved arrow) is controlled such that a weight can be lifted during the wall moves quasi-statically in the direction that the particle pushes. This engine harnesses heat from the heat reservoir (yellow region around each boxes) and convert it into mechanical work, cyclically, and thus corresponds to a perpetual-motion engine of the second kind, which is prohibited by the second-law of thermodynamics.

Szilard interprets the coupling between the location of the particle and the direction of the lever as a sort of memory faculty and points out that the coupling is the main cause that enables an amount of work to be extracted from the heat reservoir. He infers, therefore, that establishing the coupling must be accompanied by a production of entropy (dissipation of heat into the environment) which compensates for the lost heat in the reservoir. In [16], Sagawa and Ueda have proved this idea in the form of a fluctuation theorem of information exchange, generalizing the second-law of thermodynamics by taking information into account:

$$\left\langle e^{-\sigma+\Delta I}\right\rangle = 1, \tag{1}$$

where σ is the entropy production of a system X, and ΔI is the change of mutual information between the system X and another system Y, such as a demon, during a process λ_t for $0 \leq t \leq \tau$. Here the bracket indicates the ensemble average over all microscopic trajectories of X and over all states of Y. By Jensen's inequality [17], Equation (1) implies

$$\langle \sigma \rangle \geq \langle \Delta I \rangle. \tag{2}$$

This tells indeed that establishing a correlation between the two subsystems, $\langle \Delta I \rangle > 0$, accompanies an entropy production, $\langle \sigma \rangle > 0$, and expenditure of this correlation, $\langle \Delta I \rangle < 0$, serves as a source of entropy decrease, $\langle \sigma \rangle < 0$. In proving this theorem, they have assumed that the state of system Y does not evolve in time. This assumption causes no problem for simple models of measurement and feedback control. However, in biological systems, it is not unusual that both subsystems that exchange information with each other co-evolve in time. For example, transmembrane receptor proteins transmit signals through thermodynamic coupling between extracellular ligands and conformation of intracellular parts of the receptors during a dynamic allosteric transition [18,19]. In this paper, we relax the constraint that Sagawa and Ueda have assumed, and generalize the fluctuation theorem of information exchange to be applicable to more involved situations, where the two subsystems can influence each other so that the states of both systems co-evolve in time.

2. Results

2.1. Theoretical Framework

We consider a finite classical stochastic system composed of subsystems X and Y that are in contact with a heat reservoir of inverse temperature $\beta \equiv 1/(k_B T)$ where k_B is the Boltzmann constant and T is the temperature of the reservoir. We allow both systems X and Y to be driven far from equilibrium by changing external parameter λ_t during time $0 \leq t \leq \tau$ [20–22]. We assume that time evolutions of subsystems X and Y are described by a classical stochastic dynamics from $t = 0$ to $t = \tau$ along trajectories $\{x_t\}$ and $\{y_t\}$, respectively, where x_t (y_t) denotes a specific microstate of X (Y) at time t for $0 \leq t \leq \tau$ on each trajectory. Since both trajectories fluctuate, we repeat the process λ_t with appropriate initial joint probability distribution $p_0(x,y)$ over all microstates (x,y) of systems X and Y. Then the joint probability distribution $p_t(x,y)$ would evolve for $0 \leq t \leq \tau$. Let $p_t(x) := \int p_t(x,y)\,dy$ and $p_t(y) := \int p_t(x,y)\,dx$ be the corresponding marginal probability distributions. We assume

$$p_0(x,y) \neq 0 \text{ for all } (x,y) \tag{3}$$

so that we have $p_t(x,y) \neq 0$, $p_t(x) \neq 0$, and $p_t(y) \neq 0$ for all x and y during $0 \leq t \leq \tau$.

Now, the entropy production σ during process λ_t for $0 \leq t \leq \tau$ is given by

$$\sigma := \Delta s + \beta Q_b, \tag{4}$$

where Δs is the sum of changes in stochastic entropy along $\{x_t\}$ and $\{y_t\}$, and Q_b is heat dissipated into the reservoir (entropy production in the reservoir) [23,24]. In detail, we have

$$\begin{aligned}
\Delta s &:= \Delta s_x + \Delta s_y, \\
\Delta s_x &:= -\ln p_\tau(x_\tau) + \ln p_0(x_0), \\
\Delta s_y &:= -\ln p_\tau(y_\tau) + \ln p_0(y_0).
\end{aligned} \tag{5}$$

We note that the stochastic entropy $s[p_t(\circ)] := -\ln p_t(\circ)$ of microstate $\circ$ at time t can be interpreted as uncertainty of occurrence of $\circ$ at time t: The greater the probability that state $\circ$ occurs, the smaller the uncertainty of occurrence of state $\circ$.

Now we consider situations where system X exchanges information with system Y during process λ_t. By this, we mean that trajectory $\{x_t\}$ of system X evolves depending on the trajectory $\{y_t\}$ of system Y. Then, information I_t at time t between x_t and y_t is characterized by the reduction of uncertainty of x_t due to given y_t [16]:

$$\begin{aligned}
I_t(x_t, y_t) &:= s[p_t(x_t)] - s[p_t(x_t|y_t)] \\
&= \ln \frac{p_t(x_t, y_t)}{p_t(x_t) p_t(y_t)},
\end{aligned} \tag{6}$$

where $p_t(x_t|y_t)$ is the conditional probability distribution of x_t given y_t. We note that this is called the (time-dependent form of) thermodynamic coupling function [19]. The larger the value of $I_t(x_t, y_t)$ is, the more information is being shared between x_t and y_t for their occurrence. We note that $I_t(x_t, y_t)$ vanishes if x_t and y_t are independent at time t, and the average of $I_t(x_t, y_t)$ with respect to $p_t(x_t, y_t)$ over all microstates is the mutual information between the two subsystems, which is greater than or equal to zero [17].

2.2. Proof of Fluctuation Theorem of Information Exchange

Now we are ready to prove the fluctuation theorem of information exchange in this general setup. We define reverse process $\lambda'_t := \lambda_{\tau-t}$ for $0 \leq t \leq \tau$, where the external parameter is time-reversed [25,26]. Here we set the initial probability distribution $p'_0(x, y)$ for the reverse process as the final (time $t = \tau$) probability distribution for the forward process $p_\tau(x, y)$ so that we have

$$p'_0(x) = \int p'_0(x,y)\, dy = \int p_\tau(x,y)\, dy = p_\tau(x),$$
$$p'_0(y) = \int p'_0(x,y)\, dx = \int p_\tau(x,y)\, dx = p_\tau(y). \tag{7}$$

Then, by Equation (3), we have $p'_t(x,y) \neq 0$, $p'_t(x) \neq 0$, and $p'_t(y) \neq 0$ for all x and y during $0 \leq t \leq \tau$. We also consider the time-reversed conjugate for each $\{x_t\}$ and $\{y_t\}$ for $0 \leq t \leq \tau$ as follows:

$$\{x'_t\} := \{x^*_{\tau-t}\},$$
$$\{y'_t\} := \{y^*_{\tau-t}\}, \tag{8}$$

where $*$ denotes momentum reversal. The microscopic reversibility condition connects the time-reversal symmetry of the microscopic dynamics to non-equilibrium thermodynamics, and reads in this framework as follows [23,27–29]:

$$\frac{p(\{x_t\},\{y_t\}|x_0,y_0)}{p'(\{x'_t\},\{y'_t\}|x'_0,y'_0)} = e^{\beta Q_b}, \tag{9}$$

where $p(\{x_t\}, \{y_t\}|x_0, y_0)$ is the conditional joint probability distribution of paths $\{x_t\}$ and $\{y_t\}$ conditioned at initial microstates x_0 and y_0, and $p'(\{x'_t\}, \{y'_t\}|x'_0, y'_0)$ is that for the reverse process. Now we have the following:

$$\frac{p'(\{x'_t\},\{y'_t\})}{p(\{x_t\},\{y_t\})} = \frac{p'(\{x'_t\},\{y'_t\}|x'_0,y'_0)}{p(\{x_t\},\{y_t\}|x_0,y_0)} \cdot \frac{p'_0(x'_0,y'_0)}{p_0(x_0,y_0)} \tag{10}$$

$$= \frac{p'(\{x'_t\},\{y'_t\}|x'_0,y'_0)}{p(\{x_t\},\{y_t\}|x_0,y_0)} \cdot \frac{p'_0(x'_0,y'_0)}{p'_0(x'_0)p'_0(y'_0)} \cdot \frac{p_0(x_0)p_0(y_0)}{p_0(x_0,y_0)} \cdot \frac{p'_0(x'_0)}{p_0(x_0)} \cdot \frac{p'_0(y'_0)}{p_0(y_0)} \tag{11}$$

$$= \exp\{-\beta Q_b + I_\tau(x_\tau, y_\tau) - I_0(x_0, y_0) - \Delta s_x - \Delta s_y\} \tag{12}$$

$$= \exp\{-\sigma + \Delta I\}. \tag{13}$$

To obtain Equation (11) from Equation (10), we multiply Equation (10) by $\frac{p'_0(x'_0)p'_0(y'_0)}{p'_0(x'_0)p'_0(y'_0)}$ and $\frac{p_0(x_0)p_0(y_0)}{p_0(x_0)p_0(y_0)}$, which are 1. We obtain Equation (12) by applying Equations (5)–(7) and (9) consecutively to Equation (11). Finally, we set $\Delta I := I_\tau(x_\tau, y_\tau) - I_0(x_0, y_0)$, and use Equation (4) to obtain Equation (13) from Equation (12).

We note that Equation (13) generalizes the detailed fluctuation theorem in the presence of information exchange that is proved in [16]. Now we obtain the generalized version of Equation (1) by using Equation (13) as follows:

$$\left\langle e^{-\sigma+\Delta I} \right\rangle = \int e^{-\sigma+\Delta I} p(\{x_t\},\{y_t\}) d\{x_t\} d\{y_t\}$$
$$= \int p'(\{x'_t\},\{y'_t\}) d\{x'_t\} d\{y'_t\} = 1. \quad (14)$$

Here we use the fact that there is a one-to-one correspondence between the forward and the reverse paths due to the time-reversal symmetry of the underlying microscopic dynamics such that $d\{x_t\} = d\{x'_t\}$ and $d\{y_t\} = d\{y'_t\}$ [30].

2.3. Corollary

Before discussing a corollary, we remark one thing: we have used similar notation to that used by Sagawa and Ueda in [16], but there is an important difference. Most importantly, their entropy production σ_{su} reads as follows:

$$\sigma_{su} := \Delta s_{su} + \beta Q_b,$$

where $\Delta s_{su} := \Delta s_x$. In [16], system X is in contact with the heat reservoir, but system Y is not. Nor does system Y evolve over time. Thus they have considered entropy production in system X and the bath. In this paper, both systems X and Y are in contact with the reservoir, and system Y also evolves in time. Thus both subsystems X and Y as well as the heat bath contribute to the entropy production as expressed in Equations (4) and (5). Keeping in mind this difference, we apply Jensen's inequality to Equation (14) to obtain

$$\langle \sigma \rangle \geq \langle \Delta I \rangle. \quad (15)$$

It tells us that firstly, establishing correlation between X and Y accompanies entropy production, and secondly, established correlation serves as a source of entropy decrease.

Now as a corollary, we refine the generalized fluctuation theorem in Equation (14) by including energetic terms. To this end, we define local free energy $\mathcal{F}_x$ of system X at x_t and $\mathcal{F}_y$ of system Y at y_t as follows:

$$\mathcal{F}_x(x_t, t) := E_x(x_t, t) - Ts[p_t(x_t)]$$
$$\mathcal{F}_y(y_t, t) := E_y(y_t, t) - Ts[p_t(y_t)], \quad (16)$$

where E_x and E_y are internal energy of systems X and Y, respectively, and $s[p_t(\circ)] := -\ln p_t(\circ)$ is stochastic entropy [23,24]. Here T is the temperature of the heat bath and argument t indicates dependency of each terms on external parameter λ_t. During the process λ_t, work done on the systems is expressed by the first law of thermodynamics as follows:

$$W := \Delta E + Q_b, \quad (17)$$

where ΔE is the change in internal energy of the systems. If we assume that systems X and Y are weakly coupled, in that interaction energy between X and Y is negligible compared to internal energy of X and Y, we may have

$$\Delta E := \Delta E_x + \Delta E_y, \quad (18)$$

where $\Delta E_x := E_x(x_\tau, \tau) - E_x(x_0, 0)$ and $\Delta E_y := E_y(y_\tau, \tau) - E_y(y_0, 0)$ [31]. We rewrite Equation (12) by adding and subtracting the change of internal energy ΔE_x of X and ΔE_y of Y as follows:

$$\frac{p'(\{x'_t\},\{y'_t\})}{p(\{x_t\},\{y_t\})} = \exp\{-\beta(Q_b + \Delta E_x + \Delta E_y) + \Delta I + \beta\Delta E_x - \Delta s_x + \beta\Delta E_y - \Delta s_y\} \quad (19)$$
$$= \exp\{-\beta(W - \Delta\mathcal{F}_x - \Delta\mathcal{F}_y) + \Delta I\}, \quad (20)$$

where we have applied Equations (16)–(18) consecutively to Equation (19) to obtain Equation (20). Here $\Delta \mathcal{F}_x := \mathcal{F}_x(x_\tau, \tau) - \mathcal{F}_x(x_0, 0)$ and $\Delta \mathcal{F}_y := \mathcal{F}_y(y_\tau, \tau) - \mathcal{F}_y(y_0, 0)$. Now we obtain fluctuation theorem of information exchange with energetic terms as follows:

$$\left\langle e^{-\beta(W - \Delta \mathcal{F}_x - \Delta \mathcal{F}_y) + \Delta I} \right\rangle = \int e^{-\beta(W - \Delta \mathcal{F}_x - \Delta \mathcal{F}_y) + \Delta I} p(\{x_t\}, \{y_t\}) d\{x_t\} d\{y_t\}$$
$$= \int p'(\{x'_t\}, \{y'_t\}) d\{x'_t\} d\{y'_t\} = 1, \quad (21)$$

which generalizes known relations in the literature [31–36]. We note that Equation (21) holds under the weak-coupling assumption between systems X and Y during the process λ_t. By Jensen's inequality, Equation (21) implies

$$\langle W \rangle \geq \left\langle \Delta \mathcal{F}_x + \Delta \mathcal{F}_y + \frac{\Delta I}{\beta} \right\rangle. \quad (22)$$

We remark that $\langle \Delta \mathcal{F}_x \rangle + \langle \Delta \mathcal{F}_y \rangle$ in Equation (22) is the difference in non-equilibrium free energy, which is different from the change in equilibrium free energy that appears in similar relations in the literature [32–36].

3. Examples

3.1. Measurement

Let X be a device (or a demon) which measures the state of other system and Y be a measured system, both of which are in contact with a heat bath of inverse temperature β (see Figure 2a). We consider a dynamic measurement process, which is described as follows: X and Y are prepared separately in equilibrium such that X and Y are not correlated initially, i.e., $I_0(x_0, y_0) = 0$ for all x_0 and y_0. At time $t = 0$, device X is put in contact with system Y so that the coupling of X and Y occurs due to their (weak) interactions until time $t = \tau$, at which a single measurement process finishes. We note that system Y is allowed to evolve in time during the process. Since each process fluctuates, we repeat the measurement many times to obtain probability distribution $p_t(x, y)$ for $0 \leq t \leq \tau$.

A distinguished feature of the framework in this paper is that mutual information $I_t(x_t, y_t)$ in Equation (6) enables us to obtain the time-varying amount of established information during the dynamic coupling process, unlike other approaches where they either provide the amount of information at a fixed time [31,36,37] or one of the system is fixed during the coupling process [16]. For example, let us assume that the probability distribution $p_t(x_t, y_t)$ at an intermediate time t is as shown in Table 1.

Table 1. The joint probability distribution of x and y at an intermediate time t: Here we assume for simplicity that both systems X and Y have two states, 0 (left) and 1 (right).

X\Y	0 (Left)	1 (Right)
0 (Left)	1/3	1/6
1 (Right)	1/6	1/3

Then we have the following:

$$I_t(x_t = 0, y_t = 0) = \ln \frac{1/3}{(1/2) \cdot (1/2)} = \ln(4/3),$$
$$I_t(x_t = 0, y_t = 1) = \ln \frac{1/6}{(1/2) \cdot (1/2)} = \ln(2/3),$$
$$I_t(x_t = 1, y_t = 0) = \ln \frac{1/6}{(1/2) \cdot (1/2)} = \ln(2/3),$$
$$I_t(x_t = 1, y_t = 1) = \ln \frac{1/3}{(1/2) \cdot (1/2)} = \ln(4/3),$$
(23)

so that $\langle \Delta I \rangle = (1/3)\ln(4/3) + (1/6)\ln(2/3) + (1/6)\ln(2/3) + (1/3)\ln(4/3) \approx \ln(1.06)$. Thus by Equation (15) we obtain the lower bound of the average entropy production for the coupling that has been established until time t from the uncorrelated initial state, as follows: $\langle \sigma \rangle \geq \langle \Delta I \rangle \approx \ln 1.06$. If there is no measurement error at final time τ such that $p_\tau(x_\tau = 0, y_\tau = 1) = p_\tau(x_\tau = 1, y_\tau = 0) = 0$ and $p_\tau(x_\tau = 0, y_\tau = 0) = p_\tau(x_\tau = 1, y_\tau = 1) = 1/2$, then we may have $\langle \sigma \rangle \geq \langle \Delta I \rangle = \ln 2$, which is greater than $\ln 1.06$.

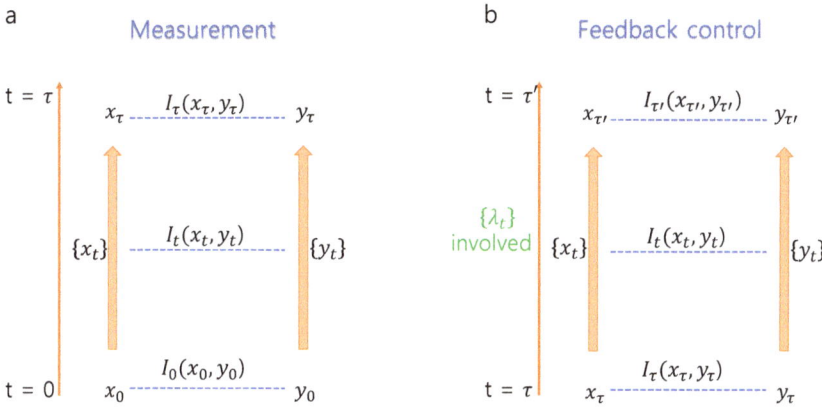

Figure 2. Measurement and feedback control: system X is, for example, a measuring device and system Y is a measured system. X and Y co-evolve, as they interact weakly, along trajectories $\{x_t\}$ and $\{y_t\}$, respectively. (a) Coupling is being established during the measurement process so that $I_t(x_t, y_t)$ for $0 \leq t \leq \tau$ may be increased (not necessarily monotonically). (b) Established correlation is being used as a source of work through external parameter λ_t so that $I_t(x_t, y_t)$ for $\tau \leq t \leq \tau'$ may be decreased (not necessarily monotonically).

3.2. Feedback Control

Unlike the case in [16], we need not to exchange subsystems X and Y to consider feedback control after the measurement. Thus we proceed continuously to feedback control immediately after each measurement process at time τ (see Figure 2b). We assume that correlation $I_\tau(x_\tau, y_\tau)$ at time τ is given by the values in Equation (23) and final correlation at later time τ' is zero, i.e., $I_{\tau'}(x_{\tau'}, y_{\tau'}) = 0$. By feedback control, we mean that external parameter λ_t for $\tau \leq t \leq \tau'$ is manipulated in a pre-determined manner [16], while systems X and Y co-evolve in time, such that the established correlation is used as a source of work while $I_t(x_t, y_t)$ for $\tau \leq t \leq \tau'$ is decreased, not necessarily monotonically. Equation (21) provides an exact relation on the energetics of this process. We rewrite its corollary, Equation (22), with respect to extractable work $W_{\text{ext}} := -W$ as follows:

$$\langle W_{\text{ext}} \rangle \leq -\left\langle \Delta \mathcal{F}_x + \Delta \mathcal{F}_y + \frac{\Delta I}{\beta} \right\rangle. \tag{24}$$

Then the extractable work on top of the conventional bound, $-\langle \Delta \mathcal{F}_x + \Delta \mathcal{F}_y \rangle$, is additionally given by $-\Delta I/\beta = \ln(1.06)$, which comes from the consumption of the established correlation.

4. Conclusions

We have proved the fluctuation theorem of information exchange, Equation (14), which holds even during the co-evolution of two systems that exchange information with each other. Equation (14) tells us that establishing correlation between two systems necessarily accompanies entropy production which is contributed by both systems and the heat reservoir, as expressed in Equations (4) and (5). We have also proved, as a corollary of Equation (14), the fluctuation theorem of information exchange with energetic terms, Equation (21), under the assumption of weak coupling between the two subsystems. Equation (21) reveals the exact relationship between non-equilibrium free energy of both sub-systems and mutual information that is established/consumed through their interactions. This more generalized framework than that in [16], enables us to apply thermodynamics of information to biological systems, where molecules generate/consume correlations through their information processing mechanisms [4–6]. Since the new framework is applicable to fully non-equilibrium situations, thermodynamic coupling during a dynamic allosteric transition, for example, may be analyzed based on this theoretical framework beyond current equilibrium thermodynamic approach [18,19].

Funding: L.J. was supported by the National Research Foundation of Korea grant funded by the Korean Government (NRF-2010-0006733, NRF-2012R1A1A2042932, NRF-2016R1D1A1B02011106), and in part by Kwangwoon University Research Grant in 2016.

Conflicts of Interest: The author declares no conflict of interest.

References

1. Hartwell, L.H.; Hopfield, J.J.; Leibler, S.; Murray, A.W. From molecular to modular cell biology. *Nature* **1999**, *402*, C47. [CrossRef] [PubMed]
2. Crofts, A.R. Life, information, entropy, and time: Vehicles for semantic inheritance. *Complexity* **2007**, *13*, 14–50. [CrossRef] [PubMed]
3. Cheong, R.; Rhee, A.; Wang, C.J.; Nemenman, I.; Levchenko, A. Information transduction capacity of noisy biochemical signaling networks. *Science* **2011**, *334*, 354–358. [CrossRef]
4. McGrath, T.; Jones, N.S.; ten Wolde, P.R.; Ouldridge, T.E. Biochemical Machines for the Interconversion of Mutual Information and Work. *Phys. Rev. Lett.* **2017**, *118*, 028101. [CrossRef] [PubMed]
5. Ouldridge, T.E.; Govern, C.C.; ten Wolde, P.R. Thermodynamics of Computational Copying in Biochemical Systems. *Phys. Rev. X* **2017**, *7*, 021004. [CrossRef]
6. Becker, N.B.; Mugler, A.; ten Wolde, P.R. Optimal Prediction by Cellular Signaling Networks. *Phys. Rev. Lett.* **2015**, *115*, 258103. [CrossRef] [PubMed]
7. Cheng, F.; Liu, C.; Shen, B.; Zhao, Z. Investigating cellular network heterogeneity and modularity in cancer: A network entropy and unbalanced motif approach. *BMC Syst. Biol.* **2016**, *10*, 65. [CrossRef]
8. Whitsett, J.A.; Guo, M.; Xu, Y.; Bao, E.L.; Wagner, M. SLICE: Determining cell differentiation and lineage based on single cell entropy. *Nucleic Acids Res.* **2016**, *45*, e54.
9. Statistical Dynamics of Spatial-Order Formation by Communicating Cells. *iScience* **2018**, *2*, 27–40. [CrossRef]
10. Maire, T.; Youk, H. Molecular-Level Tuning of Cellular Autonomy Controls the Collective Behaviors of Cell Populations. *Cell Syst.* **2015**, *1*, 349–360. [CrossRef]
11. Mehta, P.; Schwab, D.J. Energetic costs of cellular computation. *Proc. Natl. Acad. Sci. USA* **2012**, *109*, 17978–17982. [CrossRef] [PubMed]
12. Govern, C.C.; ten Wolde, P.R. Energy dissipation and noise correlations in biochemical sensing. *Phys. Rev. Lett.* **2014**, *113*, 258102. [CrossRef] [PubMed]
13. Leff, H.S.; Rex, A.F. *Maxwell's Demon: Entropy, Information, Computing*; Princeton University Press: Princeton, NJ, USA, 2014.
14. Landauer, R. Information is physical. *Phys. Today* **1991**, *44*, 23–29. [CrossRef]
15. Szilard, L. On the decrease of entropy in a thermodynamic system by the intervention of intelligent beings. *Behav. Sci.* **1964**, *9*, 301–310. [CrossRef]
16. Sagawa, T.; Ueda, M. Fluctuation theorem with information exchange: Role of correlations in stochastic thermodynamics. *Phys. Rev. Lett.* **2012**, *109*, 180602. [CrossRef] [PubMed]
17. Cover, T.M.; Thomas, J.A. *Elements of Information Theory*; John Wiley & Sons: Hoboken, NJ, USA, 2012.
18. Tsai, C.J.; Nussinov, R. A unified view of how allostery works? *PLoS Comput. Biol.* **2014**, *10*, e1003394. [CrossRef]

19. Cuendet, M.A.; Weinstein, H.; LeVine, M.V. The allostery landscape: Quantifying thermodynamic couplings in biomolecular systems. *J. Chem. Theory Comput.* **2016**, *12*, 5758–5767. [CrossRef]
20. Jarzynski, C. Equalities and inequalities: Irreversibility and the second law of thermodynamics at the nanoscale. *Annu. Rev. Condens. Matter Phys.* **2011**, *2*, 329–351. [CrossRef]
21. Seifert, U. Stochastic thermodynamics, fluctuation theorems and molecular machines. *Rep. Prog. Phys.* **2012**, *75*, 126001. [CrossRef]
22. Spinney, R.; Ford, I. Fluctuation Relations: A Pedagogical Overview. In *Nonequilibrium Statistical Physics of Small Systems*; Wiley-VCH Verlag GmbH & Co. KGaA: Weinheim, Germany, 2013; pp. 3–56.
23. Crooks, G.E. Entropy production fluctuation theorem and the nonequilibrium work relation for free energy differences. *Phys. Rev. E* **1999**, *60*, 2721–2726. [CrossRef]
24. Seifert, U. Entropy production along a stochastic trajectory and an integral fluctuation theorem. *Phys. Rev. Lett.* **2005**, *95*, 040602. [CrossRef]
25. Ponmurugan, M. Generalized detailed fluctuation theorem under nonequilibrium feedback control. *Phys. Rev. E* **2010**, *82*, 031129. [CrossRef]
26. Horowitz, J.M.; Vaikuntanathan, S. Nonequilibrium detailed fluctuation theorem for repeated discrete feedback. *Phys. Rev. E* **2010**, *82*, 061120.
27. Kurchan, J. Fluctuation theorem for stochastic dynamics. *J. Phys. A Math. Gen.* **1998**, *31*, 3719. [CrossRef]
28. Maes, C. The fluctuation theorem as a Gibbs property. *J. Stat. Phys.* **1999**, *95*, 367–392. [CrossRef]
29. Jarzynski, C. Hamiltonian derivation of a detailed fluctuation theorem. *J. Stat. Phys.* **2000**, *98*, 77–102. [CrossRef]
30. Goldstein, H.; Poole, C., Jr.; Safko, J.L. *Classical Mechanics*, 3rd ed.; Pearson: London, UK, 2001.
31. Parrondo, J.M.; Horowitz, J.M.; Sagawa, T. Thermodynamics of information. *Nat. Phys.* **2015**, *11*, 131–139. [CrossRef]
32. Kawai, R.; Parrondo, J.M.R.; den Broeck, C.V. Dissipation: The phase-space perspective. *Phys. Rev. Lett.* **2007**, *98*, 080602. [CrossRef]
33. Generalization of the second law for a transition between nonequilibrium states. *Phys. Lett. A* **2010**, *375*, 88–92. [CrossRef]
34. Generalization of the second law for a nonequilibrium initial state. *Phys. Lett. A* **2010**, *374*, 1001–1004. [CrossRef]
35. Esposito, M.; Van den Broeck, C. Second law and Landauer principle far from equilibrium. *Europhys. Lett.* **2011**, *95*, 40004. [CrossRef]
36. Sagawa, T.; Ueda, M. Generalized Jarzynski equality under nonequilibrium feedback control. *Phys. Rev. Lett.* **2010**, *104*, 090602. [CrossRef] [PubMed]
37. Horowitz, J.M.; Parrondo, J.M. Thermodynamic reversibility in feedback processes. *Europhys. Lett.* **2011**, *95*, 10005. [CrossRef]

 © 2019 by the author. Licensee MDPI, Basel, Switzerland. This article is an open access article distributed under the terms and conditions of the Creative Commons Attribution (CC BY) license (http://creativecommons.org/licenses/by/4.0/).

Article

A Note on the Sequence Related to Catalan Numbers

Jin Zhang [1] and Zhuoyu Chen [2,*]

[1] School of Information Engineering, Xi'an University, Xi'an 710127, China; zhangjin0921@xawl.edu.cn
[2] School of Mathematics, Northwest University, Xi'an 710127, China
* Correspondence: chenzymath@stumail.nwu.edu.cn

Received: 7 February 2019; Accepted: 9 March 2019; Published: 13 March 2019

Abstract: The main purpose of this paper is to find explicit expressions for two sequences and to solve two related conjectures arising from the recent study of sums of finite products of Catalan numbers by Zhang and Chen.

Keywords: new sequence; Catalan numbers; elementary and combinatorial methods; congruence; conjecture

MSC: 11B83; 11B75

1. Introduction

Let n be any non-negative integer. Then, $C_n = \frac{1}{n+1} \cdot \binom{2n}{n}$ ($n = 0, 1, 2, 3, \cdots$) are defined as the Catalan numbers. For example, the first several values of the Catalan numbers are $C_0 = 1$, $C_1 = 1$, $C_2 = 2$, $C_3 = 5$, $C_4 = 14$, $C_5 = 42$, $C_6 = 132$, $C_7 = 429$, $C_8 = 1430, \cdots$. The generating function of the sequence $\{C_n\}$ is:

$$\frac{2}{1+\sqrt{1-4x}} = \sum_{n=0}^{\infty} \frac{\binom{2n}{n}}{n+1} \cdot x^n = \sum_{n=0}^{\infty} C_n \cdot x^n. \tag{1}$$

This sequence occupies a pivotal position in combinatorial mathematics, so lots of counting problems are closely related to it. A great number of examples can be found in a study by Stanley [1]. Because of these, plenty of scholars have researched the properties of Catalan numbers and obtained a large number of vital and meaningful results. Interested readers can refer to the relevant references [2–26], which is not an exhaustive list. Very recently, Zhang and Chen [27] researched the calculation problem of the following convolution sums:

$$\sum_{a_1+a_2+\cdots+a_h=n} C_{a_1} \cdot C_{a_2} \cdot C_{a_3} \cdots C_{a_h}, \tag{2}$$

where the summation has taken over all h-dimension non-negative integer coordinates $(a_1, a_2, \cdots, a_h)$, such that the equation $a_1 + a_2 + \cdots + a_h = n$.

They first introduced two new recursive sequences, $C(h, i)$ and $D(h, i)$, and after the elementary and combinatorial methods, they proved the following two significant conclusions:

Theorem 1. *For any positive integer h, one gets the identity:*

$$\sum_{a_1+a_2+\cdots+a_{2h+1}=n} C_{a_1} \cdot C_{a_2} \cdot C_{a_3} \cdots C_{a_{2h+1}}$$
$$= \frac{1}{(2h)!} \sum_{i=0}^{h} C(h,i) \sum_{j=0}^{\min(n,i)} \frac{(n-j+h+i)! \cdot C_{n-j+h+i}}{(n-j)!} \cdot \binom{i}{j} \cdot (-4)^j,$$

where the sequence $C(h, i)$ is defined as $C(1, 0) = -2$, $C(h, h) = 1$, $C(h + 1, h) = C(h, h - 1) - (8h + 2) \cdot C(h, h)$, $C(h + 1, 0) = 8 \cdot C(h, 1) - 2 \cdot C(h, 0)$, and for all integers $1 \leq i \leq h - 1$, we acquire the recursive formula:

$$C(h+1, i) = C(h, i-1) - (8i+2) \cdot C(h, i) + (4i+4)(4i+2) \cdot C(h, i+1).$$

Theorem 2. *For any positive integer h and non-negative n, one can obtain:*

$$= \frac{1}{(2h-1)!} \sum_{i=0}^{h-1} \sum_{j=0}^{n} D(h, i+1) \cdot \binom{i+\frac{1}{2}}{j} \cdot (-4)^j \cdot \frac{(n-j+h+i)! \cdot C_{n-j+h+i}}{(n-j)!},$$

where $\binom{n+\frac{1}{2}}{i} = \left(n+\frac{1}{2}\right) \cdot \left(n-1+\frac{1}{2}\right) \cdots \left(n-i+1+\frac{1}{2}\right)/i!$, the sequence $D(k, i)$ are defined as $D(k, 0) = 0$, $D(k, k) = 1$, $D(k+1, k) = D(k, k-1) - (8k-2)$, $D(k+1, 1) = 24D(k, 2) - 6D(k, 1)$, and for all integers $1 \leq i \leq k - 1$,

$$D(k+1, i) = D(k, i-1) - (8i-2) \cdot D(k, i) + 4i(4i+2) \cdot D(k, i+1).$$

Meanwhile, through numerical observation, Zhang and Chen [27] also proposed the following two conjectures:

Conjecture 1. *Let p be a prime. Then, for any integer $0 \leq i < \frac{p+1}{2}$, we obtain the congruence:*

$$C\left(\frac{p+1}{2}, i\right) \equiv 0 \bmod p(p+1).$$

Conjecture 2. *Let p be a prime. Then, for any integer $0 \leq i < \frac{p+1}{2}$, we obtain the congruence:*

$$D\left(\frac{p+1}{2}, i\right) \equiv 0 \bmod p(p-1).$$

For easy comparison, here we list some of the values of $C(h, i)$ and $D(h, i)$ with $1 \leq h \leq 6$ and $0 \leq i \leq h$ in the following Tables 1 and 2.

Table 1. Values of $C(k, i)$.

$C(k,i)$	$i=0$	$i=1$	$i=2$	$i=3$	$i=4$	$i=5$	$i=6$
$k=1$	-2	1					
$k=2$	12	-12	1				
$k=3$	-120	180	-30	1			
$k=4$	1680	-3360	840	-56	1		
$k=5$	$-30,240$	75,600	$-25,200$	2520	-90	1	
$k=6$	665,280	$-1,995,840$	831,600	$-110,880$	5940	-132	1

Table 2. Values of $D(k, i)$.

$D(k,i)$	$i=0$	$i=1$	$i=2$	$i=3$	$i=4$	$i=5$	$i=6$
$k=1$	0	1					
$k=2$	0	-6	1				
$k=3$	0	60	-20	1			
$k=4$	0	-840	420	-42	1		
$k=5$	0	15,120	$-10,080$	1512	-72	1	
$k=6$	0	-332640	277,200	$-55,440$	3960	-110	1

Based on these two tables and a large number of numerical calculations, we found that these conjectures are not only correct, but also have generalized conclusions. Actually, they provide a simpler and clearer representation.

In this paper, by using some notes from Zhang and Chen's work [27] as well as some basic and combinatorial methods, we are going to prove the following:

Theorem 3. *Let h be a positive integer. Then, for any integer i with $0 \leq i \leq h$, we acquire the identity:*

$$C(h,i) = (-1)^{h-i} \cdot \frac{(2h)!}{(h-i)! \cdot (2i)!}.$$

Theorem 4. *Let h be a positive integer. Then, for any integer i with $1 \leq i \leq h$, we acquire the identity:*

$$D(h,i) = (-1)^{h-i} \cdot \frac{(2h-1)!}{(h-i)! \cdot (2i-1)!}.$$

Based on the above two theorems, we may instantly deduce the following two corollaries:

Corollary 1. *Let h be any positive integer. Then, for any integer $0 \leq i \leq h-1$, we gain the congruence:*

$$C(h,i) \equiv 0 \bmod 2h(2h-1).$$

Corollary 2. *Let h be any positive integer. Then, for any integer $0 \leq i \leq h-1$, we gain the congruence:*

$$D(h,i) \equiv 0 \bmod (2h-1)(2h-2).$$

Suppose that we consider p an odd prime, and that when $h = \frac{p+1}{2}$ in Corollary 1 and Corollary 2, combined with the identities $2h(2h-1) = p(p+1)$ and $(2h-1)(2h-2) = p(p-1)$, our Corollary 1 and Corollary 2 proves Conjecture 1 and Conjecture 2, respectively. Practically, they prove two more general conclusions.

Taking $n = 0$ in Theorem 1 and Theorem 2 and applying our theorems, we may instantly deduce the following two identities:

Corollary 3. *Let h be any positive integer. Then, we get the identity:*

$$\sum_{i=0}^{h}(-1)^{h-i}\binom{h+i}{2i} \cdot C_{h+i} = 1.$$

Corollary 4. *Let h be any positive integer. Then, we get the identity:*

$$\sum_{i=1}^{h}(-1)^{h-i}\binom{h+i-1}{2i-1} \cdot C_{h+i-1} = 1.$$

Some notes: If we replace $C(h,i)$ ($D(h,i)$) in Theorem 1 (Theorem 2) with the formula for $C(h,i)$ ($D(h,i)$) in our Theorem 3 (Theorem 4), then we can get a more accurate representation for convolution sums (2).

The proof of the results in this paper is uncomplicated, but guessing their specific forms is not easy.

2. Proofs of the Theorems

Actually, the recursive form of the sequence $C(h,i)$ or $D(h,i)$ is more complex, but as long as we are able to guess its accurate representation, it is not difficult to prove. First of all, combining the mathematical induction method, we are going to prove:

$$C(h,i) = (-1)^{h-i} \cdot \frac{(2h)!}{(h-i)! \cdot (2i)!}. \tag{3}$$

According to Table 1, we know that $C(1,0) = -2$, $C(1,1) = 1$, $C(2,0) = -12$, $C(2,1) = 12$, $C(2,2) = 1$, $C(3,0) = -120$, $C(3,1) = 180$, $C(3,2) = -30$, $C(3,3) = 1$. This means that (3) is correct for $h = 1, 2, 3$, and $0 \le i \le h$.

Assume that (3) is correct for integer $h = k$ and all $0 \le i \le k$. That is,

$$C(k,i) = (-1)^{k-i} \cdot \frac{(2k)!}{(k-i)! \cdot (2i)!}, \quad 0 \le i \le k. \tag{4}$$

Then, for $h = k+1$, if $i = h+1$, applying the definition of $C(h,i)$, we acquire $C(k+1, k+1) = 1$. If $i = 0$, combining the inductive hypothesis (4) and noting that $C(k+1, 0) = 8C(k,1) - 2C(k,0)$, we obtain:

$$C(k+1, 0) = 8 \cdot (-1)^{k-1} \cdot \frac{(2k)!}{(k-1)! \cdot 2!} - (-1)^k \cdot 2 \cdot \frac{(2k)!}{k!} = (-1)^{k+1} \frac{(2k+2)!}{(k+1)!}. \tag{5}$$

Suppose that $1 \le i \le k$. From (4) and the recursive properties of $C(h,i)$, we gain:

$$\begin{aligned}
C(k+1, i) &= C(k, i-1) - (8i+2) \cdot C(k,i) + (4i+4)(4i+2) \cdot C(k, i+1) \\
&= (-1)^{k-i+1} \frac{(2k)!}{(k-i+1)!(2i-2)!} - (-1)^{k-i}(8i+2) \frac{(2k)!}{(k-i)!(2i)!} \\
&\quad + (-1)^{k-i-1}(4i+4)(4i+2) \cdot \frac{(2k)!}{(k-i-1)!(2i+2)!} \\
&= (-1)^{k+1-i} \cdot \frac{(2k+2)!}{(k+1-i!) \cdot (2i)!}.
\end{aligned} \tag{6}$$

According to (5) and (6), we know that the Formula (3) is correct for $h = k+1$ and all integers $0 \le i \le k+1$. Theorem 3 can then be proved by mathematical induction.

In a similar way, we can also prove Theorem 4 by mathematical induction. Since the proof process is the same as the proof of Theorem 3, it is omitted.

3. Conclusions

The main purpose of this paper was to give two specific expressions for the sequences $C(h,i)$ and $D(h,i)$. As for some applications of our results, we proved two conjectures proposed by Zhang and Chen in [27].

As a matter of fact, our results are more general and not subject to prime conditions. Meanwhile, using our formulae for $C(h,i)$ and $D(h,i)$ in the theorems, we can simplify the variety of results that appear in Reference [27].

This paper not only enriches the research content of the Catalan numbers, but can also be regarded as a supplement and further improvement to Zhang and Chen's work in [27].

Author Contributions: All authors have equally contributed to this work. All authors read and approved the final manuscript.

Funding: This work is supported by the N. S. F. (11771351) and (11826205) of China.

Acknowledgments: The author would like to thank the referees for their very helpful and detailed comments, which have significantly improved the presentation of this paper.

Conflicts of Interest: The authors declare that there are no conflicts of interest regarding the publication of this paper.

References

1. Stanley, R.P. *Enumerative Combinatorics: Volume 2 (Cambridge Studieds in Advanced Mathematics)*; Cambridge University Press: Cambridge, UK, 1997; p. 49.
2. Chu, W.C. Further identities on Catalan numbers. *Discr. Math.* **2018**, *341*, 3159–3164. [CrossRef]

3. Anthony, J.; Polyxeni, L. A new interpretation of the Catalan numbers arising in the theory of crystals. *J. Algebra* **2018**, *504*, 85–128.
4. Liu, J.C. Congruences on sums of super Catalan numbers. *Results Math.* **2018**, *73*, 73–140. [CrossRef]
5. Kim, D.S.; Kim, T. A new approach to Catalan numbers using differential equations. *Russ. J. Math. Phys.* **2017**, *24*, 465–475. [CrossRef]
6. Kim, T.; Kim, D.S. Some identities of Catalan-Daehee polynomials arising from umbral calculus. *Appl. Comput. Math.* **2017**, *16*, 177–189.
7. Kim, T.; Kim, D.S. Differential equations associated with Catalan-Daehee numbers and their applications. *Revista de la Real Academia de Ciencias Exactas, Físicas y Naturales. Serie A. Matemáticas* **2017**, *111*, 1071–1081. [CrossRef]
8. Kim, D.S.; Kim, T. Triple symmetric identities for w-Catalan polynomials. *J. Korean Math. Soc.* **2017**, *54*, 1243–1264.
9. Kim, T.; Kwon, H.-I. Revisit symmetric identities for the λ-Catalan polynomials under the symmetry group of degree n. *Proc. Jangjeon Math. Soc.* **2016**, *19*, 711–716.
10. Kim, T. A note on Catalan numbers associated with p-adic integral on Zp. *Proc. Jangjeon Math. Soc.* **2016**, *19*, 493–501.
11. Basic, B. On quotients of values of Euler's function on the Catalan numbers. *J. Number Theory* **2016**, *169*, 160–173. [CrossRef]
12. Qi, F.; Shi, X.T.; Liu, F.F. An integral representation, complete monotonicity, and inequalities of the Catalan numbers. *Filomat* **2018**, *32*, 575–587. [CrossRef]
13. Qi, F.; Shi, X.T.; Mahmoud, M. The Catalan numbers: A generalization, an exponential representation, and some properties. *J. Comput. Anal. Appl.* **2017**, *23*, 937–944.
14. Qi, F.; Guo, B.N. Integral representations of the Catalan numbers and their applications. *Mathematics* **2017**, *5*, 40. [CrossRef]
15. Aker, K.; Gursoy, A.E. A new combinatorial identity for Catalan numbers. *Ars Comb.* **2017**, *135*, 391–398.
16. Dolgy, D.V.; Jang, G.-W.; Kim, D.S.; Kim, T. Explicit expressions for Catalan-Daehee numbers. *Proc. Jangjeon Math. Soc.* **2017**, *20*, 1–9.
17. Ma, Y.K.; Zhang, W.P. Some identities involving Fibonacci polynomials and Fibonacci numbers. *Mathematics* **2018**, *6*, 334. [CrossRef]
18. Zhang, W.P.; Lin, X. A new sequence and its some congruence properties. *Symmetry* **2018**, *10*, 359. [CrossRef]
19. Zhang, Y.X.; Chen, Z.Y. A new identity involving the Chebyshev polynomials. *Mathematics* **2018**, *6*, 244. [CrossRef]
20. Zhao, J.H.; Chen, Z.Y. Some symmetric identities involving Fubini polynomials and Euler numbers. *Symmetry* **2018**, *10*, 303.
21. Zhang, W.P. Some identities involving the Fibonacci numbers and Lucas numbers. *Fibonacci Q.* **2004**, *42*, 149–154.
22. Guariglia, E. Entropy and Fractal Antennas. *Entropy* **2016**, *18*, 84. [CrossRef]
23. Guido, R.C. Practical and useful tips on discrete wavelet transforms. *IEEE Signal Process. Mag.* **2015**, *32*, 162–166. [CrossRef]
24. Berry, M.V.; Lewis, Z.V. On the Weierstrass-Mandelbrot fractal function. *Proc. R. Soc. Lond. Ser. A* **1980**, *370*, 459–484. [CrossRef]
25. Guido, R.C.; Addison, P.; Walker, J. Introducing wavelets and time-frequency analysis. *IEEE Eng. Biol. Med. Mag.* **2009**, *28*, 13. [CrossRef] [PubMed]
26. Guariglia, E. Harmonic Sierpinski Gasket and Applications. *Entropy* **2018**, *20*, 714. [CrossRef]
27. Zhang, W.P.; Chen, L. On the Catalan numbers and some of their identities. *Symmetry* **2019**, *11*, 62. [CrossRef]

© 2019 by the authors. Licensee MDPI, Basel, Switzerland. This article is an open access article distributed under the terms and conditions of the Creative Commons Attribution (CC BY) license (http://creativecommons.org/licenses/by/4.0/).

Article

Bernoulli Polynomials and Their Some New Congruence Properties

Ran Duan and Shimeng Shen *

School of Mathematics, Northwest University, Xi'an 710127, China; duan.ran.stumail@stumail.nwu.edu.cn
* Correspondence: millieshen28@163.com

Received: 18 February 2019; Accepted: 8 March 2019; Published: 11 March 2019

Abstract: The aim of this article is to use the fundamental modus and the properties of the Euler polynomials and Bernoulli polynomials to prove some new congruences related to Bernoulli polynomials. One of them is that for any integer h or any non-negative integer n, we obtain the congruence $B_{2n+1}(2h) \equiv 0 \mod (2n+1)$, where $B_n(x)$ are Bernoulli polynomials.

Keywords: Euler polynomials; Bernoulli polynomials; elementary method; identity; congruence

MSC: 11B68; 11A07

---

1. Introduction

As usual, for the real number x, if $m \geq 0$ denotes any integer, the famous Bernoulli polynomials $B_m(x)$ (see [1–4]) and Euler polynomials $E_m(x)$ (see [2–5]) are decided by the coefficients of the series of powers:

$$\frac{z \cdot e^{zx}}{e^z - 1} = \sum_{m=0}^{\infty} \frac{B_m(x)}{m!} \cdot z^m \tag{1}$$

and:

$$\frac{2e^{zx}}{e^z + 1} = \sum_{m=0}^{\infty} \frac{E_m(x)}{m!} \cdot z^m. \tag{2}$$

If $x = 0$, then $E_m = E_m(0)$ and $B_m = B_m(0)$ are known as the mth Euler numbers and m^{th} Bernoulli numbers, respectively. For example, some values of B_m and E_m are $B_0 = 1$, $B_1 = -\frac{1}{2}$, $B_2 = \frac{1}{6}$, $B_3 = 0$, $B_4 = -\frac{1}{30}$, $B_5 = 0$, $B_6 = \frac{1}{42}$ and $E_0 = 1$, $E_1 = -\frac{1}{2}$, $E_2 = 0$, $E_3 = \frac{1}{4}$, $E_4 = 0$, $E_5 = -\frac{1}{2}$, $E_6 = 0$, etc. These polynomials and numbers occupy a very important position in number theory and combinatorics; this is not only because Bernoulli and Euler polynomials are well known, but also because they have a wide range of theoretical and applied values. Because of this, many scholars have studied the properties of these polynomials and numbers, and they also have obtained some valuable research conclusions. For instance, Zhang Wenpeng [6] studied a few combinational identities. As a continuation of the conclusion in [6], he showed that if p is a prime, one can obtain the congruence expression:

$$(-1)^{\frac{p-1}{2}} \cdot 2^{p-1} \cdot E_{p-1}\left(\frac{1}{2}\right) \equiv \begin{cases} 0 \mod p & \text{if } p \equiv 1 \mod 4; \\ -2 \mod p & \text{if } p \equiv 3 \mod 4. \end{cases}$$

Hou Yiwei and Shen Shimeng [3] proved the identity:

$$E_{2n-1} = -\frac{(2^{2n} - 1)}{n} \cdot B_{2n}.$$

As some corollaries of [3], Hou Yiwei and Shen Shimeng obtained several interesting congruences. For example, for p in an odd prime, one can obtain the expression:

$$E_{\frac{p-3}{2}} \equiv 0 \,(\text{mod } p), \text{ if } p \equiv 1 \text{ mod } 8.$$

Zhao Jianhong and Chen Zhuoyu [7] obtained the following deduction: if m is a positive integer, $k \geq 2$, one obtains the equation:

$$\sum_{a_1+a_2+\cdots+a_k=m} \frac{E_{a_1}}{(a_1)!} \cdot \frac{E_{a_2}}{(a_2)!} \cdots \frac{E_{a_k}}{(a_k)!} = \frac{2^{k-1}}{(k-1)!} \cdot \frac{1}{m!} \sum_{i=0}^{k-1} C(k-1,i) E_{m+k-1-i},$$

for which the summation is taken over all k-dimensional nonnegative integer coordinates $(a_1, a_2, \cdots, a_k)$ such that the equation $a_1 + a_2 + \cdots + a_k = m$, and the sequence $\{C(k,i)\}$ is decided as follows: for any integers $0 \leq i \leq k$, $C(k,k) = k!$, $C(k,0) = 1$,

$$C(k+1, i+1) = C(k, i+1) + (k+1)C(k,i), \text{ for all } 0 \leq i < k,$$

providing $C(k,i) = 0$, if $i > k$, and k is a positive integer.

T.Kim et al. did a good deal of research work and obtained a series of significant results; see [5,8–14]. Specifically, in [5], T. Kim found many valuable results involving Euler numbers and polynomials connected with zeta functions. Other papers in regard to the Bernoulli polynomials and Euler polynomials can be found in [15–19]; we will not go into detail here.

Here, we will make use of the properties of the Euler numbers, Euler polynomials, Bernoulli numbers, and Bernoulli polynomials to verify a special relationship between the Bernoulli polynomials and Euler polynomials. As some of the applications of our conclusions, we also deduce two unusual congruences involving the Bernoulli polynomials.

Theorem 1. *For any positive integers m and h, the following identity should be obtained, that is:*

$$2 \cdot B_{2m+1}(2h) = (2m+1) \cdot \left(E_{2m}(2h) + 2 \sum_{i=0}^{2h-1} E_{2m}(i) \right).$$

Theorem 2. *For any positive integers m and h, we derive the identity as below:*

$$B_{2m}(2h) - B_{2n} + m \left(E_{2m-1}(2h) - E_{2m-1} \right) = (2m) \cdot \sum_{i=1}^{2h} E_{2m-1}(i).$$

From these deductions, the following several corollaries can be inferred:

Corollary 1. *Let m be a non-negative integer. Thus, for any integer h, we obtain the congruence:*

$$B_{2m+1}(2h) \equiv 0 \text{ mod } (2m+1),$$

where $\frac{a}{b} \equiv 0 \text{ mod } k$ implies $(a,b) = 1$ and $k \mid a$ for any integers $b(b \neq 0)$ and a.

Corollary 2. *For any positive integer m and integer h, $2^{2m-1} \cdot (B_{2m}(2h) - B_{2m})$ must be an integer, and:*

$$2^{2m-1} \cdot (B_{2m}(2h) - B_{2m}) \equiv 0 \text{ mod } m.$$

Corollary 3. *For any integer h, let p be an odd prime; as a result, we have:*

$$B_p(2h) \equiv 0 \text{ mod } p \text{ and } B_{2p}(2h) \equiv B_{2p} \text{ mod } p.$$

Corollary 4. Let p be an odd prime. In this way, there exits an integer N with $N \equiv 1 \mod p$ such that the polynomial congruence:

$$N \cdot B_p(x) \equiv (x-2)(x-1)x \cdots (x-p+1) \equiv x \cdot \left(x^{p-1} - 1\right) \mod p.$$

Some notes: It is well known that congruences regarding Bernoulli numbers have interesting applications in number theory; in particular, for studying the class numbers of class-groups of number fields. Therefore, our corollaries will promote the further development of research in this field. Some important results in this field can also be found in [20–23]. Here, we will not list them one by one.

2. Several Lemmas

In this part, we will provide three straightforward lemmas. Henceforth, we will handle certain mathematical analysis knowledge and the properties of the Euler polynomials and Bernoulli polynomials, all of which can be discovered from [1–3]. Thus, they will not be repeated here.

Lemma 1. If $m \geq 0$ is an integer, polynomial $2^m \cdot E_m(x)$ denotes the integral coefficient polynomial of x.

Proof. First, from Definition 2 of the Euler polynomials $E_m(x)$, we have:

$$2e^{xz} = (e^z + 1) \cdot \frac{2e^{xz}}{e^z + 1} = \left(1 + \sum_{m=0}^{\infty} \frac{1}{n!} \cdot z^m\right) \left(\sum_{m=0}^{\infty} \frac{E_m(x)}{m!} \cdot z^m\right). \quad (3)$$

On the other hand, we also have:

$$2e^{xz} = 2 \cdot \sum_{m=0}^{\infty} \frac{x^m}{m!} \cdot z^m. \quad (4)$$

uniting (3) and (4), then comparing the coefficients of the power series, we obtain that:

$$2x^m = E_m(x) + \sum_{k=0}^{m} \binom{m}{k} E_k(x)$$

or identity:

$$2E_m(x) = 2x^m - \sum_{k=0}^{m-1} \binom{m}{k} E_k(x). \quad (5)$$

Note that $E_0(x) = 1$, $E_1(x) = x - \frac{1}{2}$, so from (5) and mathematical induction, we may immediately deduce that $2^m \cdot E_m(x)$ is an integral coefficient polynomial of x. □

Lemma 2. If m is a positive integer, the following equation can be obtained:

$$2^m \cdot B_m(x) = B_m(2x) - \frac{1}{2} \cdot m \cdot E_{m-1}(2x).$$

Proof. From Definitions 1 and 2 of the Euler polynomials and Bernoulli polynomials, we discover the identity as below:

$$\frac{2ze^{2xz}}{e^{2z} - 1} = \sum_{m=0}^{\infty} \frac{2^m \cdot B_m(x)}{m!} \cdot z^m = \left(\frac{z \cdot e^{2xz}}{e^z - 1} - \frac{z \cdot e^{2xz}}{e^z + 1}\right)$$

$$= \sum_{m=0}^{\infty} \frac{B_m(2x)}{m!} \cdot z^m - \frac{1}{2} \sum_{m=0}^{\infty} \frac{E_m(2x)}{m!} \cdot z^{m+1}. \quad (6)$$

Relating the coefficients of the power series in (6), we obtain:

$$2^m \cdot B_m(x) = B_m(2x) - \frac{m}{2} \cdot E_{m-1}(2x).$$

This proves Lemma 2. □

Lemma 3. *If m is a positive integer, then for any positive integer M, we will be able to obtain the identities:*

$$2^m \cdot (B_m(M) - B_m) = m \cdot \sum_{i=0}^{2M-1} E_{m-1}(i).$$

Proof. On the basis of Definition 2 of the Euler polynomials, we obtain:

$$\sum_{i=0}^{N-1} \frac{2ze^{iz}}{e^z + 1} = \sum_{m=0}^{\infty} \frac{1}{n!} \left(\sum_{i=0}^{N-1} E_m(i) \right) \cdot z^{m+1}. \tag{7}$$

In another aspect, we also obtain:

$$\sum_{i=0}^{N-1} \frac{2ze^{iz}}{e^z + 1} = \frac{2z \left(e^{Nz} - 1 \right)}{(e^z + 1)(e^z - 1)} = \frac{2ze^{Nz} - 2z}{e^{2z} - 1}$$

$$= \sum_{m=0}^{\infty} \frac{2^m \cdot B_m \left(\frac{N}{2} \right)}{m!} \cdot z^m - \sum_{m=0}^{\infty} \frac{2^m \cdot B_m}{m!} \cdot z^m. \tag{8}$$

Combining (7) and (8), then comparing the coefficients of the power series, we will obtain:

$$2^m \cdot \left(B_m \left(\frac{N}{2} \right) - B_m \right) = m \cdot \sum_{i=0}^{N-1} E_{m-1}(i). \tag{9}$$

Now, Lemma 3 follows from (9) with $N = 2M$. □

3. Proofs of the Theorems

Applying three simple lemmas in Section 2, we can easily finish the proofs of our theorems. Above all, we study Theorem 1. For any positive integer m, from Lemma 2, we have:

$$2^{2m+1} \cdot B_{2m+1}(M) = B_{2m+1}(2M) - \frac{2m+1}{2} \cdot E_{2m}(2M). \tag{10}$$

Note that $B_{2m+1} = 0$. From Lemma 3, we also have:

$$2^{2m+1} \cdot B_{2m+1}(M) = (2m+1) \cdot \sum_{i=0}^{2M-1} E_{2m}(i). \tag{11}$$

Combining (10) and (11), we have:

$$B_{2m+1}(2M) = \frac{2m+1}{2} \cdot E_{2m}(2M) + (2m+1) \cdot \sum_{i=0}^{2M-1} E_{2m}(i).$$

Afterwards, we prove Theorem 2. According to Lemma 2 with $x = M$ and $x = 0$, we have:

$$2^{2m} \cdot B_{2m}(M) = B_{2m}(2M) - m \cdot E_{2m-1}(2M) \tag{12}$$

and:

$$2^{2m} \cdot B_{2m} = B_{2m} - m \cdot E_{2m-1}. \tag{13}$$

Applying Lemma 3, we also have:

$$2^{2m} \cdot (B_{2m}(M) - B_{2m}) = (2m) \cdot \sum_{i=0}^{2M-1} E_{2m-1}(i). \tag{14}$$

Combining (12), (13), and (14), we have the identity:

$$B_{2m}(2M) - B_{2m} = m \cdot E_{2m-1}(2M) - m \cdot E_{2m-1} + 2m \cdot \sum_{i=0}^{2M-1} E_{2m-1}(i).$$

This proves Theorem 2.

From Lemma 1, we know that all $2^{2m} \cdot E_{2m}(i)$ ($i = 0, 1, \cdots, 2M$) are integers, and $(2^{2m}, 2m+1) = 1$, so on the basis of Theorem 1, we may directly deduce the congruence:

$$B_{2m+1}(2M) \equiv 0 \bmod (2m+1). \tag{15}$$

Since $B_{2m+1}(x)$ is an odd function (that is, $B_{2m+1}(-x) = -B_{2m+1}(x)$), and $B_{2m+1} = 0$, so (15) also holds for any integer M and non-negative integer m.

This completes the proof of Corollary 1.

Now, we study Corollary 2. On the basis of Lemma 1, we know that $2^{2m-1} \cdot E_{2m-1}(i)$ is an integer for all $1 \leq i \leq 2M$, so from Theorem 1, we know that $2^{2m-1} \cdot (B_{2m}(2M) - B_{2m})$ must be an integer, and it can be divided by m, that is,

$$2^{2m-1} \cdot (B_{2m}(2M) - B_{2m}) \equiv 0 \bmod m. \tag{16}$$

Note that $B_{2m}(x)$ is an even function, and if $M = 0$, after that, the left-hand side of (16) becomes zero; thus, the congruence (16) is correct for all integers M.

This completes the proof of Corollary 2.

Corollary 3 is a special case of Corollary 1 with $2m + 1 = p$ and Corollary 2 with $2m = 2p$.

Now, we prove Corollary 4. Since $B_p(x)$ is a pth rational coefficient polynomial of x and its first item is x^p, from Lemma 3, we know that the congruence equation $B_p(2x) \equiv 0 \bmod p$ has exactly p different solutions $x = 0, 1, 2, \cdots p - 1$, so there exits an integer N with $N \equiv 1 \bmod p$ satisfied with $N \cdot B_p(x)$, an integral coefficient polynomial of x. From [1] (see Theorem 5.23), we have the congruence:

$$N \cdot B_p(x) \equiv x(x-1)(x-2) \cdot (x-p+1) \bmod p.$$

This completes the proofs of our all results.

4. Conclusions

As we all know, the congruences of Bernoulli numbers have important applications in number theory; in particular, for studying the class numbers of class-groups of number fields. The main results of this paper are two theorems involving Bernoulli and Euler polynomials and numbers and four corollaries (or congruences). Two theorems gave some new equations regarding Bernoulli polynomials and Euler polynomials. As some applications of these theorems, we gave four interesting congruences involving Bernoulli polynomials. Especially, Corollaries 1 and 4 are very simple and beautiful. It is clear that Corollary 4 is a good reference for further research on Bernoulli polynomials.

Author Contributions: All authors have equally contributed to this work. All authors read and approved the final manuscript.

Funding: This work is supported by the NSF (11771351) and (11826205) of P. R. China.

Acknowledgments: The authors would like to thank the Editor and referee for their very helpful and detailed comments, which have significantly improved the presentation of this paper.

Conflicts of Interest: The authors declare that there are no conflicts of interest regarding the publication of this paper.

References

1. Apostol, T.M. *Introduction to Analytic Number Theory*; Springer: New York, NY, USA, 1976.
2. Knuth, D.E.; Buckholtz, T.J. Computation of Tangent, Euler, and Bernoulli numbers. *Math. Comput.* **1967**, *21*, 663–688. [CrossRef]
3. Hou, Y.W.; Shen, S.M. The Euler numbers and recursive properties of Dirichlet L-functions. *Adv. Differ. Equ.* **2018**, *2018*, 397. [CrossRef]
4. Liu, G.D. Identities and congruences involving higher-order Euler-Bernoulli numbers and polynomials. *Fibonacci Q.* **2001**, *39*, 279–284.
5. Kim, T. Euler numbers and polynomials associated with zeta functions. *Abstr. Appl. Anal.* **2008**, *2018*, 581582. [CrossRef]
6. Zhang, W.P. Some identities involving the Euler and the central factorial numbers. *Fibonacci Q.* **1998**, *36*, 154–157.
7. Zhao, J.H.; Chen, Z.Y. Some symmetric identities involving Fubini polynomials and Euler numbers. *Symmetry* **2018**, *10*, 303.
8. Kim, D.S.; Kim, T. Some symmetric identities for the higher-order q-Euler polynomials related to symmetry group S_3 arising from p-Adic q-fermionic integrals on $\mathbb{Z}_p$. *Filomat* **2016**, *30*, 1717–1721. [CrossRef]
9. Kim, T. Symmetry of power sum polynomials and multivariate fermionic p-Adic invariant integral on $\mathbb{Z}_p$. *Russ. J. Math. Phys.* **2009**, *16*, 93–96. [CrossRef]
10. Kim, T.; Kim, D.S.; Jang, G.W. A note on degenerate Fubini polynomials. *Proc. Jiangjeon Math. Soc.* **2017**, *20*, 521–531.
11. Kim, D.S.; Park, K.H. Identities of symmetry for Bernoulli polynomials arising from quotients of Volkenborn integrals invariant under S_3. *Appl. Math. Comput.* **2013**, *219*, 5096–5104. [CrossRef]
12. Kim, T.; Kim, D.S. An identity of symmetry for the degernerate Frobenius-Euler polynomials. *Math. Slovaca* **2018**, *68*, 239–243. [CrossRef]
13. Kim, T.; Kim, S.D.; Jang, G.W.; Kwon, J. Symmetric identities for Fubini polynomials. *Symmetry* **2018**, *10*, 219. [CrossRef]
14. Kim, D.S.; Rim, S.-H.; Kim, T. Some identities on Bernoulli and Euler polynomials arising from orthogonality of Legendre polynomials. *J. Inequal. Appl.* **2012**, *2012*, 227. [CrossRef]
15. Simsek, Y. Identities on the Changhee numbers and Apostol-type Daehee polynomials. *Adv. Stud. Contemp. Math.* **2017**, *27*, 199–212.
16. Guy, R.K. *Unsolved Problems in Number Theory*, 2nd ed.; Springer: New York, NY, USA, 1994.
17. Liu, G.D. The solution of problem for Euler numbers. *Acta Math. Sin.* **2004**, *47*, 825–828.
18. Zhang, W.P.; Xu, Z.F. On a conjecture of the Euler numbers. *J. Number Theory* **2007**, *127*, 283–291. [CrossRef]
19. Cho, B.; Park, H. Evaluating binomial convolution sums of divisor functions in terms of Euler and Bernoulli polynomials. *Int. J. Number Theory* **2018**, *14*, 509–525. [CrossRef]
20. Wagstaff, S.S., Jr. Prime divisors of the Bernoulli and Euler Numbers. *Number Theory Millenn.* **2002**, *3*, 357–374.
21. Bayad, A.; Aygunes, A. Hecke operators and generalized Bernoulli-Euler polynomials. *J. Algebra Number Theory Adv. Appl.* **2010**, *3*, 111–122.
22. Kim, D.S.; Kim, T. Some p-Adic integrals on $\mathbb{Z}_p$ Associated with trigonometric functions. *Russ. J. Math. Phys.* **2018**, *25*, 300–308. [CrossRef]
23. Powell, B.J. Advanced problem 6325. *Am. Math. Mon.* **1980**, *87*, 836.

© 2019 by the authors. Licensee MDPI, Basel, Switzerland. This article is an open access article distributed under the terms and conditions of the Creative Commons Attribution (CC BY) license (http://creativecommons.org/licenses/by/4.0/).

Article

Connection Problem for Sums of Finite Products of Legendre and Laguerre Polynomials

Taekyun Kim [1], Kyung-Won Hwang [2,*], Dae San Kim [3] and Dmitry V. Dolgy [4]

1. Department of Mathematics, Kwangwoon University, Seoul 01897, Korea; tkkim@kw.ac.kr
2. Department of Mathematics, Dong-A University, Busan 49315, Korea
3. Department of Mathematics, Sogang University, Seoul 04107, Korea; dskim@sogang.ac.kr
4. Institute of National Science, Far Eastern Federal University, Vladivostok 690950, Russia; dvdolgy@gmail.com
* Correspondence: khwang@dau.ac.kr

Received: 12 February 2019; Accepted: 26 February 2019; Published: 2 March 2019

Abstract: The purpose of this paper is to represent sums of finite products of Legendre and Laguerre polynomials in terms of several orthogonal polynomials. Indeed, by explicit computations we express each of them as linear combinations of Hermite, generalized Laguerre, Legendre, Gegenbauer and Jacobi polynomials, some of which involve terminating hypergeometric functions $_1F_1$ and $_2F_1$.

Keywords: Legendre polynomials; Laguerre polynomials; generalized Laguerre polynomials; Gegenbauer polynomials; hypergeometric functions $_1F_1$ and $_2F_1$

1. Preliminaries

Here, after fixing some notations that will be needed throughout this paper, we will review briefly some basic facts about orthogonal polynomials relevant to our discussion. As general references on orthogonal polynomials, we recommend the reader to refer to [1,2].

As is well known, the falling factorial sequence $(x)_n$ and the rising factorial sequence $\langle x \rangle_n$ are respectively defined by

$$(x)_n = x(x-1)\dots(x-n+1), \quad (n \geq 1), (x)_0 = 1, \tag{1}$$

$$\langle x \rangle_n = x(x+1)\dots(x+n-1), \quad (n \geq 1), \langle x \rangle_0 = 1. \tag{2}$$

The two factorial sequences are related by

$$(-1)^n (x)_n = \langle -x \rangle_n, \quad (-1)^n \langle x \rangle_n = (-x)_n. \tag{3}$$

$$\frac{(2n-2s)!}{(n-s)!} = \frac{2^{2n-2s}(-1)^s \langle \frac{1}{2} \rangle_n}{\langle \frac{1}{2} - n \rangle_s}, \quad (n \geq s \geq 0). \tag{4}$$

$$\frac{(2n+2s)!}{(n+s)!} = 2^{2n+2s} \langle \frac{1}{2} \rangle_n \langle n + \frac{1}{2} \rangle_s, \quad (n,s \geq 0). \tag{5}$$

$$\Gamma(n + \frac{1}{2}) = \frac{(2n)!\sqrt{\pi}}{2^{2n} n!}, \quad (n \geq 0), \tag{6}$$

$$\frac{\Gamma(x+1)}{\Gamma(x+1-n)} = (x)_n, \quad \frac{\Gamma(x+n)}{\Gamma(x)} = \langle x \rangle_n, \quad (n \geq 0), \tag{7}$$

$$B(x,y) = \int_0^1 t^{x-1}(1-t)^{y-1} dt = \frac{\Gamma(x)\Gamma(y)}{\Gamma(x+y)}, \quad (\operatorname{Re} x, \operatorname{Re} y > 0), \tag{8}$$

where $\Gamma(x)$ and $B(x,y)$ denote respectively the gamma and beta functions.

The hypergeometric function is defined by

$$_pF_q = (a_1,\ldots,a_p; b_1,\ldots,b_q; x) = \sum_{n=0}^{\infty} \frac{\langle a_1\rangle_n \cdots \langle a_p\rangle_n}{\langle b_1\rangle_n \cdots \langle b_q\rangle_n} \frac{x^n}{n!}. \tag{9}$$

Now, we are ready to recall some relevant facts about Legendre polynomials $P_n(x)$, Laguerre polynomials $L_n(x)$, Hermite polynomials $H_n(x)$, generalized (extended) Laguerre polynomials $L_n^\alpha(x)$, Gegenbauer polynomials $C_n^{(\lambda)}(x)$, and Jacobi polynomials $P_n^{(\alpha,\beta)}(x)$. All the facts stated here can also be found in [3–8].Interested readers may refer to [1,2,9–13] for full accounts of orthogonal polynomials and also to [14,15] for papers discussing relevant orthogonal polynomials.

The above-mentioned orthogonal polynomials are given, in terms of generating functions, by

$$F(t,x) = (1 - 2xt + t^2)^{-\frac{1}{2}} = \sum_{n=0}^{\infty} P_n(x) t^n, \tag{10}$$

$$G(t,x) = (1-t)^{-1} \exp\left(-\frac{xt}{1-t}\right) = \sum_{n=0}^{\infty} L_n(x) t^n, \tag{11}$$

$$e^{2xt-t^2} = \sum_{n=0}^{\infty} H_n(x) \frac{t^n}{n!}, \tag{12}$$

$$(1-t)^{-\alpha-1} \exp\left(-\frac{xt}{1-t}\right) = \sum_{n=0}^{\infty} L_n^\alpha(x) t^n, \tag{13}$$

$$\frac{1}{(1-2xt+t^2)^\lambda} = \sum_{n=0}^{\infty} C_n^{(\lambda)}(x) t^n, \quad (\lambda > -\frac{1}{2}, \lambda \neq 0, |t| < 1, |x| \leq 1), \tag{14}$$

$$\frac{\alpha+\beta}{R(1-t++R)^\alpha (1+t+R)^\beta} = \sum_{n=0}^{\infty} P_n^{(\alpha,\beta)}(x) t^n, \tag{15}$$

$$(R = \sqrt{1-2xt+t^2}, \alpha, \beta > -1).$$

In terms of explicit expressions, those orthogonal polynomials are given explicitly as follows:

$$P_n(x) = {}_2F_1\left(-n, n+1; 1; \frac{1-x}{2}\right)$$

$$= \frac{1}{2^n} \sum_{l=0}^{[\frac{n}{2}]} (-1)^l \binom{n}{l} \binom{2n-2l}{n} x^{n-2l}, \tag{16}$$

$$L_n(x) = {}_1F_1(-n; 1; x)$$

$$= \sum_{l=0}^{n} (-1)^{n-l} \binom{n}{l} \frac{1}{(n-l)!} x^{n-l}, \tag{17}$$

$$H_n(x) = n! \sum_{l=0}^{[\frac{n}{2}]} \frac{(-1)^l}{l!(n-2l)!} (2x)^{n-2l}, \tag{18}$$

$$L_n^\alpha(x) = \frac{\langle \alpha+1\rangle_n}{n!} {}_1F_1(-n; \alpha+1; x)$$

$$= \sum_{l=0}^{n} \frac{(-1)^l \binom{n+\alpha}{n-l}}{l!} x^l, \tag{19}$$

$$C_n^\lambda(x) = \binom{n+2\lambda-1}{n} {}_2F_1\left(-n, n+2\lambda; \lambda+\frac{1}{2}; \frac{1-x}{2}\right)$$
$$= \sum_{k=0}^{[\frac{n}{2}]} (-1)^k \frac{\Gamma(n-k+\lambda)}{\Gamma(\lambda)k!(n-2k)!} (2x)^{n-2k}, \tag{20}$$

$$P_n^{(\alpha,\beta)}(x) = \frac{\langle\alpha+1\rangle_n}{n!} {}_2F_1\left(-n, 1+\alpha+\beta+n; \alpha+1; \frac{1-x}{2}\right)$$
$$= \sum_{k=0}^{n} \binom{n+\alpha}{n-k}\binom{n+\beta}{k}\left(\frac{x-1}{2}\right)^k\left(\frac{x+1}{2}\right)^{n-k}. \tag{21}$$

For Legendre, Gegenbauer and Jacobi polynomials, we have Rodrigues' formulas, and for Hermite and generalized Laguerre polynomials, we have Rodrigues-type formulas.

$$H_n(x) = (-1)^n e^{x^2} \frac{d^n}{dx^n} e^{-x^2}, \tag{22}$$

$$L_n^\alpha(x) = \frac{1}{n!} x^{-\alpha} e^x \frac{d^n}{dx^n}(e^{-x}x^{n+\alpha}), \tag{23}$$

$$P_n(x) = \frac{1}{2^n n!} \frac{d^n}{dx^n}(x^2-1)^n, \tag{24}$$

$$(1-x^2)^{\lambda-\frac{1}{2}} C_n^{(\lambda)}(x) = \frac{(-2)^n}{n!} \frac{\langle\lambda\rangle_n}{\langle n+2\lambda\rangle_n} \frac{d^n}{dx^n}(1-x^2)^{n+\lambda-\frac{1}{2}}, \tag{25}$$

$$(1-x)^\alpha(1+x)^\beta P_n^{(\alpha,\beta)}(x) = \frac{(-1)^n}{2^n n!} \frac{d^n}{dx^n}(1-x)^{n+\alpha}(1+x)^{n+\beta}. \tag{26}$$

The orthogonal polynomials in Equations (22)–(26) satisfy the following orthogonality relations with respect to various weight functions.

$$\int_{-\infty}^{\infty} e^{-x^2} H_n(x) H_m(x) dx = 2^n n! \sqrt{\pi} \delta_{m,n}, \tag{27}$$

$$\int_0^\infty x^\alpha e^{-x} L_n^\alpha(x) L_m^\alpha(x) dx = \frac{1}{n!} \Gamma(\alpha+n+1) \delta_{m,n}, \tag{28}$$

$$\int_{-1}^{1} P_n(x) P_m(x) dx = \frac{2}{2n+1} \delta_{m,n}, \tag{29}$$

$$\int_{-1}^{1} (1-x^2)^{\lambda-\frac{1}{2}} C_n^{(\lambda)}(x) C_m^{(\lambda)}(x) dx = \frac{\pi 2^{1-2\lambda} \Gamma(n+2\lambda)}{n!(n+\lambda)\Gamma(\lambda)^2} \delta_{m,n}, \tag{30}$$

$$\int_{-1}^{1} (1-x)^\alpha (1+x)^\beta P_n^{(\alpha,\beta)}(x) P_m^{(\alpha,\beta)}(x) dx = \frac{2^{\alpha+\beta+1}\Gamma(n+\alpha+1)\Gamma(n+\beta+1)}{(2n+\alpha+\beta+1)\Gamma(n+\alpha+\beta+1)\Gamma(n+1)} \delta_{m,n}. \tag{31}$$

2. Introduction

In this paper, we will consider two sums of finite products

$$\gamma_{n,r}(x) = \sum_{i_1+\cdots+i_{2r+1}=n} P_{i_1}(x) P_{i_2}(x) \ldots P_{i_{2r+1}}(x), \quad (n, r \geq 0), \tag{32}$$

in terms of Legendre polynomials and

$$\varepsilon_{n,r}(x) = \sum_{i_1+\cdots+i_{r+1}=n} L_{i_1}\left(\frac{x}{r+1}\right) L_{i_2}\left(\frac{x}{r+1}\right) \ldots L_{i_{r+1}}\left(\frac{x}{r+1}\right), \quad (n, r \geq 0), \tag{33}$$

in terms of Laguerre polynomials. We represent each of them as linear combinations of Hermite, extended Laguerre, Legendre, Gegenbauer, and Jacobi polynomials (see Theorems 1 and 2). It is amusing to note here that, for some of these expressions, the coefficients involve certain terminating hypergeometric functions $_2F_1$ and $_1F_1$. These representations are obtained by carrying out explicit computations with the help of Propositions 1 and 2. We observe here that the formulas in Proposition 1 can be derived from the orthogonalities in Equation (27)–(31), Rodrigues' and Rodrigues-type formulas in Equation (22)–(26), and integration by parts.

Our study of such representation problems can be justified by the following. Firstly, the present research can be viewed as a generalization of the classical connection problems. Indeed, the classical connection problems are concerned with determining the coefficients in the expansion of a product of two polynomials in terms of any given sequence of polynomials (see [1,2]).

Secondly, studying such kinds of sums of finite products of special polynomials can be well justified also by the following example. Let us put

$$\alpha_m(x) = \sum_{k=1}^{m-1} \frac{1}{k(m-k)} B_k(x) B_{m-k}(x), \quad (m \geq 2),$$

where $B_n(x)$ are the Bernoulli polynomials. Then we can express $\alpha_m(x)$ as linear combinations of Bernoulli polynomials, for example from the Fourier series expansion of the function closely related to that. Indeed, we can show that

$$\sum_{k=1}^{m-1} \frac{1}{2k(2m-2k)} B_{2k}(x) B_{2m-2k}(x) + \frac{2}{2m-1} B_1(x) B_{2m-1}(x) \tag{34}$$
$$= \frac{1}{m} \sum_{k=1}^{m} \frac{1}{2k} \binom{2m}{2k} B_{2k} B_{2m-2k}(x) + \frac{1}{m} H_{2m-1} B_{2m}(x) + \frac{2}{2m-1} B_{2m-1} B_1(x),$$

where $H_m = \sum_{j=1}^{m} \frac{1}{j}$ are the harmonic numbers.

Further, some simple modification of this gives us the famous Faber-Pandharipande-Zagier identity and a slightly different variant of the Miki's identity by letting respectively $x = \frac{1}{2}$ and $x = 0$ in (34). We note here that all the other known derivations of F-P-Z and Miki's identity are quite involved, while our proof of Miki's and Faber-Pandharipande-Zagier identities follow from the polynomial identity (34), which in turn follows immediately the Fourier series expansion of $\alpha_m(x)$. Indeed, Miki makes use of a formula for the Fermat quotient $\frac{a^p - a}{p}$ modulo p^2, Shiratani-Yokoyama employs p-adic analysis, Gessel's proof is based on two different expressions for Stirling numbers of the second kind $S_2(n,k)$, and Dunne-Schubert exploits the asymptotic expansion of some special polynomials coming from the quantum field theory computations. For some details on these, we let the reader refer to the introduction in [16] and the papers therein.

The next two theorems are the main results of this paper.

Theorem 1. *For any nonnegative integers n and r, we have the following representation.*

$$\sum_{i_1 + i_2 + \cdots + i_{2r+1} = n} P_{i_1}(x) P_{i_2}(x) \ldots P_{i_{2r+1}}(x)$$
$$= \frac{2^r (n + r - \frac{1}{2})_{n+r}}{(2r-1)!!} \sum_{j=0}^{[\frac{n}{2}]} \frac{{}_1F_1(-j; \frac{1}{2} - n - r; -1)}{j!(n-2j)!} H_{n-2j}(x) \tag{35}$$

$$= \frac{1}{(2r-1)!! 2^{n+r}} \sum_{k=0}^{n} \frac{(-1)^k}{\Gamma(\alpha + k + 1)}$$
$$\times \sum_{l=0}^{[\frac{n-k}{2}]} \frac{(-1)^l (2n + 2r - 2l)! \Gamma(n - 2l + \alpha + 1)}{l!(n+r-l)!(n-k-2l)!} L_k^\alpha(x) \tag{36}$$

$$= \frac{2^{r-1}(n+r-\frac{1}{2})_{n+r}}{(2r-1)!!}$$

$$\times \sum_{j=0}^{[\frac{n}{2}]} \frac{(2n+1-4j)\,_2F_1(-j, j-n-\frac{1}{2}; \frac{1}{2}-n-r; 1)}{j!(n-j+\frac{1}{2})_{n-j+1}} P_{n-2j}(x) \tag{37}$$

$$= \frac{2^r \Gamma(\lambda)(n+r-\frac{1}{2})_{n+r}}{(2r-1)!!}$$

$$\times \sum_{j=0}^{[\frac{n}{2}]} \frac{(n+\lambda-2j)\,_2F_1(-j, j-n-r; \frac{1}{2}-n-r; 1)}{\Gamma(n+\lambda-j+1)j!} C_{n-2j}^{(\lambda)}(x) \tag{38}$$

$$= \frac{(-1)^n}{(2r-1)!!2^{n+r}} \sum_{k=0}^{n} \frac{\Gamma(k+\alpha+\beta+1)(-2)^k}{\Gamma(2k+\alpha+\beta+1)}$$

$$\times \sum_{l=0}^{[\frac{n-k}{2}]} \frac{(-1)^l(2n+2r-2l)!}{l!(n+r-l)!(n-k-2l)!} \tag{39}$$

$$\times\,_2F_1(2l+k-n, k+\beta+1; 2k+\alpha+\beta+2; 2) P_k^{(\alpha,\beta)}(x).$$

Here $(2r-1)!!$ is the double factorial given by

$$(2r-1)!! = (2r-1)(2r-3)\ldots 1, \quad (r \geq 1), (-1)!! = 1. \tag{40}$$

Remark 1. *An alternative expression for (36) is given by*

$$\gamma_{n,r}(x) = \frac{1}{\Gamma(\alpha+1)(2r-1)!!2^{n+r}}$$

$$\times \sum_{l=0}^{[\frac{n}{2}]} \frac{(-1)^l(2n+2r-2l)!\Gamma(n-2l+\alpha+1)}{l!(n+r-l)!(n-2l)!} \sum_{k=0}^{n-2l} \frac{\langle 2l-n \rangle_k}{\langle \alpha+1 \rangle_k} L_k^\alpha(x). \tag{41}$$

Theorem 2. *For any nonnegative integers n and r, we have the following representation.*

$$\sum_{i_1+i_2+\cdots+i_{r+1}=n} L_{i_1}\left(\frac{x}{r+1}\right) L_{i_2}\left(\frac{x}{r+1}\right) \ldots L_{i_{r+1}}\left(\frac{x}{r+1}\right)$$

$$= (n+r)! \sum_{k=0}^{n} \frac{(-\frac{1}{2})^k}{k!} \sum_{j=0}^{[\frac{n-k}{2}]} \frac{(\frac{1}{4})^j}{j!(n-k-2j)!(r+k+2j)!} H_k(x) \tag{42}$$

$$= (n+r)! \sum_{k=0}^{n} \frac{\,_2F_1(k-n, k+\alpha+1; r+k+1; 1)}{(n-k)!(r+k)!} L_k^\alpha(x) \tag{43}$$

$$=(n+r)! \sum_{k=0}^{n} 2^{k+1}(2k+1)$$

$$\times \sum_{j=0}^{[\frac{n-k}{2}]} \frac{(k+j+1)!}{j!(n-k-2j)!(r+k+2j)!(2k+2j+2)!} P_k(x) \tag{44}$$

$$=(n+r)!\Gamma(\lambda) \sum_{k=0}^{n} \left(-\frac{1}{2}\right)^k (k+\lambda)$$

$$\times \sum_{j=0}^{[\frac{n-k}{2}]} \frac{(\frac{1}{4})^j}{j!(n-k-2j)!(r+k+2j)!\Gamma(k+j+\lambda+1)} C_k^{(\lambda)}(x) \tag{45}$$

25

$$= (n+r)! \sum_{k=0}^{n} \frac{\Gamma(k+\alpha+\beta+1)(-2)^k}{\Gamma(2k+\alpha+\beta+1)}$$
$$\times \sum_{l=0}^{n-k} \frac{{}_2F_1(k-n+l, k+\beta+1; 2k+\alpha+\beta+2; 2)}{l!(n+r-l)!(n-k-l)!} P_k^{(\alpha,\beta)}(x). \tag{46}$$

Remark 2. *An alternative expression for (42) is as follows:*

$$\varepsilon_{n,r}(x) = (n+r)! \sum_{j=0}^{[\frac{n}{2}]} \frac{(\frac{1}{4})^j}{j!(n-2j)!(r+2j)!} \sum_{k=0}^{n-2j} \frac{(\frac{1}{2})^k \langle 2j-n \rangle_k}{k! \langle r+2j+1 \rangle_k} H_k(x). \tag{47}$$

Before we move on to the next section, we would like to mention some of the related previous works. In [16–18], sums of finite products of Bernoulli, Euler and Genocchi polynomials were represented as linear combinations of Bernoulli polynomials. These were derived from the Fourier series expansions for the functions closely related to those sums of finite products. In addition, in [9] the same had been done for sums of finite products of Chebyshev polynomials of the second kind and of Fibonacci polynomials.

On the other hand, in terms of all kinds of Chebyshev polynomials, sums of finite products of Chebyshev polynomials of the second, third and fourth kinds and of Fibonacci, Legendre and Laguerre polynomials were expressed in [11,12,19]. Further, by the orthogonal polynomials in Equations (16), and (18)–(21), sums of finite products of Chebyshev polynomials of the second kind and Fibonacci polynomials were represented in [13].

Finally, the reader may want to see [20,21] for some other aspects of Legendre and Laguerre polynomials.

3. Proof of Theorem 1

We will first state Propositions 1 and 2 that will be needed in showing Theorems 1 and 2.

The results in the next proposition can be derived from the orthogonalities in (27)–(31), Rodrigues' and Rodrigues-type formulas in (22)–(26), and integration by parts, as we mentioned earlier. The facts (a), (b), (c), (d) and (e) in Proposition 1 are respectively from (3.7) of [5], (2.3) of [7] (see also (2.4) of [3]), (2.3) of [6], (2.3) of [4] and (2.7) of [8].

Proposition 1. *For any polynomial $q(x) \in \mathbb{R}[x]$ of degree n, the following hold.*

(a)
$$q(x) = \sum_{k=0}^{n} C_{k,1} H_k(x), \text{ where } C_{k,1} = \frac{(-1)^k}{2^k k! \sqrt{\pi}} \int_{-\infty}^{\infty} q(x) \frac{d^k}{dx^k} e^{-x^2} dx,$$

(b)
$$q(x) = \sum_{k=0}^{n} C_{k,2} L_k^\alpha(x), \text{ where } C_{k,2} = \frac{1}{\Gamma(\alpha+k+1)} \int_0^\infty q(x) \frac{d^k}{dx^k} (e^{-x} x^{k+\alpha}) dx,$$

(c)
$$q(x) = \sum_{k=0}^{n} C_{k,3} P_k(x), \text{ where } C_{k,3} = \frac{2k+1}{2^{k+1} k!} \int_{-1}^{1} q(x) \frac{d^k}{dx^k} (x^2-1)^k dx,$$

(d)
$$q(x) = \sum_{k=0}^{n} C_{k,4} C_k^{(\lambda)}(x), \text{ where }$$
$$C_{k,4} = \frac{(k+\lambda) \Gamma(\lambda)}{(-2)^k \sqrt{\pi} \Gamma(k+\lambda+\frac{1}{2})} \int_{-1}^{1} q(x) \frac{d^k}{dx^k} (1-x^2)^{k+\lambda-\frac{1}{2}} dx,$$

(e)
$$q(x) = \sum_{k=0}^{n} C_{k,5} P_k^{(\alpha,\beta)}(x), \text{ where}$$

$$C_{k,5} = \frac{(-1)^k (2k+\alpha+\beta+1)\Gamma(k+\alpha+\beta+1)}{2^{\alpha+\beta+k+1}\Gamma(\alpha+k+1)\Gamma(\beta+k+1)}$$

$$\times \int_{-1}^{1} q(x) \frac{d^k}{dx^k}(1-x)^{k+\alpha}(1+x)^{k+\beta} dx.$$

Proposition 2. *The following proposition was stated in [16].*
For any nonnegative integers m and k, the following identities hold.

(a)
$$\int_{-\infty}^{\infty} x^m e^{-x^2} dx = \begin{cases} 0, & \text{if } m \equiv 1 \pmod 2, \\ \frac{m!\sqrt{\pi}}{(\frac{m}{2})! 2^m}, & \text{if } m \equiv 0 \pmod 2, \end{cases}$$

(b)
$$\int_{-1}^{1} x^m (1-x^2)^k dx = \begin{cases} 0, & \text{if } m \equiv 1 \pmod 2, \\ \frac{2^{2k+2} k! m! (k+\frac{m}{2}+1)!}{(\frac{m}{2})! (2k+m+2)!}, & \text{if } m \equiv 0 \pmod 2, \end{cases}$$

$$= 2^{2k+1} k! \sum_{s=0}^{m} \binom{m}{s} 2^s (-1)^{m-s} \frac{(k+s)!}{(2k+s+1)!},$$

(c)
$$\int_{-1}^{1} x^m (1-x^2)^{k+\lambda-\frac{1}{2}} dx = \begin{cases} 0, & \text{if } m \equiv 1 \pmod 2, \\ \frac{\Gamma(k+\lambda+\frac{1}{2})\Gamma(\frac{m}{2}+\frac{1}{2})}{\Gamma(k+\lambda+\frac{m}{2}+1)}, & \text{if } m \equiv 0 \pmod 2, \end{cases}$$

(d)
$$\int_{-1}^{1} x^m (1-x)^{k+\alpha}(1+x)^{k+\beta} dx = 2^{2k+\alpha+\beta+1} \sum_{s=0}^{m} \binom{m}{s} (-1)^{m-s} 2^s$$

$$\times \frac{\Gamma(k+\alpha+1)\Gamma(k+\beta+s+1)}{\Gamma(2k+\alpha+\beta+s+2)}.$$

Differentiation of (10) gives us the following lemma.

Lemma 1. *For any nonnegative integers n and r, we have the following identity.*

$$\sum_{i_1+i_2+\ldots i_{2r+1}=n} P_{i_1}(x), P_{i_2}(x), \ldots, P_{i_{2r+1}}(x) = \frac{1}{(2r-1)!!} P_{n+r}^{(r)}(x), \quad (48)$$

where the sum is over all nonnegative integers $i_1, i_2, \ldots, i_{2r+1}$, with $i_1 + i_2 + \ldots i_{2r+1} = n$.
By taking rth derivative of (16), we have

$$P_n^{(r)}(x) = \frac{1}{2^n} \sum_{l=0}^{[\frac{n-r}{2}]} (-1)^l \binom{n}{l} \binom{2n-2l}{n} (n-2l)_r x^{n-2l-r}. \quad (49)$$

Actually, we need the following particular case of (49).

$$P_{n+r}^{(r+k)}(x) = \frac{1}{2^{n+r}} \sum_{l=0}^{[\frac{n-k}{2}]} (-1)^l \binom{n+r}{l} \binom{2n+2r-2l}{n+r}$$

$$\times (n+r-2l)_{r+k} x^{n-2l-k}. \quad (50)$$

Here we are going to show (35), (36) and (38), leaving the other two (37) and (39) as exercises.

With $\gamma_{n,r}(x)$ as in (32), let us put

$$\gamma_{n,r}(x) = \sum_{k=0}^{n} C_{k,1} H_k(x). \tag{51}$$

Then, from (a) of Proposition 1, (48), (50), and by integrating by parts k times, we have

$$\begin{aligned}
C_{k,1} &= \frac{(-1)^k}{2^k k! \sqrt{\pi}} \int_{-\infty}^{\infty} \gamma_{n,r}(x) \frac{d^k}{dx^k} e^{-x^2} dx \\
&= \frac{(-1)^k}{2^k k! \sqrt{\pi} (2r-1)!!} \int_{-\infty}^{\infty} P_{n+r}^{(r)}(x) \frac{d^k}{dx^k} e^{-x^2} dx \\
&= \frac{1}{2^k k! \sqrt{\pi} (2r-1)!!} \int_{-\infty}^{\infty} P_{n+r}^{(r+k)}(x) e^{-x^2} dx \\
&= \frac{1}{2^{k+n+r} k! \sqrt{\pi} (2r-1)!!} \\
&\quad \times \sum_{l=0}^{[\frac{n-k}{2}]} (-1)^l \binom{n+r}{l} \binom{2n+2r-2l}{n+r} (n+r-2l)_{r+k} \\
&\quad \times \int_{-\infty}^{\infty} x^{n-2l-k} e^{-x^2} dx.
\end{aligned} \tag{52}$$

From (52) and making use of (a) of Proposition 2, we obtain

$$\begin{aligned}
C_{k,1} &= \frac{1}{2^{k+n+r} k! \sqrt{\pi} (2r-1)!!} \\
&\quad \times \sum_{l=0}^{[\frac{n-k}{2}]} (-1)^l \binom{n+r}{l} \binom{2n+2r-2l}{n+r} (n+r-2l)_{r+k} \\
&\quad \times \begin{cases} 0, & \text{if } k \not\equiv n \pmod{2}, \\ \frac{(n-k-2l)! \sqrt{\pi}}{2^{n-k-2l} (\frac{n-k}{2}-l)!}, & \text{if } k \equiv n \pmod{2}. \end{cases}
\end{aligned} \tag{53}$$

Now, from (51) and (53) and after some simplifications,

$$\begin{aligned}
\gamma_{n,r}(x) &= \frac{1}{2^{2n+r}(2r-1)!!} \sum_{\substack{0 \le k \le n \\ k \equiv n \pmod{2}}} \frac{1}{k!} \\
&\quad \times \sum_{l=0}^{[\frac{n-k}{2}]} \frac{(-4)^l (2n+2r-2l)!}{l!(n+r-l)!(\frac{n-k}{2}-l)!} H_k(x) \\
&= \frac{1}{2^{2n+r}(2r-1)!!} \sum_{j=0}^{[\frac{n}{2}]} \frac{1}{j!(n-2j)!} H_{n-2j}(x) \\
&\quad \times \sum_{l=0}^{j} \frac{(-4)^l (j)_l (2n+2r-2l)!}{l!(n+r-l)!} \\
&= \frac{2^r (n+r-\frac{1}{2})_{n+r}}{(2r-1)!!} \sum_{j=0}^{[\frac{n}{2}]} \frac{1}{j!(n-2j)!} H_{n-2j}(x) \\
&\quad \times \sum_{l=0}^{j} \frac{(-1)^l \langle -j \rangle_l}{l! \langle \frac{1}{2}-n-r \rangle_l} \\
&= \frac{2^r (n+r-\frac{1}{2})_{n+r}}{(2r-1)!!} \sum_{j=0}^{[\frac{n}{2}]} \frac{{}_1F_1(-j; \frac{1}{2}-n-r; -1)}{j!(n-2j)!} H_{n-2j}(x).
\end{aligned} \tag{54}$$

This shows (35) in Theorem 1.

Next, we put

$$\gamma_{n,r}(x) = \sum_{k=0}^{n} C_{k,2} L_k^\alpha(x). \tag{55}$$

Then, from (b) of Proposition 1, (48), (50) and integration by parts k times, we get

$$C_{k,2} = \frac{(-1)^k}{\Gamma(\alpha+k+1)(2r-1)!!} \int_0^\infty P_{n+r}^{(r+k)}(x) e^{-x} x^{k+\alpha} dx$$

$$= \frac{(-1)^k}{\Gamma(\alpha+k+1)(2r-1)!! \, 2^{n+r}} \sum_{l=0}^{[\frac{n-k}{2}]} (-1)^l \binom{n+r}{l} \binom{2n+2r-2l}{n+r}$$
$$\times (n+r-2l)_{r+k} \, \Gamma(n-2l+\alpha+1) \tag{56}$$

$$= \frac{(-1)^k}{\Gamma(\alpha+k+1)(2r-1)!! \, 2^{n+r}}$$
$$\times \sum_{l=0}^{[\frac{n-k}{2}]} \frac{(-1)^l (2n+2r-2l)! \, \Gamma(n-2l+\alpha+1)}{l!(n+r-l)!(n-k-2l)!}.$$

Combining (55) and (56), and changing order of summation, we immediately have

$$\gamma_{n,r}(x) = \frac{1}{(2r-1)!! \, 2^{n+r}} \sum_{k=0}^{n} \frac{(-1)^k}{\Gamma(\alpha+k+1)}$$
$$\times \sum_{l=0}^{[\frac{n-k}{2}]} \frac{(-1)^l (2n+2r-2l)! \, \Gamma(n-2l+\alpha+1)}{l!(n+r-l)!(n-k-2l)!} L_k^\alpha(x)$$
$$= \frac{1}{\Gamma(\alpha+1)(2r-1)!! \, 2^{n+r}} \sum_{l=0}^{[\frac{n}{2}]} \frac{(-1)^l (2n+2r-2l)! \Gamma(n-2l+\alpha+1)}{l!(n+r-l)!(n-2l)!}$$
$$\times \sum_{k=0}^{n-2l} \frac{\langle 2l-n\rangle_k}{\langle \alpha+1\rangle_k} L_k^\alpha(x). \tag{57}$$

This yields (36) in Theorem 1.

Finally, we let

$$\gamma_{n,r}(x) = \sum_{k=0}^{n} C_{k,4} C_k^{(\lambda)}(x) \tag{58}$$

Then, from (d) of Proposition 1, (48), (50), integration by parts k times and making use of (c) of Proposition 2, we have

$$C_{k,4} = \frac{(k+\lambda)\Gamma(\lambda)(-1)^k}{(-2)^k \sqrt{\pi} \Gamma(k+\lambda+\frac{1}{2})(2r-1)!!}$$
$$\times \int_{-1}^{1} P_{n+r}^{(r+k)}(x)(1-x^2)^{k+\lambda-\frac{1}{2}} dx$$

$$= \frac{(k+\lambda)\Gamma(\lambda)}{2^{k+n+r}\sqrt{\pi}\Gamma(k+\lambda+\frac{1}{2})(2r-1)!!} \tag{59}$$
$$\times \sum_{l=0}^{[\frac{n-k}{2}]} (-1)^l \binom{n+r}{l}\binom{2n+2r-2l}{n+r}(n+r-2l)_{r+k}$$
$$\times \begin{cases} 0, & \text{if } k \not\equiv n \pmod 2, \\ \frac{\Gamma(k+\lambda+\frac{1}{2})\Gamma(\frac{n-k+1}{2}-l)}{\Gamma(\frac{n+k}{2}+\lambda-l+1)}, & \text{if } k \equiv n \pmod 2. \end{cases}$$

From (58) and (59), exploiting (3), (4), (6) and (7), and after some simplifications, we finally derive

$$\gamma_{n,r}(x) = \frac{\Gamma(\lambda)}{\sqrt{\pi}(2r-1)!!2^{n+r}} \sum_{\substack{0 \leq k \leq n \\ k \equiv n \pmod{2}}} \frac{(k+\lambda)}{2^k}$$

$$\times \sum_{l=0}^{[\frac{n-k}{2}]} \frac{(-1)^l (2n+2r-2l)! \Gamma(\frac{n-k+1}{2} - l)}{l!(n+r-l)!(n-k-2l)! \Gamma(\frac{k+n}{2} + \lambda - l + 1)} C_k^{(\lambda)}(x)$$

$$= \frac{\Gamma(\lambda)}{\sqrt{\pi}(2r-1)!!2^{n+r}} \sum_{j=0}^{[\frac{n}{2}]} \frac{(n-2j+\lambda)}{2^{n-2j}} \qquad (60)$$

$$\times \sum_{l=0}^{j} \frac{(-1)^l (2n+2r-2l)! \Gamma(j-l+\frac{1}{2})}{l!(n+r-l)!(2j-2l)! \Gamma(n+\lambda-j-l+1)} C_{n-2j}^{(\lambda)}(x)$$

$$= \frac{\Gamma(\lambda)}{(2r-1)!! \, 2^{2n+r}} \sum_{j=0}^{[\frac{n}{2}]} \frac{(n-2j+\lambda)}{\Gamma(n+\lambda-j+1)}$$

$$\times \sum_{l=0}^{j} \frac{(-4)^l (2n+2r-2l)! (n+\lambda-j)_l}{l!(n+r-l)!(j-l)!} C_{n-2j}^{(\lambda)}(x)$$

$$= \frac{2^r \Gamma(\lambda)(n+r-\frac{1}{2})_{n+r}}{(2r-1)!!} \sum_{j=0}^{[\frac{n}{2}]} \frac{(n-2j+\lambda)}{\Gamma(n+\lambda-j+1) j!}$$

$$\times \sum_{l=0}^{j} \frac{(-j)_l (j-n-r)_l}{l! \left\langle \frac{1}{2} - n - r \right\rangle_l} C_{n-2j}^{(\lambda)}(x)$$

$$= \frac{2^r \Gamma(\lambda)(n+r-\frac{1}{2})_{n+r}}{(2r-1)!!}$$

$$\times \sum_{j=0}^{[\frac{n}{2}]} \frac{(n-2j+\lambda) \, {}_2F_1(-j, j-n-r; \frac{1}{2} - n - r; 1)}{\Gamma(n+\lambda-j+1) j!} C_{n-2j}^{(\lambda)}(x).$$

This completes the proof for (38) in Theorem 1.

4. Proof of Theorem 2

The proofs for (42), (43) and (45) are left to the reader as an exercise and we will show only (44) and (46) in Theorem 2.

The following lemma is important for our discussion in this section and can be derived by differentiating (11).

Lemma 2. *Let n, r be nonnegative integers. Then we have the following identity.*

$$\sum_{i_1+i_2+\cdots+i_{r+1}=n} L_{i_1}\left(\frac{x}{r+1}\right) L_{i_2}\left(\frac{x}{r+1}\right) \cdots L_{i_{r+1}}\left(\frac{x}{r+1}\right) = (-1)^r L_{n+r}^{(r)}(x), \qquad (61)$$

where the sum runs over all nonnegative integers $i_1, i_2, \ldots, i_{r+1}$, with $i_1 + i_2 + \cdots + i_{r+1} = n$.

From (17), it is immediate to see that the rth derivative of $L_n(x)$ is given by

$$L_n^{(r)}(x) = \sum_{l=0}^{n-r} (-1)^{n-l} \binom{n}{l} \frac{1}{(n-l-r)!} x^{n-l-r}. \qquad (62)$$

In particular, we have

$$L_{n+r}^{(r+k)}(x) = \sum_{l=0}^{n-k}(-1)^{n+r-l}\binom{n+r}{l}\frac{1}{(n-k-l)!}x^{n-k-l}. \tag{63}$$

With $\varepsilon_{n,r}(x)$ as in (33), let us set

$$\varepsilon_{n,r}(x) = \sum_{k=0}^{n} C_{k,3} P_k(x). \tag{64}$$

Then, from (c) of Proposition 1, (61), (63), by integration by parts k times and using (b) of Proposition 2, we get

$$\begin{aligned}
C_{k,3} &= \frac{(2k+1)(-1)^{r+k}}{2^{k+1}k!}\int_{-1}^{1}L_{n+r}^{(r+k)}(x)(x^2-1)^k dx \\
&= \frac{(-1)^{n+k}(2k+1)(n+r)!}{2^{k+1}k!}\sum_{l=0}^{n-k}\frac{(-1)^l}{l!(n+r-l)!(n-k-l)!} \tag{65} \\
&\quad \times \begin{cases} 0, & \text{if } l \not\equiv n-k \ (\text{mod } 2) \\ \frac{2^{2k+2}k!(n-k-l)!(\frac{n+k-l}{2}+1)!}{(\frac{n-k-l}{2})!(n+k-l+2)!}, & \text{if } l \equiv n-k \ (\text{mod } 2) \end{cases} \\
&= (-1)^{n+k}(2k+1)2^{k+1}(n+r)! \\
&\quad \times \sum_{\substack{0\le l\le n-k \\ l\equiv n-k \ (\text{mod } 2)}} \frac{(-1)^l(\frac{n+k-l}{2}+1)!}{l!(n+r-l)!(\frac{n-k-l}{2})!(n+k-l+2)!} \\
&= (n+r)!(2k+1)2^{k+1} \\
&\quad \times \sum_{j=0}^{[\frac{n-k}{2}]}\frac{(k+j+1)!}{j!(n-k-2j)!(r+k+2j)!(2k+2j+2)!}.
\end{aligned}$$

By combining (64) and (65) we get the following result.

$$\begin{aligned}
\varepsilon_{n,r}(x) &= (n+r)!\sum_{k=0}^{n}(2k+1)\,2^{k+1} \\
&\quad \times \sum_{j=0}^{[\frac{n-k}{2}]}\frac{(k+j+1)!}{j!(n-k-2j)!(r+k+2j)!(2k+2j+2)!}P_k(x).
\end{aligned} \tag{66}$$

This completes the proof for (44).
Finally, we put

$$\varepsilon_{n,r}(x) = \sum_{k=0}^{n} C_{k,5}\, P_k^{(\alpha,\beta)}(x). \tag{67}$$

Then, from (e) of Proposition 1, (61), (63), integration by parts k times and exploiting (d) of Proposition 2, we have

$$C_{k,5} = \frac{(-1)^r (2k+\alpha+\beta+1)\,\Gamma(k+\alpha+\beta+1)}{2^{\alpha+\beta+k+1}\Gamma(\alpha+k+1)\Gamma(\beta+k+1)}$$

$$\times \int_{-1}^{1} L_{n+r}^{(r+k)}(x)(1-x)^{k+\alpha}(1+x)^{k+\beta}dx$$

$$= \frac{(-1)^r (2k+\alpha+\beta+1)\Gamma(k+\alpha+\beta+1)}{2^{\alpha+\beta+k+1}\Gamma(\alpha+k+1)\Gamma(\beta+k+1)}$$

$$\times \sum_{l=0}^{n-k}(-1)^{n+r-l}\binom{n+r}{l}\frac{1}{(n-k-l)!}$$

$$\times \int_{-1}^{1} x^{n-k-l}(1-x)^{k+\alpha}(1+x)^{k+\beta}dx$$

$$= \frac{(n+r)!\,(-2)^k(2k+\alpha+\beta+1)\Gamma(k+\alpha+\beta+1)}{\Gamma(\beta+k+1)} \qquad (68)$$

$$\times \sum_{l=0}^{n-k}\frac{1}{l!\,(n+r-l)!}\sum_{s=0}^{n-k-l}\frac{(-2)^s\,\Gamma(k+\beta+s+1)}{s!\,(n-k-l-s)!\,\Gamma(2k+\alpha+\beta+s+2)}$$

$$= \frac{(n+r)!\,(-2)^k\Gamma(k+\alpha+\beta+1)}{\Gamma(2k+\alpha+\beta+1)}$$

$$\times \sum_{l=0}^{n-k}\frac{1}{l!\,(n+r-l)!\,(n-k-l)!}\sum_{s=0}^{n-k-l}\frac{2^s\,\langle k+l-n\rangle_s\,\langle k+\beta+1\rangle_s}{s!\,\langle 2k+\alpha+\beta+2\rangle_s}$$

$$= \frac{(n+r)!\,(-2)^k\Gamma(k+\alpha+\beta+1)}{\Gamma(2k+\alpha+\beta+1)}$$

$$\times \sum_{l=0}^{n-k}\frac{{}_2F_1(k+l-n,k+\beta+1;2k+\alpha+\beta+2;2)}{l!\,(n+r-l)!\,(n-k-l)!}$$

We now obtain

$$\varepsilon_{n,r}(x) = (n+r)!\sum_{k=0}^{n}\frac{(-2)^k\,\Gamma(k+\alpha+\beta+1)}{\Gamma(2k+\alpha+\beta+1)}$$

$$\times \sum_{l=0}^{n-k}\frac{{}_2F_1(k+l-n,k+\beta+1;2k+\alpha+\beta+2;2)}{l!\,(n+r-l)!\,(n-k-l)!}P_k^{(\alpha,\beta)}(x). \qquad (69)$$

This verifies (46) in Theorem 2.

5. Conclusions

Let $\gamma_{m,r}(x)$, $\varepsilon_{m,r}(x)$, and $\alpha_m(x)$ denote the following sums of finite products given by

$$\gamma_{n,r}(x) = \sum_{i_1+\cdots+i_{2r+1}=n} P_{i_1}(x)P_{i_2}(x)\ldots P_{i_{2r+1}}(x),$$

$$\varepsilon_{n,r}(x) = \sum_{i_1+\cdots+i_{r+1}=n} L_{i_1}\left(\frac{x}{r+1}\right)L_{i_2}\left(\frac{x}{r+1}\right)\ldots L_{i_{r+1}}\left(\frac{x}{r+1}\right),$$

$$\alpha_m(x) = \sum_{k=1}^{m-1}\frac{1}{k(m-k)}B_k(x)B_{m-k}(x),\;(m\geq 2),$$

where $P_n(x)$, $L_n(x)$, $B_n(x)$, $(n \geq 0)$ are respectively Legendre, Laguerre and Bernoulli polynomials. In this paper, we studied sums of finite products of Legendre polynomials $\gamma_{m,r}(x)$ and those of Laguerre polynomials $\varepsilon_{m,r}(x)$, and expressed them as linear combinations of the orthogonal polynomials $H_n(x)$, $L_n^\alpha(x)$, $P_n(x)$, $C_n^{(\lambda)}(x)$, and $P_n^{(\alpha,\beta)}(x)$. These have been done by carrying out explicit computations. In recent years, we have obtained similar results for many other special polynomials.

For example, we considered sums of finite products of Bernoulli, Euler and Genocchi polynomials and represented them in terms of Bernoulli polynomials. In addition, as for Chebyshev polynomials of the second, third, and fourth kinds, and Fibonacci, Legendre and Laguerre polynomials, we expressed them not only in terms of Bernoulli polynomials but also of Chebyshev polynomials of all kinds and Hermite, generalized Laguerre, Legendre, Gegenbauer and Jacobi polynomials.

We gave twofold justification for studying such sums of finite products of special polynomials. Firstly, it can be viewed as a generalization of the classical connection problem in which one wants to determine the connection coefficients in the expansion of a product of two polynomials in terms of any given sequence of polynomials. Secondly, from the representation of $\alpha_m(x)$ in terms of Bernoulli polynomials we can derive the famous Faber-Pandharipande-Zagier identity and a slightly different variant of the Miki's identity. We emphasized that these identities had been obtained by several different methods which are quite involved and not elementary, while our previous method used only elementary Fourier series expansions.

Along the same line of the present paper, we would like to continue to work on representing sums of finite products of some special polynomials in terms of various kinds of special polynomials and to find interesting applications of them in mathematics, science and engineering areas.

Author Contributions: All authors contributed equally to the manuscript and typed, read, and approved the final manuscript.

Funding: This work was supported by the Dong-A University research Fund.

Conflicts of Interest: The authors declare no conflict of interest.

References

1. Andrews, G.E.; Askey, R.; Roy, R. *Special Functions, Encyclopedia of Mathematics and Its Applications*; Cambridge University Press: Cambridge, UK, 1999.
2. Beals, R.; Wong, R. Special functions and orthogonal polynomials. In *Cambridge Studies in Advanced Mathematics*; Cambridge University Press: Cambridge, UK, 2016.
3. Kim, D.S.; Kim, T.; Dolgy, D.V. Some identities on Laguerre polynomials in connection with Bernoulli and Euler numbers. *Discret. Dyn. Nat. Soc.* **2012**, *2012*, 619197. [CrossRef]
4. Kim, D.S.; Kim, T.; Rim, S. Some identities involving Gegenbauer polynomials. *Adv. Differ. Equ.* **2012**, *2012*, 219. [CrossRef]
5. Kim, D.S.; Kim, T.; Rim, S.; Lee, S.H. Hermite polynomials and their applications associated with Bernoulli and Euler numbers. *Discret. Dyn. Nat. Soc.* **2012**, *2012*, 974632. [CrossRef]
6. Kim, D.S.; Rim, S.; Kim, T. Some identities on Bernoulli and Euler polynomials arising from orthogonality of Legendre polynomials. *J. Inequal. Appl.* **2012**, *2012*, 227. [CrossRef]
7. Kim, T.; Kim, D.S. Extended Laguerre polynomials associated with Hermite, Bernoulli, and Euler numbers and polynomials. *Abstr. Appl. Anal.* **2012**, *2012*, 957350. [CrossRef]
8. Kim, T.; Kim, D.S.; Dolgy, D.V. Some identities on Bernoulli and Hermite polynomials associated with Jacobi polynomials. *Discret. Dyn. Nat. Soc.* **2012**, *2012*, 584643. [CrossRef]
9. Kim, T.; Kim, D.S.; Dolgy, D.V.; Park, J.-W. Sums of finite products of Chebyshev polynomials of the second kind and of Fibonacci polynomials. *J. Inequal. Appl.* **2018**, *2018*, 148. [CrossRef] [PubMed]
10. Kim, T.; Kim, D.S.; Dolgy, D.V.; Park, J.-W. Sums of finite products of Legendre and Laguerre polynomials. *Adv. Differ. Equ.* **2018**, *2018*, 277. [CrossRef]
11. Kim, T.; Kim, D.S.; Dolgy, D.V.; Ryoo, C.S. Representing sums of finite products of Chebyshev polynomials of third and fourth kinds by Chebyshev polynomials. *Symmetry* **2018**, *10*, 258. [CrossRef]
12. Kim, T.; Kim, D.S.; Jang, G.-W.; Kwon, J. Sums of finite products of Legendre and Laguerre polynomials by Chebyshev polynomials. *Adv. Stud. Contemp. Math.* **2018**, *28*, 551–565.
13. Kim, T.; Kim, D.S.; Kwon, J.; Dolgy, D.V. Expressing sums of finite products of Chebyshev polynomials of the second kind and of Fibonacci polynomials by several orthogonal polynomials. *Mathematics* **2018**, *6*, 210. [CrossRef]

14. Cesarano, C.; Fornaro, C. A note on two-variable Chebyshev polynomials. *Georgian Math. J.* **2017**, *24*, 339–349. [CrossRef]
15. Cesarano, C.; Fornaro, C.; Vazquez, L. A note on a special class of Hermite polynomials. *Int. J. Pure Appl. Math.* **2015**, *98*, 261–273. [CrossRef]
16. Kim, T.; Kim, D.S.; Jang, L.C.; Jang, G. Sums of finite products of Genocchi functions. *Adv. Differ. Equ.* **2017**; *2017*, 268. [CrossRef]
17. Agarwal, R.P.; Kim, D.S.; Kim, T.; Kwon, J. Sums of finite products of Bernoulli functions. *Adv. Differ. Equ.* **2017**, *2017*, 237. [CrossRef]
18. Kim, T.; Kim, D.S.; Jang, G.-W.; Kwon, J. Sums of finite products of Euler functions. In *Advances in Real and Complex Analysis with Applications*; Trends in Mathematics; Springer: Berlin, Germany, 2017, pp. 243–260.
19. Kim, T.; Dolgy, D.V.; Kim, D.S. Representing sums of finite products of Chebyshev polynomials of second kind and Fibonacci polynomials in terms of Chebyshev polynomials. *Adv. Stud. Contemp. Math. (Kyungshang)* **2018**, *28*, 321–335.
20. Araci, S.; Acikgoz, M.; Bagdasaryan, A.; Sen, E. The Legendre polynomials associated with Bernoulli, Euler, Hermite and Bernstein polynomials. *Turk. J. Anal. Number Theory* **2013**, *1*, 1–3. [CrossRef]
21. Khan, W.A.; Araci, S.; Acikgoz, M. A new class of Laguerre-based Apostol type polynomials. *Cogent Math.* **2016**, *3*, 1243839. [CrossRef]

© 2019 by the authors. Licensee MDPI, Basel, Switzerland. This article is an open access article distributed under the terms and conditions of the Creative Commons Attribution (CC BY) license (http://creativecommons.org/licenses/by/4.0/).

Article

On Central Complete and Incomplete Bell Polynomials I

Taekyun Kim [1,*], Dae San Kim [2] and Gwan-Woo Jang [1]

1 Department of Mathematics, Kwangwoon University, Seoul 139-701, Korea; gwjang@kw.ac.kr
2 Department of Mathematics, Sogang University, Seoul 121-742, Korea; dskim@sogang.ac.kr
* Correspondence: tkkim@kw.ac.kr; Tel.: +82-(0)2-940-8368

Received: 27 December 2018; Accepted: 19 February 2019; Published: 22 February 2019

Abstract: In this paper, we introduce central complete and incomplete Bell polynomials which can be viewed as generalizations of central Bell polynomials and central factorial numbers of the second kind, and also as 'central' analogues for complete and incomplete Bell polynomials. Further, some properties and identities for these polynomials are investigated. In particular, we provide explicit formulas for the central complete and incomplete Bell polynomials related to central factorial numbers of the second kind.

Keywords: central incomplete Bell polynomials; central complete Bell polynomials; central complete Bell numbers

1. Introduction

In this paper, we introduce central incomplete Bell polynomials $T_{n,k}(x_1, x_2, \cdots, x_{n-k+1})$ given by

$$\frac{1}{k!}\Big(\sum_{m=1}^{\infty} \frac{1}{2^m}(x_m - (-1)^m x_m)\frac{t^m}{m!}\Big)^k = \sum_{n=k}^{\infty} T_{n,k}(x_1, x_2, \cdots, x_{n-k+1})\frac{t^n}{n!}$$

and central complete Bell polynomials $B_n^{(c)}(x|x_1, x_2, \cdots, x_n)$ given by

$$exp\Big(x\sum_{i=1}^{\infty} \frac{1}{2^i}(x_i - (-1)^i x_i)\frac{t^i}{i!}\Big) = \sum_{n=0}^{\infty} B_n^{(c)}(x|x_1, x_2, \cdots, x_n)\frac{t^n}{n!}$$

and investigate some properties and identities for these polynomials. They can be viewed as generalizations of central Bell polynomials and central factorial numbers of the second kind, and also as 'central' analogues for complete and incomplete Bell polynomials.

Here, we recall that the central factorial numbers $T(n,k)$ of the second kind and the central Bell polynomials $B_n^{(c)}(x)$ are given in terms of generating functions by

$$\frac{1}{k!}\big(e^{\frac{t}{2}} - e^{-\frac{t}{2}}\big)^k = \sum_{n=k}^{\infty} T(n,k)\frac{t^n}{n!}, \quad e^{x(e^{\frac{t}{2}} - e^{-\frac{t}{2}})} = \sum_{n=0}^{\infty} B_n^{(c)}(x)\frac{t^n}{n!},$$

so that $T_{n,k}(1,1,\cdots,1) = T(n,k)$ and $B_n^{(c)}(x|1,1,\cdots,1) = B_n^{(c)}(x)$.

The incomplete and complete Bell polynomials have applications in such diverse areas as combinatorics, probability, algebra, modules over a *-algebra (see [1,2]), quasi local algebra and analysis. Here, we recall some applications of them and related works. The incomplete Bell polynomials $B_{n,k}(x_1, x_2, \cdots, x_{n-k+1})$ (see [3,4]) arise naturally when we want to find higher-order derivatives of composite functions. Indeed, such higher-order derivatives can be expressed in terms of incomplete Bell polynomials, which is known as Faà di Bruno formula given as in the following (see [3]):

$$\frac{d^n}{dt^n} g(f(t)) = \sum_{k=0}^{n} g^{(k)}(f(t)) B_{n,k}(f'(t), f''(t), \cdots, f^{(n-k+1)}(t)).$$

For the curious history on this formula, we let the reader refer to [5].

In addition, the number of monomials appearing in $B_{n,k} = B_{n,k}(x_1, x_2, \cdots, x_{n-k+1})$ is the number of partitioning a set with n elements into k blocks and the coefficient of each monomial is the number of partitioning a set with n elements as the corresponding k blocks. For example,

$$B_{10,7} = 3150 x_2^3 x_1^4 + 2520 x_3 x_2 x_1^5 + 210 x_4 x_1^6$$

shows that there are three ways of partitioning a set with 10 elements into seven blocks, and 3150 partitions with blocks of size 2, 2, 2, 1, 1, 1, 1, 2520 partitions with blocks of size 3, 2, 1, 1, 1, 1, 1, and 210 partitions with blocks of size 4, 1,1, 1, 1, 1. This example is borrowed from [4], which gives a practical way of computing $B_{n,k}$ for any given n, k (see [4], (1.5)).

Furthermore, the incomplete Bell polynomials can be used in constructing sequences of binomial type (also called associated sequences). Indeed, for any given scalars $c_1, c_2, \cdots, c_n, \cdots$ the following form a sequence of binomial type

$$s_n(x) = \sum_{k=0}^{n} B_{n,k}(c_1, c_2, \cdots, c_{n-k+1}) x^k, \quad (n = 0, 1, 2, \cdots)$$

and, conversely, any sequence of binomial type arises in this way for some scalar sequence $c_1, c_2, \cdots, c_n, \cdots$. For these, the reader may want to look at the paper [6].

There are certain connections between incomplete Bell polynomials and combinatorial Hopf algebras such as the Hopf algebra of word symmetric functions, the Hopf algebra of symmetric functions, the Faà di Bruno algebra, etc. The details can be found in [7].

The complete Bell polynomials $B_n(x_1, x_2, \cdots, x_n)$ (see [3,8–10]) have applications to probability theory. Indeed, the nth moment $\mu_n = E[X^n]$ of the random variable X is the nth complete Bell polynomial in the first n cumulants. Namely,

$$\mu_n = B_n(\kappa_1, \kappa_2, \cdots, \kappa_n).$$

For many applications to probability theory and combinatorics, the reader can refer to the Ph. D. thesis of Port [10].

Many special numbers, like Stirling numbers of both kinds, Lah numbers and idempotent numbers, appear in many combinatorial and number theoretic identities involving complete and incomplete Bell polynomials. For these, the reader refers to [3,8].

The central factorial numbers have received less attention than Stirling numbers. However, according to [11], they are at least as important as Stirling numbers, said to be "as important as Bernoulli numbers, or even more so". A systematic treatment of these important numbers was given in [11], including their properties and applications to difference calculus, spline theory, and to approximation theory, etc. For some other related references on central factorial numbers, we let the reader refer to [1,2,12–14]. Here, we note that central Bell polynomials and central factorial numbers of the second kind are respectively 'central' analogues for Bell polynomials and Stirling numbers of the second kind. They have been studied recently in [13,15].

The complete Bell polynomials and the incomplete Bell polynomials are respectively mutivariate versions for Bell polynomials and Stirling numbers of the second kind. This paper deals with central complete and incomplete Bell polynomials which are 'central' analogues for the complete and incomplete Bell polynomials. In addition, they can be viewed as generalizations of central Bell polynomials and central factorial numbers of the second kind (see [15]). The outline of the paper is as follows. After giving an introduction to the present paper in Section 1, we review some known properties and results about Bell polynomials, and incomplete and complete Bell polynomials in Section 2. We state the new and main results of this paper in Section 3, where we introduce central incomplete and complete Bell polynomials and investigate some properties and identities for them. In particular, Theorems 1 and 3 give basic formulas for computing central incomplete Bell polynomials and central complete Bell polynomials, respectively. We remark that the number of monomials appearing in $T_{n,k}(x_1, 2x_2, \cdots, 2^{n-k}x_{n-k+1})$ is the number of partitioning a set with n elements into k blocks with odd sizes and the coefficient of each monomial is the number of partitioning a set with n elements as the corresponding k blocks with odd sizes. This is illustrated by an example. Furthermore, we give expressions for the central incomplete and complete Bell polynomials with some various special arguments and also for the connection between the two Bell polynomials. We defer more detailed study of the central incomplete and complete Bell polynomials to a later paper.

2. Preliminaries

The Stirling numbers of the second kind are given in terms of generating function by (see [3,16])

$$\frac{1}{k!}(e^t - 1)^k = \sum_{n=k}^{\infty} S_2(n,k)\frac{t^n}{n!}. \tag{1}$$

The Bell polynomials are also called Tochard polynomials or exponential polynomials and defined by (see [9,13,15,17])

$$e^{x(e^t-1)} = \sum_{n=0}^{\infty} B_n(x)\frac{t^n}{n!}. \tag{2}$$

From Equations (1) and (2), we immediately see that (see [3,18])

$$\begin{aligned} B_n(x) &= e^{-x} \sum_{k=0}^{\infty} \frac{k^n}{k!} x^k \\ &= \sum_{k=0}^{n} x^k S_2(n,k), \quad (n \geq 0). \end{aligned} \tag{3}$$

When $x = 1$, $B_n = B_n(1)$ are called Bell numbers.

The (exponential) incomplete Bell polynomials are also called (exponential) partial Bell polynomials and defined by the generating function (see [9,15])

$$\frac{1}{k!}\Big(\sum_{m=1}^{\infty} x_m \frac{t^m}{m!}\Big)^k = \sum_{n=k}^{\infty} B_{n,k}(x_1, \cdots, x_{n-k+1})\frac{t^n}{n!}, \quad (k \geq 0). \tag{4}$$

Thus, by Equation (4), we get

$$\begin{aligned} B_{n,k}(x_1, \cdots, x_{n-k+1}) &= \sum \frac{n!}{i_1! i_2! \cdots i_{n-k+1}!}\Big(\frac{x_1}{1!}\Big)^{i_1}\Big(\frac{x_2}{2!}\Big)^{i_2} \times \cdots \\ &\quad \times \Big(\frac{x_{n-k+1}}{(n-k+1)!}\Big)^{i_{n-k+1}}, \end{aligned} \tag{5}$$

where the summation runs over all integers $i_1, \cdots, i_{n-k+1} \geq 0$ such that $i_1 + i_2 + \cdots + i_{n-k+1} = k$ and $i_1 + 2i_2 + \cdots + (n-k+1)i_{n-k+1} = n$.

From (1) and (4), we easily see that

$$B_{n,k}\underbrace{(1,1,\cdots,1)}_{n-k+1-times} = S_2(n,k), \quad (n,k \geq 0). \tag{6}$$

We easily deduce from (5) the next two identities:

$$B_{n,k}(\alpha x_1, \alpha x_2, \cdots, \alpha x_{n-k+1}) = \alpha^k B_{n,k}(x_1, x_2, \cdots, x_{n-k+1}) \tag{7}$$

and

$$B_{n,k}(\alpha x_1, \alpha^2 x_2, \cdots, \alpha^{n-k+1} x_{n-k+1}) = \alpha^n B_{n,k}(x_1, x_2, \cdots, x_{n-k+1}), \tag{8}$$

where $\alpha \in \mathbb{R}$ (see [15]).

From (4), it is not difficult to note that

$$\sum_{n=k}^{\infty} B_{n,k}(x,1,0,0,\cdots,0)\frac{t^n}{n!} = \frac{1}{k!}\left(xt + \frac{t^2}{2}\right)^k$$

$$= \frac{t^k}{k!}\sum_{n=0}^{k}\binom{k}{n}\left(\frac{t}{2}\right)^n x^{k-n} \tag{9}$$

$$= \sum_{n=0}^{k}\frac{(n+k)!}{k!}\binom{k}{n}\frac{1}{2^n}x^{k-n}\frac{t^{n+k}}{(n+k)!},$$

and

$$\sum_{n=k}^{\infty} B_{n,k}(x,1,0,0,\cdots,0)\frac{t^n}{n!} = \sum_{n=0}^{\infty} B_{n+k,k}(x,1,0,\cdots,0)\frac{t^{n+k}}{(n+k)!}. \tag{10}$$

Combining (9) with (10), we have

$$B_{n+k,k}(x,1,0,\cdots,0) = \frac{(n+k)!}{k!}\binom{k}{n}\frac{1}{2^n}x^{k-n}, \quad (0 \leq n \leq k). \tag{11}$$

Replacing n by $n-k$ in (11) yields the following identity

$$B_{n,k}(x,1,0,\cdots,0) = \frac{n!}{k!}\binom{k}{n-k}x^{2k-n}\left(\frac{1}{2}\right)^{n-k}, \quad (k \leq n \leq 2k). \tag{12}$$

We recall here that the (exponential) complete Bell polynomials are defined by

$$\exp\left(\sum_{i=1}^{\infty} x_i \frac{t^i}{i!}\right) = \sum_{n=0}^{\infty} B_n(x_1, x_2, \cdots, x_n)\frac{t^n}{n!}. \tag{13}$$

Then, by (4) and (13), we get

$$B_n(x_1, x_2, \cdots, x_n) = \sum_{k=0}^{n} B_{n,k}(x_1, x_2, \cdots, x_{n-k+1}). \tag{14}$$

From (3), (6), (7) and (14), we have

$$B_n(x, x, \cdots, x) = \sum_{k=0}^{n} x^k B_{n,k}(1, 1, \cdots, 1)$$
$$= \sum_{k=0}^{n} x^k S_2(n, k) = B_n(x), \quad (n \geq 0). \tag{15}$$

We recall that the central factorial numbers of the second kind are given by (see [19,20])

$$\frac{1}{k!}\left(e^{\frac{t}{2}} - e^{-\frac{t}{2}}\right)^k = \sum_{n=k}^{\infty} T(n, k) \frac{t^n}{n!}, \tag{16}$$

where $k \geq 0$.

From (16), it is not difficult to derive the following expression

$$T(n, k) = \frac{1}{k!} \sum_{j=0}^{k} \binom{k}{j} (-1)^{k-j} \left(j - \frac{k}{2}\right)^n, \tag{17}$$

where $n, k \in \mathbb{Z}$ with $n \geq k \geq 0$, (see [16,20]).

In [20], the central Bell polynomials $B_n^{(c)}(x)$ are defined by

$$B_n^{(c)}(x) = \sum_{k=0}^{n} T(n, k) x^k, \quad (n \geq 0). \tag{18}$$

When $x = 1$, $B_n^{(c)} = B_n^{(c)}(1)$ are called the central Bell numbers.

It is not hard to derive the generating function for the central Bell polynomials from (18) as follows (see [15]):

$$e^{x\left(e^{\frac{t}{2}} - e^{-\frac{t}{2}}\right)} = \sum_{n=0}^{\infty} B_n^{(c)}(x) \frac{t^n}{n!}. \tag{19}$$

By making use of (19), the following Dobinski-like formula was obtained earlier in [15]:

$$B_n^{(c)}(x) = \sum_{l=0}^{\infty} \sum_{j=0}^{\infty} \binom{l+j}{j} (-1)^j \frac{1}{(l+j)!} \left(\frac{l}{2} - \frac{j}{2}\right)^n x^{l+j}, \tag{20}$$

where $n \geq 0$.

Motivated by (4) and (13), we will introduce central complete and incomplete Bell polynomials and investigate some properties and identities for these polynomials. Also, we present explicit formulas for the central complete and incomplete Bell polynomials related to central factorial numbers of the second kind.

3. On Central Complete and Incomplete Bell Polynomials

In view of (13), we may consider the *central incomplete Bell polynomials* which are given by

$$\frac{1}{k!}\left(\sum_{m=1}^{\infty} \frac{1}{2^m}(x_m - (-1)^m x_m) \frac{t^m}{m!}\right)^k = \sum_{n=k}^{\infty} T_{n,k}(x_1, x_2, \cdots, x_{n-k+1}) \frac{t^n}{n!}, \tag{21}$$

where $k = 0, 1, 2, 3, \cdots$.

For $n,k \geq 0$ with $n - k \equiv 0 \pmod{2}$, by (4) and (5), we get

$$T_{n,k}(x_1, x_2, \cdots, x_{n-k+1}) = \sum \frac{n!}{i_1! i_2! \cdots i_{n-k+1}!} \left(\frac{x_1}{1!}\right)^{i_1} \left(\frac{0}{2 \cdot 2!}\right)^{i_2} \\ \times \left(\frac{x_3}{2^2 \cdot 3!}\right)^{i_3} \cdots \left(\frac{x_{n-k+1}}{2^{n-k}(n-k+1)!}\right)^{i_{n-k+1}}, \tag{22}$$

where the summation is over all integers $i_1, i_2, \cdots, i_{n-k+1} \geq 0$ such that $i_1 + \cdots + i_{n-k+1} = k$ and $i_1 + 2i_2 + \cdots + (n-k+1)i_{n-k+1} = n$.

From (5) and (22), we note that

$$T_{n,k}(x_1, x_2, \cdots, x_{n-k+1}) = B_{n,k}\left(x_1, 0, \frac{x_3}{2^2}, 0, \cdots, \frac{x_{n-k+1}}{2^{n-k}}\right), \tag{23}$$

where $n, k \geq 0$ with $n - k \equiv 0 \pmod{2}$ and $n \geq k$.
Therefore, from (22) and (23), we obtain the following theorem.

Theorem 1. *For $n, k \geq 0$ with $n \geq k$ and $n - k \equiv 0 \pmod{2}$, we have*

$$T_{n,k}(x_1, x_2, \cdots, x_{n-k+1}) = B_{n,k}\left(x_1, 0, \frac{x_3}{2^2}, 0, \cdots, \frac{x_{n-k+1}}{2^{n-k}}\right) \\ = \sum \frac{n!}{i_1! i_3! \cdots i_{n-k+1}!} \left(\frac{x_1}{1!}\right)^{i_1} \left(\frac{x_3}{2^2 \cdot 3!}\right)^{i_3} \times \cdots \times \left(\frac{x_{n-k+1}}{2^{n-k}(n-k+1)!}\right)^{i_{n-k+1}}, \tag{24}$$

where the summation is over all integers $i_1, i_2, \cdots, i_{n-k+1} \geq 0$ such that $i_1 + i_3 + \cdots + i_{n-k+1} = k$ and $i_1 + 3i_3 + \cdots + (n-k+1)i_{n-k+1} = n$.

Remark 1. *Theorem 1 shows in particular that we have*

$$T_{n,k}(x_1, 2x_2, \cdots, 2^{n-k} x_{n-k+1}) = B_{n,k}(x_1, 0, x_3, 0, \cdots, x_{n-k+1}).$$

From this, we note that the number of monomials appearing in $T_{n,k}(x_1, 2x_2, \cdots, 2^{n-k} x_{n-k+1})$ is the number of partitioning a set with n elements into k blocks with odd sizes and the coefficient of each monomial is the number of partitioning a set with n elements as the corresponding k blocks with odd sizes. For example, from the example in Section 3 of [4], we have

$$T_{13,7}(x_1, 2x_2, 2^2 x_3, 2^3 x_4, 2^4 x_5, 2^5 x_6, 2^6 x_7) = 200,200 x_3^3 x_1^4 + 72,072 x_5 x_3 x_1^5 + 1716 x_7 x_1^6.$$

Thus, there are three ways of partitioning a set with 13 elements into seven blocks with odd sizes, and 200,200 partitions with blocks of size 3, 3, 3, 1, 1, 1, 1, 72,072 partitions with blocks of size 5, 3, 1, 1, 1, 1, 1, and 1716 partitions with blocks of size 7, 1, 1, 1, 1, 1, 1.

For $n, k \geq 0$ with $n \geq k$ and $n - k \equiv 0 \pmod 2$, by (21), we get

$$\sum_{n=k}^{\infty} T_{n,k}(x, x^2, x^3, \cdots, x^{n-k+1}) \frac{t^n}{n!} = \frac{1}{k!}\left(xt + \frac{x^3 t^3}{2^2 3!} + \frac{x^5 t^5}{2^4 5!} + \cdots\right)^k$$

$$= \frac{1}{k!}\left(e^{\frac{x}{2}t} - e^{-\frac{x}{2}t}\right)^k = \frac{1}{k!}e^{-\frac{kx}{2}t}\left(e^{xt} - 1\right)^k$$

$$= \frac{1}{k!}\sum_{l=0}^{k}\binom{k}{l}(-1)^{k-l}e^{(l-\frac{k}{2})xt} \quad (25)$$

$$= \frac{1}{k!}\sum_{l=0}^{k}\binom{k}{l}(-1)^{k-l}\sum_{n=0}^{\infty}\left(l - \frac{k}{2}\right)^n x^n \frac{t^n}{n!}$$

$$= \sum_{n=0}^{\infty}\left(\frac{x^n}{k!}\sum_{l=0}^{k}\binom{k}{l}(-1)^{k-l}\left(l - \frac{k}{2}\right)^n\right)\frac{t^n}{n!}.$$

Now, the next theorem follows by comparing the coefficients on both sides of (25).

Theorem 2. *For $n, k \geq 0$ with $n - k \equiv 0 \pmod 2$, we have*

$$\frac{x^n}{k!}\sum_{l=0}^{k}\binom{k}{l}(-1)^{k-l}\left(l - \frac{k}{2}\right)^n = \begin{cases} T_{n,k}(x, x^2, \cdots, x^{n-k+1}), & \text{if } n \geq k, \\ 0, & \text{if } n < k. \end{cases} \quad (26)$$

In particular,

$$\frac{1}{k!}\sum_{l=0}^{k}\binom{k}{l}(-1)^{k-l}\left(l - \frac{k}{2}\right)^n = \begin{cases} T_{n,k}(1, 1, \cdots, 1), & \text{if } n \geq k, \\ 0, & \text{if } n < k. \end{cases} \quad (27)$$

For $n, k \geq 0$ with $n - k \equiv 0 \pmod 2$ and $n \geq k$, by (17) and (27), we get

$$T_{n,k}(1, 1, \cdots, 1) = T(n, k). \quad (28)$$

Therefore, by (26)–(28) and Theorem 1, we obtain the following corollary

Corollary 1. *For $n, k \geq 0$ with $n - k \equiv 0 \pmod 2$, $n \geq k$, we have*

$$T_{n,k}(x, x^2, \cdots, x^{n-k+1}) = x^n T_{n,k}(1, 1, \cdots, 1)$$

and

$$T_{n,k}(1, 1, \cdots, 1) = T(n, k) = B_{n,k}\left(1, 0, \frac{1}{2^2}, \cdots, \frac{1}{2^{n-k}}\right)$$

$$= \sum \frac{n!}{i_1! i_3! \cdots i_{n-k+1}!}\left(\frac{1}{1!}\right)^{i_1}\left(\frac{1}{2^2 3!}\right)^{i_3}\cdots\left(\frac{1}{2^{n-k}(n-k+1)!}\right)^{i_{n-k+1}},$$

where $i_1 + i_3 + \cdots + i_{n-k+1} = k$ and $i_1 + 3i_3 + \cdots + (n-k+1)i_{n-k+1} = n$.

For $n, k \geq 0$ with $n \geq k$ and $n - k \equiv 0 \pmod 2$, we observe that

$$\sum_{n=k}^{\infty} T_{n,k}(x, 1, 0, 0, \cdots, 0)\frac{t^n}{n!} = \frac{1}{k!}(xt)^k. \quad (29)$$

Thus, we have
$$T_{n,k}(x,1,0,\cdots,0) = x^k \binom{0}{n-k}.$$

The next two identities follow easily from (24):
$$T_{n,k}(x,x,\cdots,x) = x^k T_{n,k}(1,1,\cdots,1), \qquad (30)$$

and
$$T_{n,k}(\alpha x_1, \alpha x_2, \cdots, \alpha x_{n-k+1}) = \alpha^k T_{n,k}(x_1, x_2, \cdots, x_{n-k+1}),$$

where $n,k \geq 0$ with $n - k \equiv 0 \pmod{2}$ and $n \geq k$.

Now, we observe that
$$\exp\left(x \sum_{i=1}^{\infty} \left(\tfrac{1}{2}\right)^i (x_i - (-1)^i x_i) \frac{t^i}{i!}\right)$$
$$= \sum_{k=0}^{\infty} x^k \frac{1}{k!} \left(\sum_{i=1}^{\infty} \left(\tfrac{1}{2}\right)^i (x_i - (-1)^i x_i) \frac{t^i}{i!}\right)^k$$
$$= 1 + \sum_{k=1}^{\infty} x^k \frac{1}{k!} \left(\sum_{i=1}^{\infty} \left(\tfrac{1}{2}\right)^i (x_i - (-1)^i x_i) \frac{t^i}{i!}\right)^k \qquad (31)$$
$$= 1 + \sum_{k=1}^{\infty} x^k \sum_{n=k}^{\infty} T_{n,k}(x_1, x_2, \cdots, x_{n-k+1}) \frac{t^n}{n!}$$
$$= 1 + \sum_{n=1}^{\infty} \left(\sum_{k=1}^{n} x^k T_{n,k}(x_1, x_2, \cdots, x_{n-k+1})\right) \frac{t^n}{n!}.$$

In view of (13), it is natural to define the *central complete Bell polynomials* by
$$\exp\left(x \sum_{i=1}^{\infty} \left(\tfrac{1}{2}\right)^i (x_i - (-1)^i x_i) \frac{t^i}{i!}\right) = \sum_{n=0}^{\infty} B_n^{(c)}(x|x_1, x_2, \cdots, x_n) \frac{t^n}{n!}. \qquad (32)$$

Thus, by (31) and (32), we get
$$B_n^{(c)}(x|x_1, x_2, \cdots, x_n) = \sum_{k=0}^{n} x^k T_{n,k}(x_1, x_2, \cdots, x_{n-k+1}). \qquad (33)$$

When $x = 1$, $B_n^{(c)}(1|x_1, x_2, \cdots, x_n) = B_n^{(c)}(x_1, x_2, \cdots, x_n)$ are called the *central complete Bell numbers*.

For $n \geq 0$, we have
$$B_n^{(c)}(x_1, x_2, \cdots, x_n) = \sum_{k=0}^{n} T_{n,k}(x_1, x_2, \cdots, x_{n-k+1}) \qquad (34)$$

and
$$B_0^{(c)}(x_1, x_2, \cdots, x_n) = 1.$$

By (18) and (33), we get

$$B_n^{(c)}(1,1,\cdots,1) = \sum_{k=0}^{n} T_{n,k}(1,1,\cdots,1) = \sum_{k=0}^{n} T(n,k) = B_n^{(c)}, \qquad (35)$$

and

$$B_n^{(c)}(x|1,1,\cdots,1) = \sum_{k=0}^{n} x^k T_{n,k}(1,1,\cdots,1) = \sum_{k=0}^{n} x^k T(n,k) = B_n^{(c)}(x). \qquad (36)$$

From (31), we note that

$$\exp\left(\sum_{i=1}^{\infty} \left(\frac{1}{2}\right)^i (x_i - (-1)^i x_i)\frac{t^i}{i!}\right)$$

$$= 1 + \sum_{n=1}^{\infty} \frac{1}{n!}\left(\sum_{i=1}^{\infty} \left(\frac{1}{2}\right)^i (x_i - (-1)^i x_i)\frac{t^i}{i!}\right)^n$$

$$= 1 + \frac{1}{1!}\sum_{i=1}^{\infty} \left(\frac{1}{2}\right)^i (x_i - (-1)^i x_i)\frac{t^i}{i!} + \frac{1}{2!}\left(\sum_{i=1}^{\infty} \left(\frac{1}{2}\right)^i (x_i - (-1)^i\right.$$

$$\left.\times x_i)\frac{t^i}{i!}\right)^2 + \frac{1}{3!}\left(\sum_{i=1}^{\infty} \left(\frac{1}{2}\right)^i (x_i - (-1)^i x_i)\frac{t^i}{i!}\right)^3 + \cdots \qquad (37)$$

$$= 1 + \frac{1}{1!}x_1 t + \frac{1}{2!}x_1^2 t^2 + \frac{1}{3!}\left(x_1^3 + \frac{1}{2^2}x_3\right)t^3 + \cdots$$

$$= \sum_{n=0}^{\infty} \left(\sum_{m_1+2m_2+\cdots+nm_n=n} \frac{n!}{m_1!m_2!\cdots m_n!}\left(\frac{x_1}{1!}\right)^{m_1}\left(\frac{0}{2!2}\right)^{m_2}\right.$$

$$\left.\times \left(\frac{x_3}{3!2^2}\right)^{m_3}\cdots \left(\frac{x_n(1-(-1)^n)}{n!2^n}\right)^{m_n}\right)\frac{t^n}{n!}.$$

Now, for $n \in \mathbb{N}$ with $n \equiv 1 \pmod{2}$, by (32), (34) and (37), we get

$$B_n^{(c)}(x_1, x_2, \cdots, x_n) = \sum_{k=0}^{n} T_{n,k}(x_1, x_2, \cdots, x_{n-k+1})$$

$$= \sum_{m_1+3m_3+\cdots+nm_n=n} \frac{n!}{m_1!m_3!\cdots m_n!}\left(\frac{x_1}{1!}\right)^{m_1}\left(\frac{x_3}{3!2^2}\right)^{m_3}\cdots \left(\frac{x_n}{n!2^{n-1}}\right)^{m_n}. \qquad (38)$$

Therefore, Equation (38) yields the following theorem.

Theorem 3. *For $n \in \mathbb{N}$ with $n \equiv 1 \pmod{2}$, we have*

$$B_n^{(c)}(x_1, x_2, \cdots, x_n) = \sum_{k=0}^{n} T_{n,k}(x_1, x_2, \cdots, x_{n-k+1})$$

$$= \sum_{m_1+3m_3+\cdots+nm_n=n} \frac{n!}{m_1!m_3!\cdots m_n!}\left(\frac{x_1}{1!}\right)^{m_1}\left(\frac{x_3}{3!2^2}\right)^{m_3}\cdots \left(\frac{x_n}{n!2^{n-1}}\right)^{m_n}.$$

Example 1. *Here, we illustrate Theorem 3 with the following example:*

$$B_5^{(c)}(x_1, 2x_2, 2^2 x_3, 2^3 x_4, 2^4 x_5) = \frac{5!}{0!0!1!}\left(\frac{x_1}{1!}\right)^0\left(\frac{x_3}{3!}\right)^0\left(\frac{x_5}{5!}\right)^1 + \frac{5!}{2!1!0!}\left(\frac{x_1}{1!}\right)^2\left(\frac{x_3}{3!}\right)^1\left(\frac{x_5}{5!}\right)^0$$

$$+ \frac{5!}{5!0!0!}\left(\frac{x_1}{1!}\right)^5\left(\frac{x_3}{3!}\right)^0\left(\frac{x_5}{5!}\right)^0 = x_1^5 + 10 x_1^2 x_3 + x_5,$$

$$T_{5,1}(x_1, 2x_2, 2^2x_3, 2^3x_4, 2^4x_5) = \frac{5!}{0!0!1!}\left(\frac{x_1}{1!}\right)^0\left(\frac{x_3}{3!}\right)^0\left(\frac{x_5}{5!}\right)^1 = x_5,$$

$$T_{5,3}(x_1, 2x_2, 2^2x_3) = \frac{5!}{2!1!}\left(\frac{x_1}{1!}\right)^2\left(\frac{x_3}{3!}\right)^1 = 10x_1^2 x_3, \quad T_{5,5}(x_1) = \frac{5!}{5!}\left(\frac{x_1}{1!}\right)^5 = x_1^5,$$

$$T_{5,0}(x_1, 2x_2, 2^2x_3, 2^3x_4, 2^4x_5, 2^5x_6) = 0, \quad T_{5,2}(x_1, 2x_2, 2^2x_3, 2^3x_4) = 0, \quad T_{5,4}(x_1, 2x_2) = 0.$$

On the one hand, we have

$$\exp\left(x\sum_{i=1}^{\infty}\left(\frac{1}{2}\right)^i(1-(-1)^i)\frac{t^i}{i!}\right) = 1 + \sum_{k=1}^{\infty}\frac{x^k}{k!}\left(\sum_{n=k}^{\infty}\left(\frac{1}{2}\right)^i(1-(-1)^i)\frac{t^i}{i!}\right)^k$$

$$= 1 + \sum_{k=1}^{\infty} x^k \sum_{n=k}^{\infty} T_{n,k}(1,1,\cdots,1)\frac{t^n}{n!} \tag{39}$$

$$= 1 + \sum_{n=1}^{\infty}\left(\sum_{k=1}^{n} x^k T_{n,k}(1,1,\cdots,1)\right)\frac{t^n}{n!}.$$

On the other hand, from (19), we have

$$\exp\left(x\sum_{i=1}^{\infty}\left(\frac{1}{2}\right)^i(1-(-1)^i)\frac{t^i}{i!}\right) = \exp\left(x\left(t + \frac{1}{2^2}t^3 + \frac{1}{2^4}t^5 + \cdots\right)\right)$$

$$= \exp\left(x\left(e^{\frac{t}{2}} - e^{-\frac{t}{2}}\right)\right) = \sum_{n=0}^{\infty} B_n^{(c)}(x)\frac{t^n}{n!}. \tag{40}$$

Therefore, by (39) and (40), we obtain the following theorem.

Theorem 4. *For $n, k \geq 0$ with $n \geq k$, we have*

$$\sum_{k=0}^{n} x^k T_{n,k}(1,1,\cdots,1) = B_n^{(c)}(x).$$

We note from Theorem 4 the next identities:

$$\sum_{k=0}^{n} x^k T_{n,k}(1,1,\cdots,1) = \sum_{k=0}^{n} T_{n,k}(x,x,\cdots,x) = B_n^{(c)}(x,x,\cdots,x). \tag{41}$$

Thus, Theorem 4 and (41) together give us the following corollary.

Corollary 2. *For $n \geq 0$, we have*

$$B_n^{(c)}(x,x,\cdots,x) = B_n^{(c)}(x).$$

The Stirling numbers of the first kind are given in terms of the generating function by (see [3,21])

$$\frac{1}{k!}\left(\log(1+t)\right)^k = \sum_{n=k}^{\infty} S_1(n,k)\frac{t^n}{n!}, \quad (k \geq 0). \tag{42}$$

In order to get the following result and using (42), we first observe that

$$\frac{1}{k!}\left(\log\left(1+\frac{x}{1-\frac{x}{2}}\right)\right)^k = \sum_{l=k}^{\infty} S_1(l,k)\frac{1}{l!}\left(\frac{x}{1-\frac{x}{2}}\right)^l$$

$$= \sum_{l=k}^{\infty} S_1(l,k)\frac{x^l}{l!}\left(1-\frac{x}{2}\right)^{-l}$$

$$= \sum_{l=k}^{\infty} \frac{1}{l!}S_1(l,k) \sum_{n=l}^{\infty} \binom{n-1}{l-1}(\frac{1}{2})^{n-l} x^n \qquad (43)$$

$$= \sum_{n=k}^{\infty} \left(\sum_{l=k}^{n} \frac{1}{l!}S_1(l,k)\binom{n-1}{l-1}(\frac{1}{2})^{n-l}\right) x^n.$$

The following equation can be derived from (21) and (43):

$$\sum_{n=k}^{\infty} T_{n,k}(0!,1!,2!,\cdots,(n-k)!)\frac{t^n}{n!}$$

$$= \frac{1}{k!}\left(t+\left(\frac{1}{2}\right)^2\frac{t^3}{3}+\left(\frac{1}{2}\right)^4\frac{t^5}{5}+\left(\frac{1}{2}\right)^6\frac{t^7}{7}+\cdots\right)^k$$

$$= \frac{1}{k!}\left(\log\left(1+\frac{t}{2}\right)-\log\left(1-\frac{t}{2}\right)\right)^k = \frac{1}{k!}\left(\log\left(\frac{1+\frac{t}{2}}{1-\frac{t}{2}}\right)\right)^k \qquad (44)$$

$$= \frac{1}{k!}\left(\log\left(1+\frac{t}{1-\frac{t}{2}}\right)\right)^k = \sum_{n=k}^{\infty}\left(\sum_{l=k}^{n}\frac{S_1(l,k)}{l!}\binom{n-1}{l-1}(\frac{1}{2})^{n-l}\right)t^n.$$

Now, we obtain the following theorem by comparing the coefficients on both sides of (44).

Theorem 5. *For $n,k \geq 0$ with $n \geq k$, we have*

$$T_{n,k}(0!,1!,2!,\cdots,(n-k)!) = n!\sum_{l=k}^{n}\frac{S_1(l,k)}{l!}\binom{n-1}{l-1}(\frac{1}{2})^{n-l}.$$

4. Conclusions

In this paper, we introduced central complete and incomplete Bell polynomials which can be viewed as generalizations of central Bell polynomials and central factorial numbers of the second kind, and also as 'central' analogues for complete and incomplete Bell polynomials. As examples and recalling some relevant works, we reminded the reader that the incomplete and complete Bell polynomials appearing in a Faà di Bruno formula, which encode integer partition information, can be used in constructing sequences of binomial type, have connections with combinatorial Hopf algebras, have applications in probability theory and arise in many combinatorial and number theoretic identities. One additional thing we want to mention here is that the Faà di Bruno formula has been proved to be very useful in finding explicit expressions for many special numbers arising from many different families of linear and nonlinear differential equations having generating functions of some special numbers and polynomials as solutions (see [22]).

The main results of the present paper are stated in Section 3, in which we introduced central incomplete and complete Bell polynomials and investigated some properties and identities. In particular, in Theorems 1 and 3, we gave basic formulas for computing central incomplete Bell polynomials and central complete Bell polynomials, respectively. We remarked that the number of monomials appearing in $T_{n,k}(x_1,2x_2,\cdots,2^{n-k}x_{n-k+1})$ is the number of partitioning n into k odd parts and the coefficient of each monomial is the number of partitioning n as the corresponding k odd parts. This was illustrated by an example. Furthermore, we gave expressions for the central incomplete and complete Bell polynomials with some various special arguments and also for the connection between the two Bell polynomials. In the near future, we hope to find some further properties, identities and various applications for central complete and incomplete Bell polynomials.

Author Contributions: T.K. and D.S.K. conceived the framework and structured the whole paper; T.K. wrote the paper; D.S.K. and G.-W.J. checked the results of the paper; T.K. and D.S.K. completed the revision of the article.

Funding: This research received no external funding.

Acknowledgments: This paper is dedicated to the 70th birthday of Gradimir V. Milovanovic. In addition, we would like to express our sincere condolences on the death of Simsek's mother. The authors thank the referees for their helpful suggestions and comments which improved the original manuscript greatly.

Conflicts of Interest: The authors declare no conflict of interest.

References

1. Bagarello, F.; Trapani, C.; Triolo, S. Representable states on quasilocal quasi*-algebras. *J. Math. Phys.* **2011**, *52*, 013510. [CrossRef]
2. Trapani, C.; Triolo, S. Representations of modules over a *-algebra and related seminorms. *Stud. Math.* **2008**, *184*, 133–148. [CrossRef]
3. Comtet, L. *Advanced Combinatorics: The Art of Finite and Infinite Expansions*; D. Reidel Publishing Co.: Dordrecht, The Netherlands, 1974.
4. Cvijović, D. New identities for the partial Bell polynomials. *Appl. Math. Lett.* **2011**, *24*, 1544–1547. [CrossRef]
5. Johnson, W.P. The curious history of Faà di Bruno's formula. *Am. Math. Mon.* **2002**, *109*, 217–234.
6. Mihoubi, M. Bell polynomials and binomial type sequences. *Discret. Math.* **2008**, *308*, 2450–2459. [CrossRef]
7. Aboud, A.; Bultel, J.-P.; Chouria, A.; Luque, J.-G.; Mallet, O. Word Bell polynomials. *Sém. Lothar. Combin.* **2017**, *75*, B75h.
8. Connon, D.F.; Various applications of the (exponential) complete Bell polynomials. *arXiv* **2010**, arXiv:1001.2835.
9. Kolbig, K.S. The complete Bell polynomials for certain arguments in terms of Stirling numbers of the first kind. *J. Comput. Appl. Math.* **1994**, *51*, 113–116. [CrossRef]
10. Port, D. Polynomial Maps with Applications to Combinatorics and Probability Theory. Ph.D. Thesis, Massachusetts Institute of Technology, Cambridge, MA, USA, February 1994.
11. Butzer, P.L.; Schmidt, M.; Stark, E.L.; Vogt, L. Central factorial numbers, their main properties and some applications. *Numer. Funct. Anal. Optim.* **1989**, *10*, 419–488. [CrossRef]
12. Charalambides, C.A. Central factorial numbers and related expansions. *Fibonacci Quart.* **1981**, *19*, 451–456.
13. Kim, T. A note on central factorial numbers. *Proc. Jangjeon. Math. Soc.* **2018**, *21*, 575–588.
14. Zhang, W. Some identities involving the Euler and the central factorial numbers. *Fibonacci Quart.* **1998**, *36*, 154–157.
15. Kim, T.; Kim, D.S. A note on central Bell numbers and polynomials. *Russ. J. Math. Phys.* **2019**, *26*, in press.
16. Kim, D.S.; Kwon, J.; Dolgy, D.V.; Kim, T. On central Fubini polynomials associated with central factorial numbers of the second kind. *Proc. Jangjeon Math. Soc.* **2018**, *21*, 589–598.
17. Bouroubi, S.; Abbas, M. New identities for Bell's polynomials: New approaches. *Rostock. Math. Kolloq.* **2006**, *61*, 49–55.
18. Carlitz, L. Some remarks on the Bell numbers. *Fibonacci Quart.* **1980**, *18*, 66–73.
19. Carlitz, L.; Riordan, J. The divided central differences of zero. *Can. J. Math.* **1963**, *15*, 94–100. [CrossRef]
20. Kim, T.; Kim, D.S. On λ-Bell polynomials associated with umbral calculus. *Russ. J. Math. Phys.* **2017**, *24*, 69–78. [CrossRef]
21. Kim, D.S.; Kim, T. Some identities of Bell polynomials. *Sci. China Math.* **2015**, *58*, 2095–2104. [CrossRef]
22. Kim, T.; Kim, D.S. Differential equations associated with degenerate Changhee numbers of the second kind. *Rev. R. Acad. Cienc. Exactas Fís. Nat. Ser. A Math.* **2018**, doi:10.1007/s13398-018-0576-y. [CrossRef]

© 2019 by the authors. Licensee MDPI, Basel, Switzerland. This article is an open access article distributed under the terms and conditions of the Creative Commons Attribution (CC BY) license (http://creativecommons.org/licenses/by/4.0/).

Article

New Families of Three-Variable Polynomials Coupled with Well-Known Polynomials and Numbers

Can Kızılateş [1,*], Bayram Çekim [2], Naim Tuğlu [2] and Taekyun Kim [3]

1. Faculty of Art and Science, Department of Mathematics, Zonguldak Bülent Ecevit University, Zonguldak 67100, Turkey
2. Faculty of Science, Department of Mathematics, Gazi University, Teknikokullar, Ankara 06500, Turkey; bayramcekim@gazi.edu.tr (B.Ç.); naimtuglu@gazi.edu.tr (N.T.)
3. Department of Mathematics, Kwangwoon University, Seoul 139-701, Korea; tkkim@kw.ac.kr
* Correspondence: cankizilates@gmail.com

Received: 27 December 2018; Accepted: 3 February 2019; Published: 20 February 2019

Abstract: In this paper, firstly the definitions of the families of three-variable polynomials with the new generalized polynomials related to the generating functions of the famous polynomials and numbers in literature are given. Then, the explicit representation and partial differential equations for new polynomials are derived. The special cases of our polynomials are given in tables. In the last section, the interesting applications of these polynomials are found.

Keywords: Fibonacci polynomials; Lucas polynomials; trivariate Fibonacci polynomials; trivariate Lucas polynomials; generating functions

MSC: 11B39; 11B37; 05A19

1. Introduction

In literature, the Fibonacci and Lucas numbers have been studied extensively and some authors tried to enhance and derive some directions to mathematical calculations using these special numbers [1–3]. By favour of the Fibonacci and Lucas numbers, one of these directions verges on the tribonacci and the tribonacci-Lucas numbers. In fact, M. Feinberg in 1963 has introduced the tribonacci numbers and then derived some properties for these numbers in [4–7]. Elia in [4] has given and investigated the tribonacci-Lucas numbers. The tribonacci numbers T_n for any integer $n > 2$ are defined via the following recurrence relation

$$T_n = T_{n-1} + T_{n-2} + T_{n-3}, \tag{1}$$

with the initial values $T_0 = 0$, $T_1 = 1$, and $T_2 = 1$. Similarly, by way of the initial values $K_0 = 3$, $K_1 = 1$, and $K_2 = 3$, the tribonacci-Lucas numbers K_n are given by the recurrence relation

$$K_n = K_{n-1} + K_{n-2} + K_{n-3}. \tag{2}$$

By dint of the above extensions, the tribonacci and tribonacci-Lucas numbers are introduced with the help of the following generating functions, respectively:

$$\sum_{n=0}^{\infty} T_n t^n = \frac{t}{1 - t - t^2 - t^3}, \tag{3}$$

and

$$\sum_{n=0}^{\infty} K_n t^n = \frac{3 - 2t - t^2}{1 - t - t^2 - t^3}. \tag{4}$$

Moreover, some authors define a large class of polynomials by using the Fibonacci and the tribonacci numbers [6–9]. Firstly, the well-known Fibonacci polynomials are defined via the recurrence relation
$$F_{n+1}(x) = xF_n(x) + F_{n-1}(x),$$
with $F_0(x) = 0$, $F_1(x) = 1$. The well-known Lucas polynomials are defined with the help of the recurrence relation
$$L_{n+1}(x) = xL_n(x) + L_{n-1}(x),$$
with $L_0(x) = 2$, $L_1(x) = x$.

Fibonacci or Fibonacci-like polynomials have been studied by many mathematicians for many years. Recently, in [10], Kim et al. kept in mind the sums of finite products of Fibonacci polynomials and of Chebyshev polynomials of the second kind and obtained Fourier series expansions of functions related to them. In [11], Kim et al. studied the convolved Fibonacci numbers by using the generating functions of them and gave some new identities for the convolved Fibonacci numbers. In [12], Wang and Zhang studied some sums of powers Fibonacci polynomials and Lucas polynomials. In [13], Wu and Zhang obtained the several new identities involving the Fibonacci polynomials and Lucas polynomials.

Afterwards, by giving the Pell and Jacobsthal polynomials, in 1973, Hoggatt and Bicknell [6] introduced the tribonacci polynomials. The tribonacci polynomials are defined by the recurrence relation for $n \geq 0$,
$$t_{n+3}(x) = x^2 t_{n+2}(x) + x t_{n+1}(x) + t_n(x), \tag{5}$$
where $t_0(x) = 0$, $t_1(x) = 1$, and $t_2(x) = x^2$. The tribonacci-Lucas polynomials are defined by the recurrence relation for $n \geq 0$,
$$k_{n+3}(x) = x^2 k_{n+2}(x) + x k_{n+1}(x) + k_n(x), \tag{6}$$
where $k_0(x) = 3$, $k_1(x) = x^2$, and $k_2(x) = x^4 + 2x$, respectively. Here we note that $t_n(1) = T_n$ which is the tribonacci numbers and $k_n(1) = K_n$ which is the tribonacci-Lucas numbers. Also for these polynomials, we have the generating function as follows:
$$\sum_{n=0}^{\infty} t_n(x) t^n = \frac{t}{1 - x^2 t - xt^2 - t^3}, \tag{7}$$
and
$$\sum_{n=0}^{\infty} k_n(x) t^n = \frac{3 - 2x^2 t - xt^2}{1 - x^2 t - xt^2 - t^3}. \tag{8}$$

On the other hand, some authors try to define the second and third variables of these polynomials with the help of these numbers. For example [8], for integer $n > 2$, the recurrence relations of the trivariate Fibonacci and Lucas polynomials are as follows:
$$H_n(x,y,z) = xH_{n-1}(x,y,z) + yH_{n-2}(x,y,z) + zH_{n-3}(x,y,z), \tag{9}$$
with $H_0(x,y,z) = 0$, $H_1(x,y,z) = 1$, $H_2(x,y,z) = x$ and
$$K_n(x,y,z) = xK_{n-1}(x,y,z) + yK_{n-2}(x,y,z) + zK_{n-3}(x,y,z), \tag{10}$$
with $K_0(x,y,z) = 3$, $K_1(x,y,z) = x$, $K_2(x,y,z) = x^2 + 2y$, respectively. Also for these, we have the generating functions as follows:
$$\sum_{n=0}^{\infty} H_n(x,y,z) t^n = \frac{t}{1 - xt - yt^2 - zt^3}, \tag{11}$$

and

$$\sum_{n=0}^{\infty} K_n(x,y,z)t^n = \frac{3-2xt-yt^2}{1-xt-yt^2-zt^3}. \tag{12}$$

After that, Ozdemir and Simsek [14] give the family of two-variable polynomials, reducing some well-known polynomials and obtaining some properties of these polynomials. In light of these polynomials, we introduce the families of three-variable polynomials with the new generalized polynomials reduced to the generating functions of the famous polynomials and numbers in literature. Then, we obtain the explicit representations and partial differential equations for new polynomials. The special cases of our polynomials are given in tables. Also the last section, we give the interesting applications of these polynomials.

2. The New Generalized Polynomials: Definitions and Properties

Now, we introduce the original and wide generating functions reduce the well-known polynomials and the well-known numbers such as the trivariate Fibonacci and Lucas polynomials, the tribonacci and the tribonacci-Lucas polynomials, the tribonacci and the tribonacci-Lucas numbers, and so on.

Firstly, some properties of these functions are investigated. Then, in the case of the new generating function, we give some properties the particular well-known polynomials as tables.

Via the following generating functions, a new original and wide family of three-variable polynomials denoted by $S_j := S_j(x,y,z;k,m,n,c)$ is defined as follows:

$$T := M(t;x,y,z;k,m,n,c) = \sum_{j=0}^{\infty} S_j t^j = \frac{1}{1-x^k t - y^m t^{m+n} - z^c t^{m+n+c}}, \tag{13}$$

where $k,m,n,c \in \mathbb{N} - \{0\}$, and $\left| x^k t + y^m t^{m+n} + z^c t^{m+n+c} \right| < 1$. Now we derive the explicit representation for polynomials S_j. By means of Taylor series of the generating function of the right hand side of (13), we can write

$$T = \sum_{j=0}^{\infty} S_j t^j = \sum_{j=0}^{\infty} \left(x^k t + y^m t^{m+n} + z^c t^{m+n+c} \right)^j.$$

After that, using the binomial expansion and taking $j+s$ instead of j, we get

$$T = \sum_{j=0}^{\infty} \sum_{s=0}^{\infty} \binom{j+s}{s} \left(x^k t \right)^j \left(t^{m+n} \right)^s \left(y^m + z^c t^c \right)^s.$$

Lastly, using the expansion of $(y^m + z^c t^c)^s$, taking $u+s$ instead of s, taking $j-(m+n+c)u$ instead of j and taking $j-(m+n)s$ instead of j, respectively, we have

$$T = \sum_{j=0}^{\infty} \sum_{s=0}^{\lfloor \frac{j}{n+m} \rfloor} \sum_{u=0}^{\lfloor \frac{j-(m+n)s}{n+m+c} \rfloor} \binom{j-(n+m-1)s-(n+m+c-1)u}{s+u}\binom{s+u}{u} \left(x^k\right)^{j-(n+m)(s+u)-cu} z^{cu} y^{ms} t^j.$$

Thus after the equalization of coefficients of t^j, we obtain

$$S_j = \sum_{s=0}^{\lfloor \frac{j}{n+m} \rfloor} \sum_{u=0}^{\lfloor \frac{j-(m+n)s}{n+m+c} \rfloor} \binom{j-(n+m-1)s-(n+m+c-1)u}{s+u}\binom{s+u}{u} \left(x^k\right)^{j-(n+m)(s+u)-cu} z^{cu} y^{ms}. \tag{14}$$

Note that for $z = 0$, our polynomials reduces to the polynomials Equation (4) [14].

49

Remark 1. As a similar to Theorem 2.3 in [14], we can write the following relation

$$S_j(2x, -1, 0; 1, 1, 1, c) = \sum_{r=0}^{j} P_{j-r}(x) P_r(x),$$

where $P_r(x)$ are the Legendre polynomials.

To obtain other wide family of well-known polynomials, we define the second new generating function for the family of the polynomials $W_j := W_j(x, y, z; k, m, n, c)$ as follows

$$R := R(t; x, y, z; k, m, n, c) = \sum M(t; x, y, z; k, m, n, c) t^n$$

$$= \frac{t^n}{1 - x^k t - y^m t^{m+n} - z^c t^{m+n+c}}$$

$$= \sum_{j=0}^{\infty} W_j t^j, \tag{15}$$

where $k, m, n, c \in \mathbb{N} - \{0\}$, and $\left| x^k t + y^m t^{m+n} + z^c t^{m+n+c} \right| < 1$. Similarly for $z = 0$, our polynomials reduces to the polynomials in (5) in [14]. Now we give some special case. Firstly taking $k = m = n = c = 1$ in (15), we give the generating function

$$\frac{t}{1 - xt - yt^2 - zt^3} = \sum_{j=0}^{\infty} W_j(x, y, z; 1, 1, 1) t^j,$$

where $W_j(x, y, z; 1, 1, 1) = H_j(x, y, z)$, which are trivariate Fibonacci polynomials in (11). Secondly, writing $k = m = n = c = 1$ and $x \to x^2, y \to x, z \to 1$, we have the generating function

$$\frac{t}{1 - x^2 t - xt^2 - t^3} = \sum_{j=0}^{\infty} W_j(x^2, x, 1; 1, 1, 1, 1) t^j,$$

where $W_j(x^2, x, 1; 1, 1, 1, 1) = t_j(x)$ which are the tribonacci polynomials in (7). In the above generating function, for $x = 1$, we find the generating function of the tribonacci numbers in (3). Now, we give other special cases as the following table related to (15).

Now, we define a new family of the polynomials denoted by $K_j := K_j(x, y, z; k, m, n, c)$ via the generating function

$$\sum_{j=0}^{\infty} K_j t^j = \frac{\alpha(t; x, y) - \beta(t; x, y) t^n}{1 - x^k t - y^m t^{m+n} - z^c t^{m+n+c}}, \tag{16}$$

where $k, m, n, c \in \mathbb{N} - \{0\}$, $\alpha(t; x, y)$ and $\beta(t; x, y)$ are arbitrary polynomials depending on t, x, y and $\left| x^k t + y^m t^{m+n} + z^c t^{m+n+c} \right| < 1$. Thirdly, via (16), we give

$$3M(t; x, y, z; 1, 1, 1) - 2xR(t; x, y, z; 1, 1, 1, 1) - ytR(t; x, y, z; 1, 1, 1, 1) = \frac{3 - 2xt - yt^2}{1 - xt - yt^2 - zt^3}$$

$$= \sum_{j=0}^{\infty} K_j(x, y, z) t^j,$$

where $K_j(x, y, z)$ are the trivariate Lucas polynomials in (12). Due to the last equation, we have the polynomial representation

$$3S_j(x, y, z; 1, 1, 1) - 2xW_j(x, y, z; 1, 1, 1, 1) - ytW_j(x, y, z; 1, 1, 1, 1) = K_j(x, y, z). \tag{17}$$

In (17) substituting $x \to x^2, y \to x, z \to 1$, we get

$$3S_j(x^2, x, 1; 1, 1, 1, 1) - 2xW_j(x^2, x, 1; 1, 1, 1, 1) - ytW_j(x^2, x, 1; 1, 1, 1, 1) = k_j(x),$$

where $k_j(x)$ are the tribonacci-Lucas polynomials in (6). In the above representation, for $x = 1$, we find the generating function of the tribonacci-Lucas numbers in (4).

Now, we give other special cases as Table 1 and Table 2 related to (15) and (16) respectively.

Table 1. Special cases of W_j.

x	y	z	k	m	n	c	Special Case
x	y	z	1	1	1	1	Trivariate Fibonacci Polynomials [8]
x^2	x	1	1	1	1	1	tribonacci Polynomials [8]
x	y	0	1	1	1	c	Bivariate Fibonacci Polynomials [9]
x	1	0	1	p	1	c	Fibonacci p–Polynomials [9]
$2x$	1	0	1	p	1	c	Pell p–Polynomials [9]
x	1	0	1	1	1	c	Fibonacci Polynomials [9]
$2x$	1	0	1	1	1	c	Pell Polynomials [9]
1	$2y$	0	k	1	1	c	Jacobsthal Polynomials [9]
$3x$	-2	0	1	1	1	c	Fermat Polynomials [15]
x	-2	0	1	1	1	c	First kind of Fermat–Horadam Polynomials [16]
x	$-\alpha$	0	1	1	1	c	Second kind of Dickson Polynomials [17]
$x+2$	-1	0	1	1	1	c	Morgan–Voyce Polynomials [18]
$x+1$	$-x$	0	1	1	1	c	Delannoy Polynomials [19]
$h(x)$	1	0	1	1	1	c	$h(x)$–Fibonacci Polynomials [2]
$p(x)$	$q(x)$	0	1	1	1	c	(p,q)–Fibonacci Polynomials [15]
1	1	0	k	1	1	c	Fibonacci Numbers [9]
2	1	0	1	1	1	c	Pell Numbers [9]
1	2	0	k	1	1	c	Jacobsthal Numbers [9]

Table 2. Special cases of K_j

α	β	x	y	z	k	m	n	c	Special Case
3	$2x+yt$	x	y	z	1	1	1	1	Trivariate Lucas Polynomials [8]
3	$2x^2+xt$	x^2	x	1	1	1	1	1	tribonacci-Lucas Polynomials [8]
2	xz	x	y	0	1	1	1	c	Bivariate Lucas Polynomials [9]
$p+1$	px	x	1	0	1	p	1	c	Lucas p–Polynomials [9]
0	-1	$2x$	1	0	1	p	1	c	Pell Lucas p–Polynomials [9]
2	x	x	1	0	1	1	1	c	Lucas Polynomials [9]
2	$2x$	$2x$	1	0	1	1	1	c	Pell Lucas Polynomials [9]
2	1	1	$2y$	0	k	1	1	c	Jacobsthal Lucas Polynomials [9]
2	$3x$	$3x$	-2	0	1	1	1	c	Fermat Lucas Polynomials [15]
2	x	x	-2	0	1	1	1	c	Second kind of Fermat–Horadam P. [16]
2	x	x	$-\alpha$	0	1	1	1	c	First kind of Dickson Polynomials [17]
2	$x+2$	$x+2$	-1	0	1	1	1	c	Morgan–Voyce Polynomials [18]
2	$x+1$	$x+1$	$-x$	0	1	1	1	c	Corona Polynomials [19]
2	$h(x)$	$h(x)$	1	0	1	1	1	c	$h(x)$–Lucas Polynomials [2]
2	$p(x)$	$p(x)$	$q(x)$	0	1	1	1	c	(p,q)–Lucas Polynomials [15]
2	1	1	1	0	k	1	1	c	Lucas Numbers [9]
2	2	2	1	0	1	1	1	c	Pell–Lucas Numbers [9]
2	1	1	2	0	k	1	1	c	Jacobsthal–Lucas Numbers [9]
t	t	2	2	-1	1	1	1	1	Squares of Fibonacci Numbers [1]

3. Partial Differential Equations for Polynomials in (13)

With the help of the derivatives of these generating functions with regard to some variable and algebraic arrangements, we derive some partial differential equations for new polynomials. Taking the derivative with regard to x, y, z, t of the generating function in (13), respectively, they hold

$$\frac{\partial}{\partial x} M = kx^{k-1} t M^2, \tag{18}$$

$$\frac{\partial}{\partial y} M = my^{m-1} t^{n+m} M^2, \tag{19}$$

$$\frac{\partial}{\partial z} M = cz^{c-1} t^{n+m+c} M^2, \tag{20}$$

$$\frac{\partial}{\partial t} M = \left(x^k + y^m(n+m)t^{n+m-1} + z^c(n+m+c)t^{n+m+c-1} \right) M^2. \tag{21}$$

From (13) and (18), we get the following theorem.

Theorem 1. *For $j \geq 0$, we have the first relation as follows:*

$$\frac{\partial}{\partial x} S_j = kx^{k-1} \sum_{l=0}^{j-1} S_{j-l-1} S_l.$$

Combining (13) and (19), we have the next theorem.

Theorem 2. *For $j \geq m+n$, we have the second relation as follows:*

$$\frac{\partial}{\partial y} S_j = \sum_{l=0}^{j-m-n} my^{m-1} S_{j-m-n-l} S_l.$$

With the help of considering (13) and (20), we get the next result.

Theorem 3. *For $j \geq m+n+c$, we have the third relation as follows:*

$$\frac{\partial}{\partial z} S_j = cz^{c-1} \sum_{l=0}^{j-m-n-c} S_{j-m-n-c-l} S_l.$$

Lastly, by means of (13) and (21), we get the following result.

Theorem 4.

(i) For $m+n-1 \leq j \leq m+n+c-1$, then we obtain

$$(j+1) S_{j+1} = x^k \sum_{l=0}^{j} S_{j-l} S_l + y^m (n+m) \sum_{l=0}^{j-m-n+1} S_l S_{j-m-n-l+1}.$$

(ii) For $j \leq m+n-1$, then we derive

$$(j+1) S_{j+1} = x^k \sum_{l=0}^{j} S_{j-l} S_l.$$

(iii) For $j \geq m + n + c - 1$, then we get

$$(j+1)S_{j+1} = x^k \sum_{l=0}^{j} S_{j-l}S_l + y^m(n+m) \sum_{l=0}^{j-m-n+1} S_l S_{j-m-n-l+1}$$
$$+ z^c(n+m+c) \sum_{l=0}^{j-m-n-c+1} S_l S_{j-m-n-c-l+1}.$$

After that, using the partial differential equations in (18)–(21), we get the new partial differential equation for S_j.

Theorem 5. *For $j \geq 0$, we have*

$$jS_j = \frac{x}{k}\frac{\partial}{\partial x}S_j + \left(\frac{n+m}{m}\right) y \frac{\partial}{\partial y}S_j + \left(\frac{n+m+c}{c}\right) z \frac{\partial}{\partial z}S_j.$$

Proof. Combining (18)–(21), we get

$$\frac{\partial}{\partial t}M - \frac{x}{kt}\frac{\partial}{\partial x}M = \left(\frac{n+m}{m}\right)\frac{y}{t}\frac{\partial}{\partial y}M + \left(\frac{n+m+c}{c}\right)\frac{z}{t}\frac{\partial}{\partial z}M.$$

In the above, using (13), we get the desired result. □

4. Some Applications of Generating Functions

In this section, by using these functions, some identities connected with these polynomials are derived. Furthermore, in the special case, we show that these identities reduce to the well-known sum identities connected with the well-known numbers in literature.

Case 1. Taking $t = \frac{1}{a}$ in (15) for $|a| > 1$, we get the following equation

$$\sum_{j=0}^{\infty} \frac{W_j}{a^j} = \frac{a^{m+c}}{a^{m+n+c} - x^k a^{m+n+c-1} - y^m a^c - z^c}. \tag{22}$$

(i) Substituting $a = 2$, $x \to x^2$, $y \to x$, $z \to 1$ and $k = m = n = c = 1$ in (22), we obtain the relation for the tribonacci polynomials as

$$\sum_{j=0}^{\infty} \frac{t_j(x)}{2^j} = \frac{4}{7 - 4x^2 - 2x}. \tag{23}$$

Writing $x = 1$ in (23), we have

$$\sum_{j=0}^{\infty} \frac{T_j}{2^j} = 4,$$

where T_j are the tribonacci numbers.

(ii) Taking $a = 10$, $x \to x^2$, $y \to x$, $z \to 1$ and $k = m = n = c = 1$ in (22), we get

$$\sum_{j=0}^{\infty} \frac{T_j(x)}{10^{j+2}} = \frac{1}{999 - 100x^2 - 10x}, \tag{24}$$

and writing $x = 1$ (24), we get for the tribonacci numbers

$$\sum_{j=0}^{\infty} \frac{T_j}{10^{j+2}} = \frac{1}{889}.$$

(iii) Substituting $x \to x$, $y \to 1$, $z \to 0$, $a = 2$, and $k = m = n = c = 1$ into (22), we get for the Fibonacci polynomials

$$\sum_{j=0}^{\infty} \frac{F_j(x)}{2^j} = \frac{2}{3 - 2x}, \tag{25}$$

which was given in [14]. Then taking $x = 1$ in (25), we have for Fibonacci numbers

$$\sum_{j=0}^{\infty} \frac{F_j}{2^j} = 2,$$

which was given in [14].

(iv) Substituting $x \to x$, $y \to 1$, $z \to 0$, $a = 3$, and $k = m = n = c = 1$ in (22), we get for the Fibonacci polynomials

$$\sum_{j=0}^{\infty} \frac{F_j(x)}{3^{j+1}} = \frac{1}{8 - 3x}. \tag{26}$$

Taking $x = 1$ in (26), we get for the Fibonacci numbers

$$\sum_{j=0}^{\infty} \frac{F_j}{3^{j+1}} = \frac{1}{5} = \frac{1}{F_5},$$

was given in page 424 in [1].

(v) Substituting $x \to x$, $y \to 1$, $z \to 0$, $a = 8$, and $k = m = n = c = 1$ in (22), we get for the Fibonacci polynomials

$$\sum_{j=0}^{\infty} \frac{F_j(x)}{8^{j+1}} = \frac{1}{63 - 3x}. \tag{27}$$

Taking $x = 1$ in (27), we get for the Fibonacci numbers

$$\sum_{j=0}^{\infty} \frac{F_j}{8^{j+1}} = \frac{1}{55} = \frac{1}{F_{10}},$$

was given in page 424 in [1].

(vi) Substituting $x \to x$, $y \to 1$, $z \to 0$, $a = -10$, and $k = m = n = c = 1$ in (22), we get for the Fibonacci polynomials

$$\sum_{j=0}^{\infty} \frac{F_j(x)}{(-10)^{j+1}} = \frac{1}{99 + 10x}. \tag{28}$$

Taking $x = 1$ in (28), we get for the Fibonacci numbers

$$\sum_{j=0}^{\infty} \frac{F_j}{(-10)^{j+1}} = \frac{1}{109},$$

was given in page 427 in [1].

(vii) Substituting $x \to 2x$, $y \to 1$, $z \to 0$, $a = 3$, and $k = m = n = c = 1$ in (22), we get

$$\sum_{j=0}^{\infty} \frac{P_j(x)}{3^{j+1}} = \frac{1}{8 - 6x}, \tag{29}$$

where $P_j(x)$ are the Pell polynomials. Then taking $x = 1$ in (29), we have

$$\sum_{j=0}^{\infty} \frac{P_j}{3^{j+1}} = \frac{1}{2},$$

where P_j are the Pell numbers.

(viii) Substituting $x \to 1, y \to 2y, z \to 0, a = 3$, and $k = m = n = c = 1$ in (22), we get

$$\sum_{s=0}^{\infty} \frac{J_s(x)}{3^{s+1}} = \frac{1}{6 - 2y}, \tag{30}$$

where $J_s(x)$ are the Jacobsthal polynomials. Then taking $y = 1$ in (30), we have

$$\sum_{s=0}^{\infty} \frac{J_s}{3^{s+1}} = \frac{1}{4},$$

where J_s are the Jacobsthal numbers.

Case 2. Taking $t = \frac{1}{a}$ in (16) for $|a| > 1$, we get the following equation

$$\sum_{j=0}^{\infty} \frac{K_j}{a^j} = \frac{a^{m+n+c}\alpha(t;x,y) - a^{m+c}\beta(t;x,y)}{a^{m+n+c} - x^k a^{m+n+c-1} - y^m a^c - z^c}. \tag{31}$$

(i) Substituting $x \to x^2, y \to x, z \to 1, a = 2$, and $k = m = n = c = 1, \alpha(t;x,y) = 3, \beta(t;x,y) = 2x^2 + xt$ in (31), we get

$$\sum_{j=0}^{\infty} \frac{k_j(x)}{2^j} = \frac{24 - 8x^4 - 2x^2}{7 - 4x^2 - 2x}, \tag{32}$$

where $k_j(x)$ are the tribonacci-Lucas polynomials. Then taking $x = 1$ in (32), we have

$$\sum_{j=0}^{\infty} \frac{k_j}{2^{j+1}} = 7,$$

where $k_j(x)$ are the tribonacci-Lucas numbers.

(ii) Substituting $x \to x, y \to 1, z \to 0, a = 2$, and $k = m = n = c = 1, \alpha(t;x,y) = 2, \beta(t;x,y) = x$ in (31), we get

$$\sum_{j=0}^{\infty} \frac{L_j(x)}{2^{j+1}} = \frac{4 - x}{3 - 2x}, \tag{33}$$

where $L_j(x)$ are the Lucas polynomials. Then taking $x = 1$ in (33), we have

$$\sum_{j=0}^{\infty} \frac{L_j}{2^{j+1}} = 3,$$

where L_j are the Lucas numbers.

(iii) Substituting $x \to x, y \to 1, z \to 0, a = 10$, and $k = m = n = c = 1, \alpha(t;x,y) = 2, \beta(t;x,y) = x$ in (31), we get

$$\sum_{j=0}^{\infty} \frac{L_j(x)}{10^j} = \frac{200 - 10x}{99 - 10x}, \tag{34}$$

where $L_j(x)$ are the Lucas polynomials. Then taking $x = 1$ in (34), for L_j are the Lucas numbers, we have

$$\sum_{j=0}^{\infty} \frac{L_j}{10^{j+1}} = \frac{19}{89} = \frac{L_6 - L_1}{F_{11}}$$

was given in page 427 in [1].

(iv) Substituting $x \to x, y \to 1, z \to 0, a = 3$, and $k = m = n = c = 1, \alpha(t;x,y) = 2, \beta(t;x,y) = x$ in (31), we get

$$\sum_{j=0}^{\infty} \frac{L_j(x)}{3^{j+1}} = \frac{6-x}{8-3x}, \tag{35}$$

and taking $x = 1$ in (35), we have

$$\sum_{j=0}^{\infty} \frac{L_j}{3^{j+1}} = 1.$$

(v) Substituting $x \to x, y \to 1, z \to 0, a = 8$, and $k = m = n = c = 1, \alpha(t;x,y) = 2, \beta(t;x,y) = x$ in (31), we get

$$\sum_{j=0}^{\infty} \frac{L_j(x)}{8^{j+1}} = \frac{16-x}{63-8x}, \tag{36}$$

and taking $x = 1$ in (36), we have

$$\sum_{j=0}^{\infty} \frac{L_j}{8^{j+1}} = \frac{3}{11} = \frac{L_2}{L_5}.$$

(vi) Substituting $x \to x, y \to 1, z \to 0, a = -10$, and $k = m = n = c = 1, \alpha(t;x,y) = 2, \beta(t;x,y) = x$ in (31), we get

$$\sum_{j=0}^{\infty} \frac{L_j(x)}{(-10)^{j+1}} = \frac{-20-x}{99+10x}. \tag{37}$$

Taking $x = 1$ in (37), we have

$$\sum_{j=0}^{\infty} \frac{L_j}{(-10)^{j+1}} = \frac{-21}{109},$$

was given in page 427 in [1].

(vii) Substituting $x \to 2x, y \to 1, z \to 0, a = 5$, and $k = m = n = c = 1, \alpha(t;x,y) = 2, \beta(t;x,y) = 2x$ in (31), we get

$$\sum_{j=0}^{\infty} \frac{Q_j(x)}{5^{j+1}} = \frac{5-x}{12-5x}, \tag{38}$$

where $Q_j(x)$ are the Pell Lucas polynomials. Then taking $x = 1$ in (38), we have

$$\sum_{j=0}^{\infty} \frac{Q_j}{5^{j+1}} = \frac{4}{7},$$

where Q_j are the Pell Lucas numbers.

(viii) Substituting $x \to 1, y \to 2y, z \to 0, a = 3$, and $k = m = n = c = 1, \alpha(t;x,y) = 2, \beta(t;x,y) = 1$ in (31), we get

$$\sum_{s=0}^{\infty} \frac{j_s(y)}{3^{s+1}} = \frac{5}{6-2y}, \tag{39}$$

where $j_s(y)$ are the Jocabsthal Lucas polynomials. Then taking $y = 1$ in (39), we have

$$\sum_{s=0}^{\infty} \frac{j_s}{3^{s+1}} = \frac{5}{4},$$

where j_s is Jocabsthal Lucas number.

(ix) Substituting $x \to 2, y \to 2, z \to -1, a = 4,$ and $k = m = n = c = 1, \alpha(t;x,y) = t, \beta(t;x,y) = t$ in (31), for the square of Fibonacci numbers F_j, we get

$$\sum_{j=0}^{\infty} \frac{F_j^2}{4^j} = \frac{12}{25},$$

was given in page 439 in [1].

Let us give Tables 3 and 4 containing the obtained formulas for simplify reading.

Table 3. Special cases of Equation (22) for $k = m = n = c = 1$.

a	x	y	z	Formulas
2	x^2	x	1	$\sum_{j=0}^{\infty} \frac{t_j(x)}{2^j} = \frac{4}{7-4x^2-2x}$
2	1	1	1	$\sum_{j=0}^{\infty} \frac{T_j}{2^j} = 4$
10	x^2	x	1	$\sum_{j=0}^{\infty} \frac{T_j(x)}{10^{j+2}} = \frac{1}{999-100x^2-10x}$
10	1	1	1	$\sum_{j=0}^{\infty} \frac{T_j}{10^{j+2}} = \frac{1}{889}$
2	x	1	0	$\sum_{j=0}^{\infty} \frac{F_j(x)}{2^j} = \frac{2}{3-2x}$
2	1	1	0	$\sum_{j=0}^{\infty} \frac{F_j}{2^j} = 2$
3	x	1	0	$\sum_{j=0}^{\infty} \frac{F_j(x)}{3^{j+1}} = \frac{1}{8-3x}$
3	1	1	0	$\sum_{j=0}^{\infty} \frac{F_j}{3^{j+1}} = \frac{1}{5} = \frac{1}{F_5}$
8	x	1	0	$\sum_{j=0}^{\infty} \frac{F_j(x)}{8^{j+1}} = \frac{1}{63-3x}$
8	1	1	0	$\sum_{j=0}^{\infty} \frac{F_j}{8^{j+1}} = \frac{1}{55} = \frac{1}{F_{10}}$
-10	x	1	0	$\sum_{j=0}^{\infty} \frac{F_j(x)}{(-10)^{j+1}} = \frac{1}{99+10x}$
-10	1	1	0	$\sum_{j=0}^{\infty} \frac{F_j}{(-10)^{j+1}} = \frac{1}{109}$
3	2x	1	0	$\sum_{j=0}^{\infty} \frac{P_j(x)}{3^{j+1}} = \frac{1}{8-6x}$
3	2	1	0	$\sum_{j=0}^{\infty} \frac{P_j}{3^{j+1}} = \frac{1}{2}$
3	1	2y	0	$\sum_{s=0}^{\infty} \frac{j_s(x)}{3^{s+1}} = \frac{1}{6-2y}$
3	1	2	0	$\sum_{s=0}^{\infty} \frac{j_s}{3^{s+1}} = \frac{1}{4}$

Table 4. Special cases of Equation (31) for $k = m = n = c = 1$.

a	x	y	z	α	β	Formulas
2	x^2	x	1	3	$2x^2 + xt$	$\sum_{j=0}^{\infty} \frac{k_j(x)}{2^j} = \frac{24-8x^4-2x^2}{7-4x^2-2x}$
2	1	1	1	3	$2+t$	$\sum_{j=0}^{\infty} \frac{k_j}{2^{j+1}} = 7$
2	x	1	0	2	x	$\sum_{j=0}^{\infty} \frac{L_j(x)}{2^{j+1}} = \frac{4-x}{3-2x}$
2	1	1	0	2	1	$\sum_{j=0}^{\infty} \frac{L_j}{2^{j+1}} = 3$
10	x	1	0	2	x	$\sum_{j=0}^{\infty} \frac{L_j(x)}{10^j} = \frac{200-10x}{99-10x}$
10	1	1	0	2	1	$\sum_{j=0}^{\infty} \frac{L_j}{10^{j+1}} = \frac{19}{89} = \frac{L_6 - L_1}{F_{11}}$
3	x	1	0	2	x	$\sum_{j=0}^{\infty} \frac{L_j(x)}{3^{j+1}} = \frac{6-x}{8-3x}$
3	1	1	0	2	1	$\sum_{j=0}^{\infty} \frac{L_j}{3^{j+1}} = 1$
8	x	1	0	2	x	$\sum_{j=0}^{\infty} \frac{L_j(x)}{8^{j+1}} = \frac{16-x}{63-8x}$
8	1	1	0	2	1	$\sum_{j=0}^{\infty} \frac{L_j}{8^{j+1}} = \frac{3}{11} = \frac{L_2}{L_5}$
-10	x	1	0	2	x	$\sum_{j=0}^{\infty} \frac{L_j(x)}{(-10)^{j+1}} = \frac{-20-x}{99+10x}$
-10	1	1	0	2	1	$\sum_{j=0}^{\infty} \frac{L_j}{(-10)^{j+1}} = \frac{-21}{109}$
5	$2x$	1	0	2	$2x$	$\sum_{j=0}^{\infty} \frac{Q_j(x)}{5^{j+1}} = \frac{5-x}{12-5x}$
5	2	1	0	2	2	$\sum_{j=0}^{\infty} \frac{Q_j}{5^{j+1}} = \frac{4}{7}$
3	1	$2y$	0	2	1	$\sum_{s=0}^{\infty} \frac{j_s(y)}{3^{s+1}} = \frac{5}{6-2y}$
3	1	2	0	2	1	$\sum_{s=0}^{\infty} \frac{j_s}{3^{s+1}} = \frac{5}{4}$
4	2	2	-1	1/4	1/4	$\sum_{j=0}^{\infty} \frac{F_j^2}{4^j} = \frac{12}{25}$

5. Conclusions

In the present paper, we considered the families of three-variable polynomials with the generalized polynomials reduce to generating function of the polynomials and numbers in the literature. In Section 2, we gave special polynomials and numbers as the tables related to (15) and (16). Then we obtained the explicit representations and partial differential equations for new polynomials. In the last section, we gave the interesting sum identities related to the well-known numbers and polynomials in the literature.

For all of the resuts, if the appropriate values given in the tables are taken, many infinite sums including various polynomials are obtained.

In recent years, some authors use the well-known polynomials and numbers in the applications of ordinary and fractional differential equations and difference equations (for example [20–23]). Therefore, our new families of three variables polynomials could been used for future works of some application areas such as mathematical modelling, physics, engineering, and applied sciences.

Author Contributions: All authors contributed equally to this work. All authors read and approved the final manuscript.

Funding: This research received no external funding.

Acknowledgments: The authors would like to express their sincere gratitude to the referees for their valuable comments which have significantly improved the presentation of this paper.

Conflicts of Interest: The authors declare no conflict of interest.

References

1. Koshy, T. *Fibonacci and Lucas Numbers with Applications*; John Wiley and Sons Inc.: New York, NY, USA, 2001.
2. Nalli, A.; Haukkanen, P. On generalized Fibonacci and Lucas polynomials. *Chaos Solitons Fractals* **2009**, *42*, 3179–3186. [CrossRef]
3. Vajda, S. *Fibonacci and Lucas Numbers, and the Golden Section, Theory and Applications*; Ellis Horwood Limited: Chichester, UK, 1989.
4. Elia, M. Derived sequences, the tribonacci recurrence and cubic forms. *Fibonacci Q.* **2001**, *39*, 107–109.
5. Feinberg, M. Fibonacci-tribonacci. *Fibonacci Q.* **1963**, *3*, 70–74.
6. Hoggatt, V.E., Jr. Bicknell, M. Generalized Fibonacci polynomials. *Fibonacci Q.* **1973**, *11*, 457–465.
7. Tan, M.; Zhang, Y. A note on bivariate and trivariate Fibonacci polynomials. *Southeast Asian Bull. Math.* **2005**, *29*, 975–990.
8. Kocer, E.G.; Gedikli, H. Trivariate Fibonacci and Lucas polynomials. *Konuralp J. Math.* **2016**, *4*, 247–254.
9. Tuglu, N.; Kocer, E.G.; Stakhov, A. Bivariate Fibonacci like p-polynomials. *Appl. Math. Comput.* **2011**, *217*, 10239–10246. [CrossRef]
10. Kim, T.; Kim, D.S.; Dolgy, D.V.; Park, J.W. Sums of finite products of Chebyshev polynomials of the second kind and of Fibonacci polynomials. *J. Inequal. Appl.* **2018**, *148*. [CrossRef] [PubMed]
11. Kim, T.; Dolgy, D.V.; Kim, D.S.; Seo, J.J. Convoled Fibonacci numbers and their applications. *arXiv* **2017**, arXiv:1607.06380.
12. Wang, T.; Zhang, W. Some identities involving Fibonacci, Lucas polynomials and their applications. *Bull. Math. Soc. Sci. Math. Roum.* **2012**, *55*, 95–103.
13. Wu, Z.; Zhang, W. Several identities involving the Fibonacci polynomials and Lucas polynomials. *J. Inequal. Appl.* **2013**, *2013*, 14. [CrossRef]
14. Ozdemir, G.; Simsek, Y. Generating functions for two-variable polynomials related to a family of Fibonacci type polynomials and numbers. *Filomat* **2016**, *30*, 969–975. [CrossRef]
15. Lee, G.Y.; Asci, M. Some properties of the (p,q)–Fibonacci and (p,q)–Lucas polynomials. *J. Appl. Math.* **2012**, *2012*, 264842. [CrossRef]
16. Horadam, A.F. Chebyshev and Fermat polynomials for diagonal functions. *Fibonacci Q.* **1979**, *17*, 328–333.
17. Lidl, R.; Mullen, G.; Tumwald. G. Dickson Polynomials. *Pitman Monographs and Surveys in Pure and Applied Mathematics*; Longman Scientific and Technical: Essex, UK, 1993; Volume 65,
18. Shannon, A.G.; Horadam, A.F. Some relationships among Vieta, Morgan-Voyce and Jacobsthal polynomials. In *Applications of Fibonacci Numbers*; Howard, F.T., Ed.; Kluwer Academic Publishers: Dordrecht, The Netherlands, 1999; pp. 307–323.
19. Cheon, G.-S.; Kim, H.; Shapiro, L.W. A generalization of Lucas polynomial sequence. *Discret. Appl. Math.* **2009**, *157*, 920–927. [CrossRef]
20. Bulut, H; Zhang, Pandir, Y.; Baskonus, H.M. Symmetrical hyperbolic Fibonacci function solutions of generalized Fisher equation with fractional order. *AIP Conf. Proc.* **2013**, *1558*, 1914–1918.
21. Mirzaee, F.; Hoseini, S.F. Solving singularly perturbed differential-difference equations arising in science and engineering whit Fibonacci polynomials. *Results Phys.* **2013**, *3*, 134–141. [CrossRef]
22. Mirzaee, F.; Hoseini, S.F. Solving systems of linear Fredholm integro-differential equations with Fibonacci polynomials. *Ain. Shams Eng.* **2014**, *5*, 271–283. [CrossRef]
23. Kurt, A.; Yalinbash, S.; Sezer, M. Fibonacci-collocation method for solving high-order linear Fredholm integro-differential-difference equations. *Int. J. Math. Math. Sci.* **2013**, *2013*, 486013. [CrossRef]

© 2019 by the authors. Licensee MDPI, Basel, Switzerland. This article is an open access article distributed under the terms and conditions of the Creative Commons Attribution (CC BY) license (http://creativecommons.org/licenses/by/4.0/).

Article

A Modified PML Acoustic Wave Equation

Dojin Kim

Department of Mathematics, Pusan National University, Busan 46241, Korea; kimdojin@pusan.ac.kr

Received: 3 January 2019; Accepted: 30 January 2019; Published: 2 February 2019

Abstract: In this paper, we consider a two-dimensional acoustic wave equation in an unbounded domain and introduce a modified model of the classical un-split perfectly matched layer (PML). We apply a regularization technique to a lower order regularity term employed in the auxiliary variable in the classical PML model. In addition, we propose a staggered finite difference method for discretizing the regularized system. The regularized system and numerical solution are analyzed in terms of the well-posedness and stability with the standard Galerkin method and *von Neumann* stability analysis, respectively. In particular, the existence and uniqueness of the solution for the regularized system are proved and the Courant-Friedrichs-Lewy (CFL) condition of the staggered finite difference method is determined. To support the theoretical results, we demonstrate a non-reflection property of acoustic waves in the layers.

Keywords: well-posedness; stability; acoustic wave equation; perfectly matched layer

1. Introduction

It is quite important to effectively truncate an unbounded domain in wave propagation simulations in open space, where the perfectly matched layer (PML) methods that surround the domain of interest with thin artificial absorbing layers are popularly used in easy and effective ways. After the method was introduced by J. P. Bérenger [1], which involves splitting a field into two nonphysical electromagnetic fields, many studies were conducted regarding the PML method and its modified reformulations in many different wave-type equations. These include Maxwell's equations [2,3], elastodynamics [4,5], linearized Euler equations [6–9], Helmholtz equations [10], and other types of wave equations [10–12]. Most PML models by the splitting technique, named a split PML method, yield a hyperbolic system of first order partial differential equations [1,6,13–15]. It is known that the split PML models demonstrate excellent overall performance from the viewpoint of applications. However, it was pointed out in [7,16,17] that Bérenger's split, as well as other split models, transform Maxwell's equations from being strongly hyperbolic into weakly hyperbolic. These transforms imply a transition from strong to weak well-posedness in the Cauchy problem and may lead to ill-posedness under certain low-order damping functions in PML layers [18]. The authors of [6,19] mention that the use of artificial dissipation is necessary to stabilize the numerical scheme of such formulations for long-time simulations.

The resulting concerns about the well-posedness and stability of the split PML models have prompted the development of other PMLs. Some examples of such developments, without splitting the fields, include un-split PML models using convolution integrals [20,21] and auxiliary variables [17,22,23]. In contrast to the split PML models, it is known that the un-split PML wave equations are more effective at time discretization [22] and does not make the use of additional memory for the nonphysical field variables. However, it has also been found that the un-split PML models are susceptible to developing gradual instabilities in long-time simulations [10,19]. To overcome this instability issue, various studies are reported: a low-pass filter inside the absorbing layer [6], selective damping coefficients [24], a new layer by regularizing the damping terms [8], a change of variable [25], etc. These issues are the motivation for the mathematical study of the well-posedness and stability for the un-split PML acoustic wave model in various sound speed. A time-domain analysis of PML acoustic wave equation with a constant sound speed is presented with a time-dependent point source in two dimensions using the Cagniard-de-Hoop method [25,26], which includes the time-stability and error estimates. However, it is not easy to extend the analysis to general initial value problems in variable sound speed, because those include not only straight propagating but also evanescent waves [27]. There is another approach to demonstrating the well-posedness and stability by investigating the eigenvalues of the Cauchy hyperbolic problems for the PML wave equations [4,7,12,16–18,28]. This approach gives a restricted result when the original formulation of the PML wave equation is considered in a bounded domain, in which the solutions should be affected by boundary conditions.

Alternatively, energy techniques are used to analyze the issue of stability for the PML wave equations by presenting the energy behavior for the solution in each model [12,16,29]. In general, the restriction of the PML equations to the computational domain coincides with the original problem [12], so that damping terms are required to vanish identically in the computational region. As the constant damping function can be considered as the Heaviside function, the equation $(\partial_t + \sigma_x)\partial_x = \partial_x(\partial_t + \sigma_x)$ used in [12,16,29] is not valid at the interface between the domain of interest and the layers for the constant damping case from a discontinuity. However, all these approaches only provide its well-posedness, the stability has not been clearly proved in finite PML acoustic wave equations with variable sound speed.

The main contribution of this manuscript is not only to introduce a regularized system of the second order PML acoustic wave equation that exhibits well-posedness without losing the non-reflection property of PMLs, but also to demonstrate its numerical stability. To construct the system, we adopt a regularization technique for the term $\nabla \cdot \vec{q}$ that has a lower regularity, which is introduced in [8], to regularize the PML model for the Maxwell equation, where $\vec{q}$ is the auxiliary variable (see (2)). The standard Galerkin approximation and energy estimation of the solution are used to show the well-posedness of the regularized system. A concrete energy estimate yields the boundedness of the solution (see Theorem 1) together with the existence and uniqueness of the solution under the regularity assumption of the damping terms $\sigma_x, \sigma_y \in L^\infty(\Omega)$ (see Theorem 2). As a numerical scheme for the regularized system, a family of finite difference schemes using half-step staggered grids in space and time is used. All spatial and temporal derivatives are discretized with central finite differences that maintain the second order approximation in both space and time, respectively. A concrete *von Neumann* stability analysis for the numerical scheme indicates that the scheme is stable under the Courant-Friedrichs-Lewy (CFL) condition between the temporal and spatial grids (see Theorem 3). The novel features of this study include the good performance of the solution that present not only the well-posedness and stability but also the non-reflection property of the wave propagation compared to the classical PML model; even the regularized system does not possess PMLs in the original wave equation. This novelty is numerically illustrated in Section 4.

The remainder of the manuscript is organized as follows. Section 2 describes a regularized system for the un-split PML model of the acoustic wave equation and also contains the well-posedness of its solution based on the energy estimation. In Section 3, we develop a staggered finite difference scheme for the regularized system and determine the CFL condition for the numerical stability. In Section 4, several numerical results are presented to support our theoretical analysis and demonstrate the efficiency of the regularized system. Finally, some discussions are given in Section 5.

2. Regularized System

The aim of this section is to introduce a modified PML system using a regularization technique in a classical PML model for the acoustic wave equation. For the sake of argument, we let $H^1(\Omega) := \{\varphi : \varphi, \partial_x \varphi, \partial_y \varphi \in L^2(\Omega)\}$ and $H^{-1}(\Omega)$ be the Sobolev space and dual space of $H_0^1(\Omega)$, respectively.

The target problem we consider with here is a general second order acoustic wave equation with a variable sound speed $c(\mathbf{x}) > 0$ described by

$$u_{tt}(\mathbf{x},t) - c^2(\mathbf{x})\Delta u(\mathbf{x},t) = 0 \quad \forall (\mathbf{x},t) \in \mathbb{R}^2 \times (0,T]$$

with initial conditions $u(\cdot,0) = f$ and $u_t(\cdot,0) = 0$, where $supp(f) \subset \Omega_0$ with a domain $\Omega_0 \subset\subset [-a,a] \times [-b,b] \subset \mathbb{R}^2$. Here, $T > 0$ and the sound speed $c(\mathbf{x})$ is bounded by

$$0 < c_* \leq c(\mathbf{x}) \leq c^* < \infty. \tag{1}$$

Let the domain $\Omega := [-a - L_x, a + L_x] \times [-b - L_y, b + L_y]$ consist of the computational domain $[-a,a] \times [-b,b]$ surrounded by PML layers, where $a, b, L_x, L_y > 0$. Using a complex coordinate stretch, we consider the following system of the PML wave equation which is introduced in [28]: find $(u, \vec{q})$ satisfying

$$\begin{cases} \frac{1}{c^2} u_{tt}(\mathbf{x},t) + \alpha(\mathbf{x}) u_t(\mathbf{x},t) + \beta(\mathbf{x}) u(\mathbf{x},t) - \nabla \cdot \vec{q}(\mathbf{x},t) - \Delta u(\mathbf{x},t) = 0 & \forall (\mathbf{x},t) \in \Omega \times (0,T], \\ \vec{q}_t(\mathbf{x},t) + A(\mathbf{x}) \vec{q}(\mathbf{x},t) + B(\mathbf{x}) \nabla u(\mathbf{x},t) = 0 & \forall (\mathbf{x},t) \in \Omega \times (0,T], \end{cases} \tag{2}$$

with the initial conditions

$$u(\cdot, 0) := u_0 = f, \quad u_t(\cdot, 0) := u_1 = 0, \quad \vec{q}(\cdot, 0) := \vec{q}_0 = \vec{0},$$

and the zero Dirichlet boundary condition $u(\mathbf{x}, \cdot)|_{\partial \Omega} = 0$, where

$$\alpha(\mathbf{x}) := \frac{\sigma_x + \sigma_y}{c^2}, \quad \beta(\mathbf{x}) := \frac{\sigma_x \sigma_y}{c^2}, \quad A(\mathbf{x}) =: \begin{bmatrix} \sigma_x & 0 \\ 0 & \sigma_y \end{bmatrix}, \quad B(\mathbf{x}) := \begin{bmatrix} \sigma_x - \sigma_y & 0 \\ 0 & \sigma_y - \sigma_x \end{bmatrix}.$$

Here, the damping terms $\sigma_x := \sigma_x(x)$ and $\sigma_y := \sigma_y(y)$ are assumed to be nonnegative C^0 functions which vanish in the computational domain in the sense of the analytical continuation of the PML.

Please note that a weak solution $(u, \vec{q})$ of (2) is in $H_0^1(\Omega) \times \mathbb{L}^2(\Omega)$, i.e., $\nabla \cdot \vec{q} \in H^{-1}(\Omega)$, which regularity is not enough to show the existence. In order to provide regularity on the term by an operator, we introduce a mollifier ρ_ϵ. Let $\rho \in C^\infty(\mathbb{R}^2)$ with $supp(\rho) \subseteq B_1(0)$ satisfying $\int_{\mathbb{R}^2} \rho(\mathbf{x}) d\mathbf{x} = 1$. Then, for $\epsilon > 0$, one can define a mollifier $\rho_\epsilon(\mathbf{x})$ on $\mathbb{R}^2$ by

$$\rho_\epsilon(\mathbf{x}) = \epsilon^{-2} \rho\left(\frac{|\mathbf{x}|}{\epsilon}\right) \text{ and satisfies } \int_{\mathbb{R}^2} \rho_\epsilon(\mathbf{x}) d\mathbf{x} = 1 \text{ with } supp(\rho_\epsilon) \subseteq \overline{B_\epsilon(0)}.$$

Remark 1. Let $\mathcal{R} := -\Delta + I$ be the Riesz map from $H_0^1(\Omega) \to H^{-1}(\Omega)$. Then, we consider the operator $\delta_\epsilon : H^{-1}(\Omega) \to L^2(\Omega)$ given by

$$\delta_\epsilon(\varphi) = \mathcal{R} \circ \delta_\epsilon^* \circ \mathcal{R}^{-1}(\varphi) \text{ for all } \varphi \in H^{-1}(\Omega), \tag{3}$$

where $\delta_\epsilon^* : H_0^1(\Omega) \to H_0^1(\Omega) \cap H^2(\Omega)$ is a linear bounded operator such that $\delta_\epsilon^* \to \mathbf{1}$, the identity operator in $H_0^1(\Omega)$, as $\epsilon \to 0$ in the strong operator topology (see, for detail, Theorem 3 on page 7 in [30]). Then, we obtain

$$\delta_\epsilon \to \mathbf{1} \text{ as } \epsilon \to 0 \text{ in the strong operator topology}$$

and $\|\delta_\epsilon(\varphi)\|_{L^2(\Omega)} \leq C_{\delta_\epsilon} \|\varphi\|_{H^{-1}(\Omega)}$ for some $C_{\delta_\epsilon} > 0$. Furthermore, by the isometry of $\mathcal{R}$,

$$\|\delta_\epsilon(\varphi) - \varphi\|_{H^{-1}(\Omega)} = \|\delta_\epsilon^* u - u\|_{H_0^1(\Omega)} \to 0 \text{ as } \epsilon \to 0$$

for $u \in H_0^1(\Omega)$ such that $\mathcal{R}(u) = \varphi$. Please note that δ_ε is a linear and bounded operator from $H^{-1}(\Omega)$ to $L^2(\Omega)$.

Now, following [8,31], we introduce a regularized system of the classical PML model (2) by using δ_ε in the term $\nabla \cdot \vec{q}$, which is given by

$$\begin{cases} \frac{1}{c^2} u_{tt}(\mathbf{x},t) + \alpha(\mathbf{x}) u_t(\mathbf{x},t) + \beta(\mathbf{x}) u(\mathbf{x},t) - \delta_\varepsilon \nabla \cdot \vec{q}(\mathbf{x},t) - \Delta u(\mathbf{x},t) = 0, \\ \vec{q}_t(\mathbf{x},t) + A(\mathbf{x}) \vec{q}(\mathbf{x},t) + B(\mathbf{x}) \nabla u(\mathbf{x},t) = 0, \end{cases} \quad (4)$$

with initial and boundary conditions

$$u(\cdot,0) := u_0 = f, \quad u_t(\cdot,0) := u_1 = 0, \quad \vec{q}(\cdot,0) := \vec{q}_0 = \vec{0}, \quad u(\mathbf{x},\cdot)|_{\partial\Omega} = 0.$$

The remainder of this section details the analysis of the well-posedness of the solution to the regularized system (4) based on the energy estimation under the assumption that the dampings σ_x and σ_y are in $L^\infty(\Omega)$.

2.1. Energy Estimate of Weak Solution

We assume that the damping functions σ_x, σ_y satisfy $\sigma_x, \sigma_y \in L^\infty(\Omega)$, which implies that

$$\begin{aligned} \|\alpha\|_\infty &= \|\sigma_x + \sigma_y\|_\infty < \infty, & \|\beta\|_\infty &\leq \|\sigma_x \sigma_y\|_\infty < \infty, \\ \|A\|_2 &:= \max\{\|\sigma_x\|_\infty, \|\sigma_y\|_\infty\} < \infty, & \|B\|_2 &\leq \sqrt{2}(\|\sigma_x\|_\infty + \|\sigma_y\|_\infty) < \infty \end{aligned} \quad (5)$$

under the condition of $c(\mathbf{x}) = 1$ in the layers of the PML model (2), where $\|\cdot\|_\infty$ denotes the L^∞-norm. Under these assumptions, the aim of this subsection is to provide an energy estimation of the weak solution of (4) in the sense that

$$u \in L^2(0,T; H_0^1(\Omega)), \quad \vec{q} \in L^2(0,T; \mathbb{L}^2(\Omega)) \quad (6)$$

with

$$u_t \in L^2(0,T; L^2(\Omega)), \quad u_{tt} \in L^2(0,T; H^{-1}(\Omega)), \quad \vec{q}_t \in L^2(0,T; \mathbb{L}^2(\Omega)),$$

which satisfies

$$\begin{cases} \left\langle \frac{1}{c^2} u_{tt}, w \right\rangle + (\alpha u_t, w) + (\beta u, w) - (\delta_\varepsilon \nabla \cdot \vec{q}, w) + (\nabla u, \nabla w) = 0, \\ (\vec{q}_t, \vec{v}) + (A\vec{q}, \vec{v}) + (B \nabla u, \vec{v}) = 0 \end{cases} \quad (7)$$

for each $w \in H_0^1(\Omega), \vec{v} \in \mathbb{L}^2(\Omega)$, and almost everywhere $0 \leq t \leq T$ and the initial data satisfy

$$(u(0), w) = (u_0, w), \quad < u_t(0), w > = (u_1, w), \quad \text{and } (\vec{q}(0), \vec{v}) = (\vec{q}_0, \vec{v}) \quad (8)$$

for each $w \in H_0^1(\Omega), \vec{v} \in \mathbb{L}^2(\Omega)$. Here, $< \cdot, \cdot >$ denotes the duality pairing between $H^{-1}(\Omega)$ and $H_0^1(\Omega)$, and $(\cdot, \cdot)$ is the inner product in $L^2(\Omega)$. In addition, the time derivatives are understood in a distributional sense.

Remark 2. We note that $u \in C([0,T]; L^2(\Omega))$, $u_t \in C([0,T]; H^{-1}(\Omega))$, and $\vec{q} \in C([0,T]; \mathbb{L}^2(\Omega))$. (see Theorem 2, Chapter 5.9.2 [32] for detail). Consequently, the equalities in (7), (8) make sense.

To investigate the weak solution of (4) that satisfies (7) and (8), we use the standard Galerkin approximation and estimate the energy of the solution, which will be used to show the well-posedness of the regularized system (4) in the subsequent subsection. Let $\{w_j | j \in \mathbb{N}\}$ be an c^{-2}-weighted orthonormal basis in $L^2(\Omega)$, i.e., $(c^{-2}w_j, w_k) = \delta_{jk}$, where the Kronecker delta is given by $\delta_{jk} = \begin{cases} 0 & \text{if } j \neq k, \\ 1 & \text{if } j = k \end{cases}$ of the eigenfunctions of the eigenvalue problem

$$\begin{cases} c^2 \Delta w = \lambda w & \text{in } \Omega, \quad \lambda \in \mathbb{C}, \\ w = 0 & \text{on } \partial \Omega. \end{cases}$$

Let $\mathcal{U}_k$ be the subspace generated by the orthonormal system $\{w_1, w_2, \cdots, w_k\}$ of $L^2(\Omega)$. Then, one can see that $\mathcal{U}_k$ also becomes the c^{-2}-weighted orthogonal basis of $H_0^1(\Omega)$ in the sense that

$$\left(c^{-2}w_j, w_k\right) + (\nabla w_j, \nabla w_k) = 0 \quad \text{if } j \neq k.$$

Let us also denote $\mathcal{Q}_k$, which is the space generated by the smooth functions $\{\vec{v}_1, \vec{v}_2, \cdots, \vec{v}_k\}$ such that $\{\vec{v}_k : k \in \mathbb{N}\}$ is an orthonormal basis of $\mathbb{L}^2(\Omega)$. We now construct approximate solutions $\left(u^k, \vec{q}^k\right), k = 1, 2, 3, \cdots$, in the form

$$u^k(t) = \sum_{j=1}^{k} g_j^k(t) w_j, \quad \vec{q}^k(t) = \sum_{j=1}^{k} h_j^k(t) \vec{v}_j, \tag{9}$$

whose coefficients $g_j^k(t), h_j^k(t), j = 1, 2, \cdots, k$, are chosen so that

$$g_j^k(0) = (u_0, w_j), \quad (g_j^k)_t(0) = (u_1, w_j), \quad h_j^k(0) = (\vec{q}_0, \vec{v}_j)$$

and

$$\begin{cases} \left(\dfrac{1}{c^2} u_{tt}^k, w_j\right) + \left(\alpha u_t^k + \beta u^k - \delta_\varepsilon \nabla \cdot \vec{q}^k, w_j\right) + \left(\nabla u^k, \nabla w_j\right) = 0, \\ \left(\vec{q}_t^k, \vec{v}_j\right) + \left(A\vec{q}^k, \vec{v}_j\right) + \left(B\nabla u^k, \vec{v}_j\right) = 0 \end{cases} \tag{10}$$

are satisfied for all $w_j \in \mathcal{U}_k, \vec{v}_j \in \mathcal{Q}_k, j = 1, \cdots, k$. For each integer $k = 1, 2, \cdots$, the standard theory of ordinary differential equations guarantees the existence of the approximation $\left(u^k(t), \vec{q}^k(t)\right)$ satisfying (9) and (10).

The following theorem gives a uniform bound of energy of the approximate solutions (9), which allows us to send $k \to \infty$.

Theorem 1. *There exists a constant $C_T > 0$ that depends only on $\sigma_x, \sigma_y, \Omega$, and T such that for $k \geq 1$*

$$\max_{0 \leq t \leq T} E_k(t) + \left\|u_{tt}^k\right\|_{L^2(0,T;H^{-1}(\Omega))} + \left\|\vec{q}_t^k\right\|_{L^2(0,T;\mathbb{L}^2(\Omega))} \leq C_T \left(\|u_0\|_{H_0^1(\Omega)}^2 + \|u_1\|_{L^2(\Omega)}^2 + \|\vec{q}_0\|_{\mathbb{L}^2(\Omega)}^2\right),$$

where the energy $E_k(t)$ is defined by

$$E_k(t) = \|\tfrac{1}{c} u_t^k(t)\|_{L^2(\Omega)}^2 + \|\nabla u^k(t)\|_{L^2(\Omega)}^2 + \|\vec{q}^k(t)\|_{L^2(\Omega)}^2.$$

Proof. Please note that $u_t^k \in \mathcal{U}^k$ and $\vec{q}^k \in \mathcal{Q}^k$. Hence, we apply u_t^k and $\vec{q}^k$ in the first and second equations of (10), respectively, to obtain

$$\begin{cases} \left(\dfrac{1}{c^2} u_{tt}^k, u_t^k\right) + \left(\alpha u_t^k + \beta u^k - \delta_\varepsilon \nabla \cdot \vec{q}^k, u_t^k\right) + \left(\nabla u^k, \nabla u_t^k\right) = 0, \\ \left(\vec{q}_t^k, \vec{q}^k\right) + \left(A\vec{q}^k, \vec{q}^k\right) + \left(B\nabla u^k, \vec{q}^k\right) = 0 \end{cases}$$

for almost everywhere $0 \leq t \leq T$. Combining the two equations with the equality $\left(\tfrac{1}{c^2} u_{tt}^k, u_t^k\right) = \tfrac{d}{dt}\left(\tfrac{1}{2}\|\tfrac{1}{c} u_t^k\|_{L^2(\Omega)}^2\right)$, we obtain

$$\frac{1}{2}\frac{d}{dt}E_k + F_k^1 + F_k^2 = 0,$$

where

$$F_k^1 = \left(\alpha u_t^k, u_t^k\right) + \left(\beta u^k, u_t^k\right) - \left(\delta_\varepsilon \nabla \cdot \vec{q}^k, u_t^k\right), \quad F_k^2 = \left(A\vec{q}^k, \vec{q}^k\right) + \left(B\nabla u^k, \vec{q}^k\right).$$

Based on the linear bounded operator $\varphi \longmapsto \delta_\varepsilon(\varphi)$, Hölder's inequality, assumptions for σ_x, σ_y, and Poincaré inequality, it can be noted that $E_k(t)$ satisfies the inequality

$$\frac{dE_k}{dt} \leq C_k^\varepsilon E_k \quad \text{for a suitable constant } C_k^\varepsilon > 0.$$

Furthermore, by applying Gronwall's inequality, Poincaré inequality, and (1) in the above equation, one can obtain

$$\max_{0 \leq t \leq T} \left(\|u^k(t)\|_{H_0^1(\Omega)}^2 + \|u_t^k(t)\|_{L^2(\Omega)}^2 + \|\vec{q}^k\|_{\mathbb{L}^2(\Omega)}^2\right) \leq C\left(\|u_0\|_{H_0^1(\Omega)}^2 + \|u_1\|_{L^2(\Omega)}^2 + \|\vec{q}_0\|_{\mathbb{L}^2(\Omega)}^2\right) \quad (11)$$

for some $C > 0$.

Fix any $w \in H_0^1(\Omega)$ with $\|w\|_{H_0^1(\Omega)} \leq 1$ and $\vec{v} \in \mathbb{L}^2(\Omega)$ with $\|\vec{v}\|_{\mathbb{L}^2(\Omega)} \leq 1$, and write $w = w^1 + w^2$ and $\vec{v} = \vec{v}^1 + \vec{v}^2$, where

$$w^1 \in \text{span}\{w_j\}_{j=1}^k, \quad \left(\frac{1}{c^2}w^2, w_j\right) = 0 \text{ for } j = 1, \cdots, k$$

and

$$\vec{v}^1 \in \text{span}\{\vec{v}_j\}_{j=1}^k, \quad \left(\vec{v}^2, \vec{v}_j\right) = 0 \text{ for } j = 1, \cdots, k.$$

From (9) and (10), we have

$$\left\langle \frac{1}{c^2}u_{tt}^k, w \right\rangle = \left(\frac{1}{c^2}u_{tt}^k, w\right) = \left(\frac{1}{c^2}u_{tt}^k, w^1\right)$$
$$= -(\alpha u_t^k + \beta u^k, w^1) - (\delta_\varepsilon \nabla \cdot \vec{q}^k, w^1) + (\nabla u^k, \nabla w^1),$$
$$\left(\vec{q}_t^k, \vec{v}\right) = \left(\vec{q}_t^k, \vec{v}^1\right) = -\left(A\vec{q}^k, \vec{v}^1\right) - \left(B\nabla u^k, \vec{v}^1\right).$$

Thus, we have

$$\left|\left\langle u_{tt}^k, w \right\rangle\right| + \left|\left(\vec{q}_t^k, \vec{v}\right)\right| \leq C\left(\|u^k\|_{H_0^1(\Omega)} + \|u_t^k\|_{L^2(\Omega)} + \|\vec{q}^k\|_{\mathbb{L}^2(\Omega)}\right).$$

Consequently, we obtain

$$\int_0^T \left(\|u_{tt}^k\|_{H^{-1}(\Omega)} + \|\vec{q}_t\|_{\mathbb{L}^2(\Omega)}\right) dt \leq C \int_0^T \left(\|u^k\|_{H_0^1(\Omega)}^2 + \|u_t^k\|_{L^2(\Omega)}^2 + \|\vec{q}^k\|_{\mathbb{L}^2(\Omega)}^2\right) dt \quad (12)$$
$$\leq C_T \left(\|u_0\|_{H_0^1(\Omega)}^2 + \|u_1\|_{L^2(\Omega)}^2 + \|\vec{q}_0\|_{\mathbb{L}^2(\Omega)}^2\right).$$

The proof is carried out by combining (11) and (12). □

2.2. Existence and Uniqueness

In this subsection, we will discuss the well-posedness of the regularized system by demonstrating the existence and uniqueness of the solution (6) based on the result of Theorem 1.

Theorem 2. *(Existence and Uniqueness) Assume that the initial data $(u_0, u_1, \vec{q}_0)$ are in $H_0^1(\Omega) \times L^2(\Omega) \times \mathbb{L}^2(\Omega)$. Then, the system (4) has a unique weak solution provided by $\sigma_x, \sigma_y \in L^\infty(\Omega)$.*

Proof. The energy estimates of Theorem 1 and the standard Galerkin method enable the existence of a weak solution using the fact that $\nabla \cdot : \mathbb{L}^2(\Omega) \to H^{-1}(\Omega)$ and $\delta_\epsilon : H^{-1}(\Omega) \to L^2(\Omega)$ are continuous almost everywhere $t \in [0, T]$ (see [31] for detail proof of uniqueness). □

Remark 3. *The most important concern in the proof is the estimation of the term $\delta_\epsilon \nabla \cdot \vec{q}$ in the regularized system, which has roles of a convolution, improving the stability of the system from the regularization of the term from $H^{-1}(\Omega)$ to $\mathbb{L}^2(\Omega)$.*

3. Numerical Scheme

The aim of this section is to introduce a staggered finite difference method for discretizing the regularized system and to find a stability condition for the numerical scheme. For the staggered finite difference method, we use a family of finite difference schemes [33] with half-step staggered grids in space and time. All spatial derivatives are discretized with the centered finite differences over two or three cells, which guarantees a second order approximation in space. For the time discretization, we also use the centered finite differences for the first and second order time derivatives on a uniform mesh, which is also of the second order approximation in time. Based on the standard von Neumann stability analysis technique, we analyze the stability of the numerical scheme and obtain its CFL condition.

3.1. Staggered Finite Differences

Let $\triangle t > 0$ denote the time step size and $\triangle x > 0$ and $\triangle y > 0$ denote the spatial mesh sizes in the x and y directions, respectively. In addition, we also introduce the time step $t_n = n\triangle t$ and the spatial nodes $x_i = i\triangle x$ and $y_j = j\triangle y$ for $n \in \mathbb{N} \cup \{0\}$ and $i, j \in \mathbb{Z}$. We also define staggered nodes in the time direction and the x and y directions, respectively, as $t_{n\pm\frac{1}{2}} = t_n \pm \frac{1}{2}\triangle t$, $x_{i\pm\frac{1}{2}} = x_i \pm \frac{1}{2}\triangle x$, and $y_{j\pm\frac{1}{2}} = y_j \pm \frac{1}{2}\triangle y$ for $n, i, j \in \mathbb{N}$. To simplify the notation, we denote $u_{i,j}^n := u(t_n, x_i, y_j)$ and $q_{\alpha_{i+\frac{1}{2},j+\frac{1}{2}}}^{n+\frac{1}{2}} := q_\alpha(t_{n+\frac{1}{2}}, x_{i+\frac{1}{2}}, y_{j+\frac{1}{2}})$ for $\vec{q} = (q_x, q_y)$, $\alpha = x, y$. For the discretization of the regularization defined in Remark 1 for the regularized system, the smooth function $\rho_\epsilon(x, y)$ chosen in the following examples is constant on a rectangle centered at zero,

$$\rho_\epsilon(x, y) = \rho_{\epsilon 1}(x)\rho_{\epsilon 2}(y), \qquad (13)$$

where

$$\rho_{\epsilon_k}(\xi) = \begin{cases} \frac{1}{\epsilon_k} & \text{if } \xi \in [-\frac{\epsilon_k}{2}, \frac{\epsilon_k}{2}], \quad k = 1, 2, \\ 0 & \text{elsewhere.} \end{cases}$$

For a given two-dimensional finite difference grid with spatial sizes $\triangle x$ and $\triangle y$, a possible choice of ϵ_k is $\epsilon_1 = n_x \triangle x$ and $\epsilon_2 = n_y \triangle y$ with $n_x, n_y \in \mathbb{N}$. For instance, with $n_x = n_y = 1$ and the usual integration formula (see Chapter 3 in [34]), we discretize the regularized term $\delta_\epsilon(v)_{i,j} := (\rho_\epsilon * v)_{i,j}$, using the 9-point central difference formula, as follows:

$$(\rho_\epsilon * v)_{i,j} = \frac{1}{16}\left(4v_{i,j} + 2v_{i\pm 1,j} + 2v_{i,j\pm 1} + v_{i\pm 1,j+1} + v_{i\pm 1,j-1}\right).$$

Let us now introduce new notations

$$A_{i+\frac{1}{2}}^{x\pm} := 1 \pm \frac{\triangle t}{2}\sigma_{x_{i+\frac{1}{2}}}, \quad A_{j+\frac{1}{2}}^{y\pm} := 1 \pm \frac{\triangle t}{2}\sigma_{y_{j+\frac{1}{2}}},$$

and for $k = i, j$, $\alpha = x, y$,

$$A_{i,j}^{xy\pm} := 1 \pm \frac{\triangle t}{2}(\sigma_{x_i} + \sigma_{y_j}), \quad \sigma_{\alpha_k} := \sigma_\alpha(\alpha_k), \quad \sigma_{\alpha_{k+\frac{1}{2}}} := \sigma_\alpha(\alpha_{k+\frac{1}{2}}).$$

Based on these notations, the staggered finite difference scheme for discretizing the regularized system is defined in the following steps.

Step 1. Compute $\left(qx_{i+\frac{1}{2},j+\frac{1}{2}}^{n+\frac{1}{2}}, qy_{i+\frac{1}{2},j+\frac{1}{2}}^{n+\frac{1}{2}}\right)$,

$$A_{i+\frac{1}{2}}^{x+} qx_{i+\frac{1}{2},j+\frac{1}{2}}^{n+\frac{1}{2}} = A_{i+\frac{1}{2}}^{x-} qx_{i+\frac{1}{2},j+\frac{1}{2}}^{n-\frac{1}{2}} - \Delta t (\sigma_{x_{i+\frac{1}{2}}} - \sigma_{y_{j+\frac{1}{2}}}) \tilde{\partial}_x u_{i+\frac{1}{2},j+\frac{1}{2}}^{n},$$

$$A_{j+\frac{1}{2}}^{y+} qy_{i+\frac{1}{2},j+\frac{1}{2}}^{n+\frac{1}{2}} = A_{j+\frac{1}{2}}^{y-} qy_{i+\frac{1}{2},j+\frac{1}{2}}^{n-\frac{1}{2}} - \Delta t (\sigma_{y_{j+\frac{1}{2}}} - \sigma_{x_{i+\frac{1}{2}}}) \tilde{\partial}_y u_{i+\frac{1}{2},j+\frac{1}{2}}^{n},$$

where the cell averages $\tilde{\partial}_x u_{i+\frac{1}{2},j+\frac{1}{2}}^n$ and $\tilde{\partial}_y u_{i+\frac{1}{2},j+\frac{1}{2}}^n$ are defined as

$$\tilde{\partial}_x u_{i+\frac{1}{2},j+\frac{1}{2}}^n = \frac{u_{i+1,j+1}^n - u_{i,j+1}^n + u_{i+1,j}^n - u_{i,j}^n}{2\Delta x}, \quad \tilde{\partial}_y u_{i+\frac{1}{2},j+\frac{1}{2}}^n = \frac{u_{i+1,j+1}^n - u_{i+1,j}^n + u_{i,j+1}^n - u_{i,j}^n}{2\Delta y}.$$

The definition of the cell averages allows us to compute the regularized term in (3)

$$(\delta_\varepsilon \partial_x qx)_{i,j}^n := (\rho_\varepsilon * \partial_x qx)_{i,j}^n, \quad (\delta_\varepsilon \partial_y qy)_{i,j}^n := (\rho_\varepsilon * \partial_x qy)_{i,j}^n$$

for $\partial_x q_{x_{i,j}}^n = \frac{1}{2}\left(\tilde{\partial}_x q_{x_{i,j}}^{n+\frac{1}{2}} + \tilde{\partial}_x q_{x_{i,j}}^{n-\frac{1}{2}}\right)$ and $\partial_y q_{y_{i,j}}^n = \frac{1}{2}\left(\tilde{\partial}_y q_{y_{i,j}}^{n+\frac{1}{2}} + \tilde{\partial}_y q_{y_{i,j}}^{n-\frac{1}{2}}\right)$, where the cell averages of the derivatives of the function $(q_{x_{i,j}}^{n\pm\frac{1}{2}}, q_{y_{i,j}}^{n\pm\frac{1}{2}})$ are defined as

$$\tilde{\partial}_x q_{x_{i,j}}^{n\pm\frac{1}{2}} = \frac{1}{2\Delta x}\left(q x_{i+\frac{1}{2},j+\frac{1}{2}}^{n\pm\frac{1}{2}} - q x_{i-\frac{1}{2},j+\frac{1}{2}}^{n\pm\frac{1}{2}} + q x_{i+\frac{1}{2},j-\frac{1}{2}}^{n\pm\frac{1}{2}} - q x_{i-\frac{1}{2},j-\frac{1}{2}}^{n\pm\frac{1}{2}}\right),$$

$$\tilde{\partial}_y q_{y_{i,j}}^{n\pm\frac{1}{2}} = \frac{1}{2\Delta y}\left(q y_{i+\frac{1}{2},j+\frac{1}{2}}^{n\pm\frac{1}{2}} - q y_{i+\frac{1}{2},j-\frac{1}{2}}^{n\pm\frac{1}{2}} + q y_{i-\frac{1}{2},j+\frac{1}{2}}^{n\pm\frac{1}{2}} - q y_{i-\frac{1}{2},j-\frac{1}{2}}^{n\pm\frac{1}{2}}\right).$$

Step 2. Compute $u_{i,j}^{n+1}$,

$$A_{i,j}^{xy+} u_{i,j}^{n+1} = 2u_{i,j}^n - A_{i,j}^{xy-} u_{i,j}^{n-1} + \Delta t^2 \left(-\sigma_{i,j}^{xy} u_{i,j}^n + c_{i,j}^2 \left((\delta_\varepsilon \partial_x qx)_{i,j}^n + (\delta_\varepsilon \partial_y qy)_{i,j}^n\right) + c_{i,j}^2 \Delta_n u_{i,j}^n\right), \quad (14)$$

where

$$\sigma_{i,j}^{xy} = \sigma_{x_i} \sigma_{y_j}, \quad c_{i,j} = c(x_i, y_j), \quad \Delta_n u_{i,j}^n = \frac{u_{i+1,j}^n - 2u_{i,j}^n + u_{i-1,j}^n}{\Delta x^2} + \frac{u_{i,j+1}^n - 2u_{i,j}^n + u_{i,j-1}^n}{\Delta y^2}.$$

3.2. Stability Analysis

To obtain the stability condition of the staggered finite difference scheme defined above, we restrict our concern to the constant damping case with $\sigma_x = \sigma_y = \sigma_0 \geq 0$ for simplicity in our analysis. The stability condition for the scheme in the computational domain is as follows.

Remark 4. *The CFL condition of scheme* (13)–(14) *in the computational area (i.e., $\sigma_x = \sigma_y = 0$) is*

$$c\frac{\Delta t}{h} \leq \frac{1}{\sqrt{2}}$$

for $\Delta x = \Delta y = h$ from the standard von Neumann stability analysis technique.

Generally the stability condition for the staggered finite difference scheme developed in Section 3.1 can be obtained as follows.

Theorem 3. *Assume that $\sigma_x = \sigma_y = \sigma_0 > 0$ and the sound speed c are constants. Then, the discrete scheme (13)–(14) is stable if the CFL condition*

$$c \Delta t \leq \frac{h}{\sqrt{2}} \frac{1}{(1 + \frac{\sigma_0^2 h^2}{8c^2})^{1/2}} \qquad (15)$$

is satisfied for $\Delta x = \Delta y = h$.

To prove Theorem 3 and use the technique of the standard *von Neumann* stability analysis, we recall the definition of the simple *von Neumann* polynomial and some of its properties as follows.

Definition 1. *A polynomial is a simple von Neumann polynomial if all its roots, r, lie on the unit disk ($|B(0,r)| < 1$) and its roots on the unit circle are simple roots.*

The following theorem demonstrates that a simple *von Neumann* polynomial can be a sufficient stability condition.

Theorem 4. *A sufficient stability condition is that ϕ be a simple von Neumann polynomial, where ϕ is the characteristic polynomial (see [35] for the proof).*

With Theorem 4, the stability condition for a polynomial is presented in the following.

Theorem 5. *Let ϕ be a polynomial of degree p written as*

$$\phi(z) = c_0 + c_1 z + \cdots + c_p z^p,$$

where $c_0, c_1, \cdots, c_p \in \mathbb{C}$ and $c_p \neq 0$. The polynomial ϕ is a simple von Neumann polynomial if and only if ϕ^0 is a simple von Neumann polynomial and $|\phi(0)| \leq |\bar{\phi}(0)|$, where ϕ^0 is defined as

$$\phi^0(z) = \frac{\bar{\phi}(0)\phi(z) - \phi(0)\bar{\phi}(z)}{z},$$

and the conjugate polynomial $\bar{\phi}$ is defined as

$$\bar{\phi}(z) = \bar{c}_p + \bar{c}_{p-1} z + \cdots + \bar{c}_0 z^p,$$

where $\bar{c}$ is the complex conjugate of c. The main ingredient in the proof of the theorem is Rouché's theorem; the proof is detailed in [36].

Now, we can computationally verify the stability condition (15) in Theorem 3 using Theorems 4 and 5.

Proof of Theorem 3. Assume that $\sigma_x = \sigma_y = \sigma_0$ in scheme (13)–(14) and we rewrite the scheme as the second order central difference scheme of the variables u and $\vec{q}$.

$$\frac{1}{c_{i,j}^2}\left[\frac{u_{i,j}^{n+1} - 2u_{i,j}^n + u_{i,j}^{n-1}}{\Delta t^2} + 2\sigma_0 \frac{u_{i,j}^{n+1} - u_{i,j}^{n-1}}{2\Delta t} + \sigma_0^2 u_{i,j}^n\right] \qquad (16)$$

$$= \frac{u_{i+1,j}^n - 2u_{i,j}^n + u_{i-1,j}^n}{\Delta x^2} + \frac{u_{i,j+1}^n - 2u_{i,j}^n + u_{i,j-1}^n}{\Delta y^2} + (\rho_\epsilon * \partial_x q_x)_{i,j}^n + (\rho_\epsilon * \partial_y q_y)_{i,j}^n,$$

$$\frac{\vec{q}_{i+\frac{1}{2},j+\frac{1}{2}}^{n+\frac{1}{2}} - \vec{q}_{i+\frac{1}{2},j+\frac{1}{2}}^{n-\frac{1}{2}}}{\Delta t} + \sigma_0 \frac{\vec{q}_{i+\frac{1}{2},j+\frac{1}{2}}^{n+\frac{1}{2}} + \vec{q}_{i+\frac{1}{2},j+\frac{1}{2}}^{n-\frac{1}{2}}}{2} = \vec{0}. \qquad (17)$$

By von Neumann analysis, we can assume a spatial dependence of the following form in the field quantities:

$$u_{i,j}^{n+1} = \hat{u}^{n+1}(k_x,k_y)e^{ik_xx_i+ik_yy_j}, \quad u_{i,j}^n = \hat{u}^n(k_x,k_y)e^{ik_xx_i+ik_yy_j},$$

$$\tilde{q}_{i+\frac{1}{2},j+\frac{1}{2}}^{n+\frac{1}{2}} = \hat{q}_{i+\frac{1}{2},j+\frac{1}{2}}^{n+\frac{1}{2}}(k_x,k_y)e^{ik_xx_{i+\frac{1}{2}}+ik_yy_{j+\frac{1}{2}}},$$

where k_x, k_y, is the component of the wave vector $\vec{k}$, i.e., $\vec{k} = (k_x,k_y)^T$, and the wave number is $k = \sqrt{k_x^2 + k_y^2}$. Then, we have the system $\left[\hat{u}^{n+1},\hat{u}^n,\hat{q}_x^{n+\frac{1}{2}},\hat{q}_y^{n+\frac{1}{2}}\right]^T = G\left[\hat{u}^n,\hat{u}^{n-1},\hat{q}_x^{n-\frac{1}{2}},\hat{q}_y^{n-\frac{1}{2}}\right]^T$, where the amplification matrix G of scheme (16), (17) is given by

$$G = \begin{bmatrix} -\frac{c_1}{c_2} & -\frac{c_0}{c_2} & C_{\hat{q}_x} & C_{\hat{q}_y} \\ 1 & 0 & 0 & 0 \\ 0 & 0 & \eta & 0 \\ 0 & 0 & 0 & \eta \end{bmatrix},$$

where $C_{\hat{q}_x}$ and $C_{\hat{q}_y}$ satisfy $c_2\hat{u}^{n+1} + c_1\hat{u}^n + c_0\hat{u}^{n-1} = C_{\hat{q}_x}\hat{q}_x^{n-\frac{1}{2}} + C_{\hat{q}_y}\hat{q}_y^{n-\frac{1}{2}}$ with $c_0 = \frac{1}{\Delta t^2} - \frac{\sigma_0}{\Delta t}$, $c_1 = -\frac{2}{\Delta t^2} - 2c^2\frac{\cos(k_x\Delta x)-1}{\Delta x^2} - 2c^2\frac{\cos(k_y\Delta y)-1}{\Delta y^2} + \sigma_0^2$, $c_2 = \frac{1}{\Delta t^2} + \frac{\sigma_0}{\Delta t}$, and $\eta = \frac{1-\frac{\Delta t}{2}\sigma_0}{1+\frac{\Delta t}{2}\sigma_0}$. Then, it is noted that the characteristic function of G is given by

$$\phi(G) = \left(G^2 + \frac{c_1}{c_2}G + \frac{c_0}{c_2}\right)(G-\eta)^2.$$

Please note that $|\eta| < 1$ by the assumption. It can be observed from Theorem 5 that $\phi(G)$ is a simple von Neumann polynomial if and only if $|c_1| \leq |c_0 + c_2|$, i.e.,

$$\left|\frac{2}{\Delta t^2} + 2c^2\frac{\cos(k_xh)+\cos(k_yh)-2}{h^2} - \sigma_0^2\right| \leq \frac{2}{\Delta t^2}, \quad \text{for } h = \Delta x = \Delta y.$$

This inequality gives the CFL condition (15), which completes the proof. □

Remark 5. *From the proof of Theorem 3, we notice that the characteristic function ϕ of the amplification matrix G does not depend on any quantity related to the regularized term. That is, the staggered finite difference scheme corresponding to the classical PML model (2) with a constant damping in the layers is stable under the CFL condition (15).*

4. Numerical Result

The aim of this section is to provide numerical evidence of the well-posedness of the regularized system and the non-reflection properties of the acoustic wave in the layers of the classical PML model. For the discussion of the non-reflection properties, we demonstrate the behavior of the maximum error at t_n defined as the maximum of the differences between the numerical solution and a reference solution in the computational domain $\Omega_0 := [0,1] \times [0,1]$. Here, the reference solution is taken in the same computational domain instead of the layers with an additional large domain, for example, 15 times wider in the x and y directions in our experiment, causing the wave in the computational domain to be unaffected by the wave propagating from outside in the chosen long-time step. Furthermore, we use the energy method introduced in [37] and numerically examine the well-posedness or stability of the model (4) by observing the long-time behavior of the acoustic wave energy defined by

$$\mathcal{E}(t) = \frac{1}{2}\int_{\Omega_0}\left(\frac{1}{c^2}u_t(t)^2 + \nabla u(t)\cdot\nabla u(t)\right)dx. \tag{18}$$

For the numerical simulation, we use the same initial condition defined by (4) and, in the absorbing layer, the damping function of the form given by

$$\sigma_{x_k}(x_k) = \begin{cases} 0, & |x_k| < a_k = 1, \\ \sigma_0 \left(\dfrac{|x_k - 1|}{L} \right)^\beta, & 1 \leq |x_k| \leq 1 + L, \end{cases} \quad (19)$$

where $\beta = 0, 1, 2$, σ_0 is a given constant and L denotes the thickness of the layers.

For the comparisons of non-reflection property, we first demonstrate the maximum error for both Formulas (2) and (4) with two sets of thickness and damping as $(L, \sigma_0) = (0.25, 30)$ and $(L, \sigma_0) = (0.1875, 30)$. The numerical results are displayed in Figure 1. The classical PML has slightly smaller errors than the modified one in both cases, as shown in Figure 1, but it can be observed that these errors of the modified one can be reduced by simply increasing small amounts of thickness or damping such as $L = 0.27$ or $\sigma_0 = 35$.

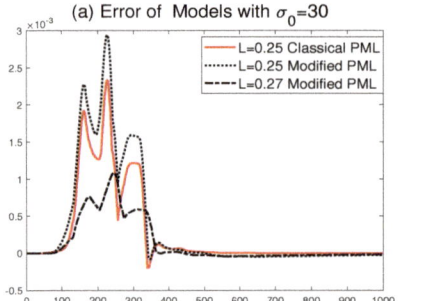
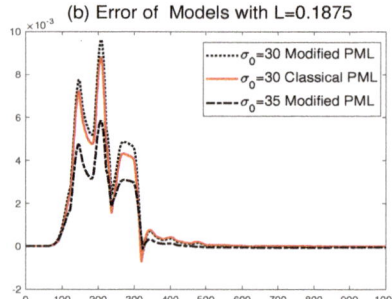

Figure 1. Comparison of errors: (**a**) a fixed damping $\sigma_0 = 30$, (**b**) a thickness $L = 0.1875$ ($\beta = 2$)

To see the influence of absorbing property by incidence angle, we demonstrate both formulas with different positions of source function. The resulted differences between reference and computed values of the solution during simulation at one point within the computational domain are plotted in Figure 2. The errors of the classical PML have relatively smaller than the modified one and both formulas have slightly better absorbing property when the angle of incidence to the interface between the computational domain and the layers is bigger.

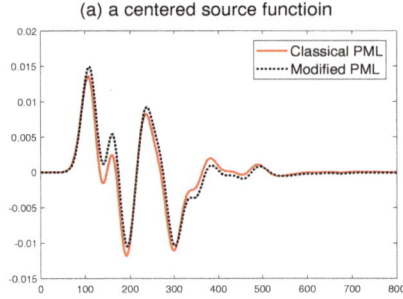
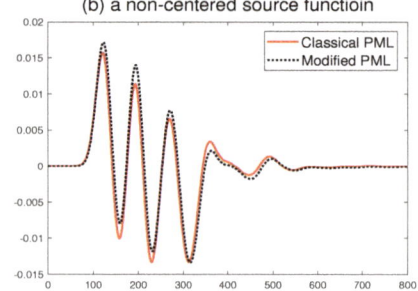

Figure 2. Comparison of the difference at a point from different positions of source function with $\sigma_0 = 35$ and $L = 0.1$

Next, to investigate the energy $\mathcal{E}(t)$ behavior, we choose a time step size Δt of $\Delta x/3$, which satisfies the CFL condition (15) to guarantee the stability of the staggered finite difference scheme (see Remark 4). Here, the first order backward and second order central finite differences in time and space, respectively, are used to discretize the energy $\mathcal{E}(t_n)$ of (18) at each time step t_n. We investigate the behavior of the energy for a long-time simulation at time $t_n = 10{,}000$ according to the thickness of the layers and magnitude of the damping. The numerical results are displayed in Figure 3: (a) the energy with various dampings $\sigma_0 = 40, 50, 50, 60, 70$ for a fixed thickness $L = 0.0625$ and (b) the energy with various thicknesses $L = 0.0625, 0.1, 0.125, 0.15$ for a fixed damping $\sigma_0 = 50$. The results indicate that the numerical stability of the modified formula is consistently stable in the long-time simulation regardless of the magnitudes of damping and thickness of the layer. This provides proof of the well-posedness of the developed system and numerical stability for the finite difference method.

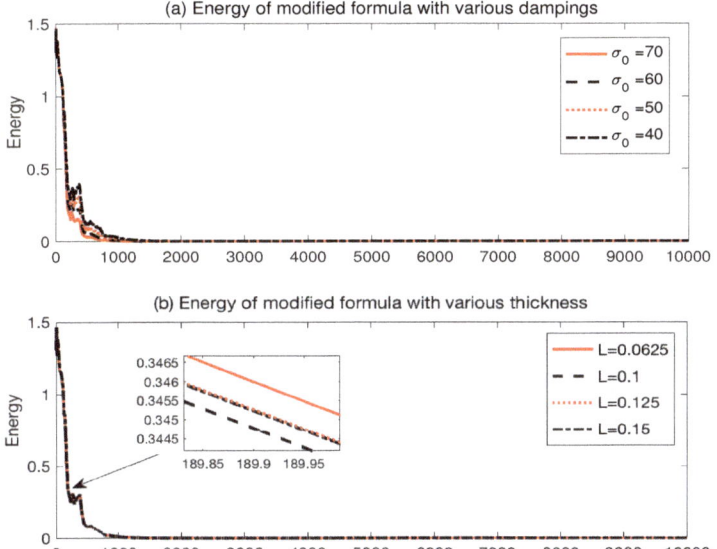

Figure 3. $\mathcal{E}(t)$ with (a) various damping values $\sigma_0 = 40, 50, 60, 70$ for a fixed thickness $L = 0.0625$ ($\beta = 0$), (b) various thickness $L = 0.0625, 0.1, 0.125, 0.15$ for a fixed damping $\sigma_0 = 50$ ($\beta = 0$).

Lastly, in order to illustrate this visual investigation, we consider the damping $\beta = 2$ and display the snap shots of the wave propagation at times $t_n = 1, 30, 60, 100, 130, 150, 200, 300, 500$ with $\sigma_0 = 35, L = 0.25$ in Figure 4. One can see that the regularized system displays a good property of non-reflection in the layers, which is the purpose of building the layers. It is remarkable that from a mathematical point of view, the analytical well-posedness without losing the non-reflection property in the layers of that the classical PML model.

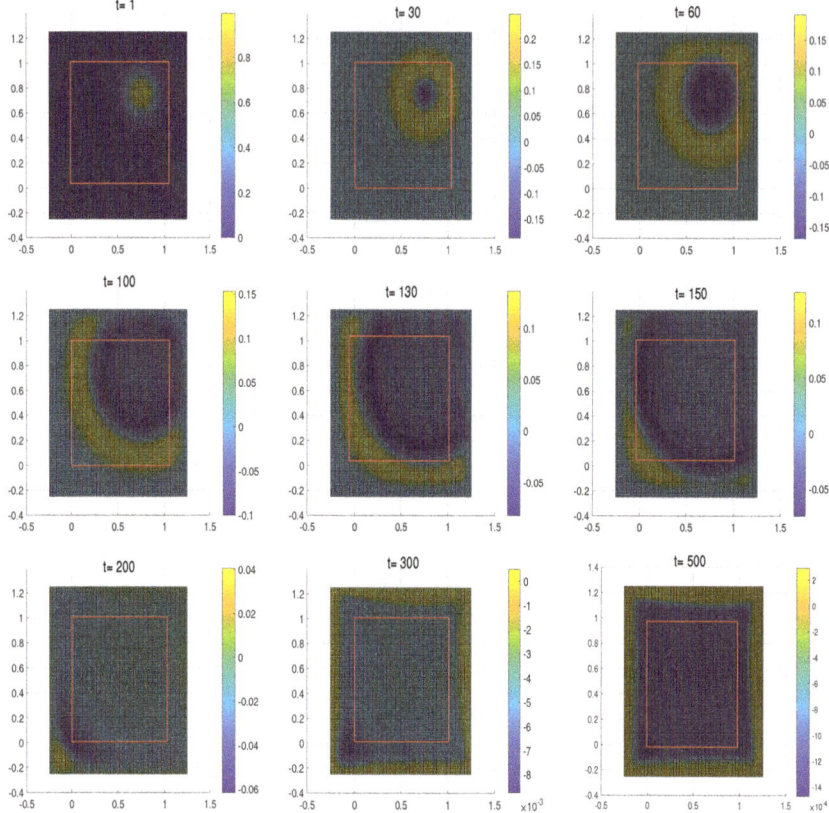

Figure 4. Snap shots of the regularized system at time $t_n = 1, 30, 60, 100, 130, 150, 200, 300, 500$ with $\sigma_0 = 35, \beta = 2, L = 0.25$ (Red rectangular box represents the computational domain.)

5. Discussion

We have introduced a new and efficient formulation related to the acoustic wave equation based on the regularization of the un-split PML wave equation. By regularizing the lower order regularity term in the original equation and the standard *von Neumann* stability analysis, we have achieved well-posedness as well as numerical stability of the solution in the new formulation. We summarize the main novelty and results of this study as follows: (1) We have proved the analytical well-posedness of our formulation without any restriction of damping terms; (2) a staggered finite difference scheme for the formulation is introduced and numerical stability is also analyzed; (3) several numerical tests are exhibited to show the numerical stability and a non-reflection property.

Funding: This research received no external funding.

Conflicts of Interest: The authors declare no conflict of interest. The funders had no role in the design of the study; in the collection, analyses, or interpretation of data; in the writing of the manuscript, or in the decision to publish the results.

Abbreviations

The following abbreviations are used in this manuscript:

PML   Perfectly Matched Layers
CFL   Courant-Friedrichs-Lewy

References

1. Bérenger, J.P. A perfectly matched layer for the absorption of electromagnetic waves. *J. Comput. Phys.* **1994**, *114*, 185–200. [CrossRef]
2. Chew, W.C.; Weedon, W.H. A 3D Perfectly matched medium from modified Maxwell's equations with stretched coordinates. *Microw. Opt. Technol. Lett.* **1994**, *7*, 599–604. [CrossRef]
3. Sjögreen, B.; Petersson, N.A. Perfectly matched layers for Maxwell's equations in second order formulation. *J. Comput. Phys.* **2005**, *209*, 19–46. [CrossRef]
4. Collino, F.; Tsogka, C. Application of the PML absorbing layer model to the linear elastodynamic problem in anisotropic heterogeneous medias. *Geophysics* **2001**, *88*, 43–73.
5. Chew, W.C.; Liu, Q.H. Perfectly matched layers for elastodynamics: A new absorbing boundary condition. *J. Comput. Acoust.* **1996**, *4*, 341–359. [CrossRef]
6. Hu, F.Q. On absorbing boundary conditions for linearized Euler equations by a perfectly matched layer. *J. Comput. Phys.* **1996**, *129*, 201–219. [CrossRef]
7. Hesthaven, J.S. On the analysis and construction of perfectly matched layers for the linearized Euler equations. *J. Comput. Phys.* **1998**, *142*, 129–147. [CrossRef]
8. Lions, J.-L.; Métral, J.; Vacus, O. Well-posed absorbing layer for hyperbolic problems. *Numer. Math.* **2002**, *92*, 535–562. [CrossRef]
9. Nataf, F. A new approach to perfectly matched layers for the linearized Euler system. *J. Comput. Phys.* **2006**, *214*, 757–772. [CrossRef]
10. Turkei, E.; Yefet, A. Absorbing PML boundary layers for wave-like equations. *Appl. Num. Math.* **1998**, *27*, 533–557. [CrossRef]
11. Barucq, H.; Diaz, J.; Tlemcani, M. New absorbing layers conditions for short water waves. *J. Comput. Phys.* **2010**, *229*, 58–72. [CrossRef]
12. Appelö, D.; Hagstrom, T.; Kress, G. Perfectly matched layer for hyperbolic systems: General formulation, well-posedness, and stability. *J. Appl. Math.* **2006**, *67*, 1–23. [CrossRef]
13. Hu, F.Q. A stable perfectly matched layer for linearized Euler equations in unsplit physical variables. *J. Comput. Phys.* **2001**, *173*, 455–480. [CrossRef]
14. Cohen, G.C. *Higher-Order Numerical Methods for Transient Wave Equations*; Springer: Berlin, Germany, 2002.
15. Zhao, L.; Cangellaris, A.C. A general approach for the development of unsplit-field time-domain implementations of perfectly matched layers for FDTD grid truncation. *IEEE Microw. Guided Lett.* **1996**, *6*, 209–211. [CrossRef]
16. Bécache, E.; Joly, P. On the analysis of Bérenger's Perfectly Matched Layers for Maxwell's equations. *Math. Model. Numer. Anal.* **2002**, *36*, 87–120. [CrossRef]
17. Abarbanel, S.; Gottlieb, D. A mathematical analysis of the PML method. *J. Comput. Phys.* **1997**, *134*, 357–363. [CrossRef]
18. Halpern, L.; Petit-Bergez, S.; Rauch, J. The analysis of matched layers. *Conflu. Math.* **2011**, *3*, 159–236. [CrossRef]
19. Abarbanel, S.; Qasimov, H.; Tsynkov, S. Long-time performance of unsplit PMLs with explicit second order schemes. *J. Sci. Comput.* **2009**, *41*, 1–12. [CrossRef]
20. Roden, J.A.; Gedney, S.D. Convolution PML (CPML): An efficient FDTD implementation of the CFS–PML for arbitrary media. *Microw. Opt. Technol. Lett.* **2000**, *27*, 334–339. [CrossRef]
21. Rylander, T.; Jin, J. Perfectly matched layer for the time domain finite element method. *J. Comput. Phys.* **2004**, *200*, 238–250. [CrossRef]
22. Komatitsch, D.; Tromp, J. A perfectly matched layer absorbing boundary condition for the second-order seismic wave equation. *Geophys. J. Int.* **2003**, *154*, 146–153. [CrossRef]

23. Appelö, D.; Kress, G. Application of a perfectly matched layer to the nonlinear wave equation. *Wave Motion* **2007**, *44*, 531–548. [CrossRef]
24. Tam, C.K.W.; Auriault, L.; Cambuli, F. Perfectly matched layer as absorbing condition for the linearized Euler equations in open and ducted domains. *J. Comput. Phys.* **1998**, *114*, 213–234. [CrossRef]
25. Diaz, J.; Joly, P. A time domain analysis of PML models in acoustics. *Comput. Methods Appl. Mech. Eng.* **2006**, *195*, 3820–3853. [CrossRef]
26. Johnson, S. *Notes on Perfectly Matched Layers*; Technical Report; Massachusetts Institute of Technology: Cambridge, MA, USA, 2010.
27. Hagstrom, T.; Hariharan, S.I. A formulation of asymptotic and exact boundary conditions using local operators. *Appl. Numer. Math.* **1998**, *27*, 403–416. [CrossRef]
28. Grote, M.J.; Sim, I. Efficient PML for the wave equation. *arXiv* **2010**, arXiv:1001.0319v1.
29. Bécache, E.; Petropoulos, P.G.; Gedney, S.D. On the long-time behavior of unsplit perfectly matched layers. *IEEE Trans. Antennas Propag.* **2004**, *52*, 1335–1342. [CrossRef]
30. Petersen, B.E. *Introduction to the Fourier Transform and Pseudo-Differential Operators*; Series: Monographs and studies in mathematics; Pitman Advanced Pub. Program: Boston, MA, USA, 1983.
31. Kim, D. The Variable Speed Wave Equation and Perfectly Matched Layers. Ph.D. Thesis, Oregon State University, Corvallis, OR, USA, 2015.
32. Evans, L.C. *Partial Differential Equations*, 2nd ed.; Graduate Series in Mathematics; Springer: Berlin, Germany, 2010.
33. LeVeque, R.J. *Finite Difference Methods for Ordinary and Partial Differential Equations*; Society for Industrial and Applied Mathematics: Philadelphia, PA, USA, 2007.
34. Trucco, E.; Verri, A. *Introductory Techniques for 3-D Computer Vision*; Prentice Hall: Upper Saddle River, NJ, USA, 1998.
35. Bidégaray-Fesquet, B. Stability of FD-TD schemes for Maxwell-Debye and Maxwell-Lorentz equations. *SIAM J. Numer. Anal.* **2008**, *46*, 2551–2566. [CrossRef]
36. Miller, J.J.H. On the location of zeros of certain classes of polynomials with applications to numerical analysis. *J. Inst. Math. Appl.* **1971**, *8*, 397–406. [CrossRef]
37. Kaltenbacher, B.; Kaltenbacher, M.; Sim, I. A modified and stable version of a Perfectly Matched Layer technique for the 3-d second order wave equation in time domain with an application to aeroacoustics. *J. Comput. Phys.* **2013**, *35*, 407–422. [CrossRef]

© 2019 by the authors. Licensee MDPI, Basel, Switzerland. This article is an open access article distributed under the terms and conditions of the Creative Commons Attribution (CC BY) license (http://creativecommons.org/licenses/by/4.0/).

Article

On the Catalan Numbers and Some of Their Identities

Wenpeng Zhang [1,2] and Li Chen [2,*]

1. School of Mathematics and Statistics, Kashgar University, Xinjiang 844006, China; wpzhang888@163.com
2. School of Mathematics, Northwest University, Xi'an 710127, Shaanxi, China
* Correspondence: cl1228@stumail.nwu.edu.cn

Received: 5 December 2018; Accepted: 4 January 2019; Published: 8 January 2019

Abstract: The main purpose of this paper is using the elementary and combinatorial methods to study the properties of the Catalan numbers, and give two new identities for them. In order to do this, we first introduce two new recursive sequences, then with the help of these sequences, we obtained the identities for the convolution involving the Catalan numbers.

Keywords: catalan numbers; elementary and combinatorial methods; recursive sequence; convolution sums

JEL Classification: 11B83; 11B75

---

1. Introduction

For any non-negative integer n, the famous Catalan numbers C_n are defined as $C_n = \frac{1}{n+1} \cdot \binom{2n}{n}$. For example, the first several Catalan numbers are $C_0 = 1, C_1 = 1, C_2 = 2, C_3 = 5, C_4 = 14, C_5 = 42, C_6 = 132, C_7 = 429, C_8 = 1430, \cdots$. The Catalan numbers C_n satisfy the recursive formula

$$C_n = \sum_{i=1}^{n} C_{i-1} \cdot C_i.$$

The generating function of the Catalan numbers C_n is

$$\frac{2}{1+\sqrt{1-4x}} = \sum_{n=0}^{\infty} \frac{\binom{2n}{n}}{n+1} \cdot x^n = \sum_{n=0}^{\infty} C_n \cdot x^n. \tag{1}$$

These numbers occupy a pivotal position in combinatorial mathematics, as many counting problems are closely related to Catalan numbers, and some famous examples can be found in R. P. Stanley [1]. Many papers related to the Catalan numbers and other special sequences can also be found in references [1–20], especially the works of T. Kim et al. give a series of new identities for the Catalan numbers, see [9–14], these are important results in the related field.

The main purpose of this paper is to consider the calculating problem of the following convolution sums involving the Catalan numbers:

$$\sum_{a_1+a_2+\cdots+a_h=n} C_{a_1} \cdot C_{a_2} \cdot C_{a_3} \cdots C_{a_h}, \tag{2}$$

where the summation is taken over all h-dimension non-negative integer coordinates $(a_1, a_2, \cdots, a_h)$ such that the equation $a_1 + a_2 + \cdots + a_h = n$.

About the convolution sums (2), it seems that none had studied it yet, at least we have not seen any related results before. We think this problem is meaningful. The reason is based on the following two aspects: First, it can reveal the profound properties of the Catalan numbers themselves. Second, for the other sequences, such as Fibonacci numbers, Fubini numbers, and Euler numbers, etc. (see [21–23]), there are corresponding results, so the Catalan numbers should have a corresponding identity. In this paper, we use the elementary and combinatorial methods to answer this question. That is, we shall prove the following:

Theorem 1. *For any positive integer h, we have the identity*

$$\sum_{a_1+a_2+\cdots+a_{2h+1}=n} C_{a_1} \cdot C_{a_2} \cdot C_{a_3} \cdots C_{a_{2h+1}}$$

$$= \frac{1}{(2h)!} \sum_{i=0}^{h} C(h,i) \sum_{j=0}^{\min(n,i)} \frac{(n-j+h+i)! \cdot C_{n-j+h+i}}{(n-j)!} \cdot \binom{i}{j} \cdot (-4)^j,$$

where $C(h,i)$ are defined as $C(1,0) = -2$, $C(h,h) = 1$, $C(h+1,h) = C(h,h-1) - (8h+2) \cdot C(h,h)$, $C(h+1,0) = 8 \cdot C(h,1) - 2 \cdot C(h,0)$, and for all integers $1 \leq i \leq h-1$, we have the recursive formula

$$C(h+1,i) = C(h,i-1) - (8i+2) \cdot C(h,i) + (4i+4)(4i+2) \cdot C(h,i+1).$$

Theorem 2. *For any positive integer h and non-negative n, we have*

$$\sum_{a_1+a_2+\cdots+a_{2h}=n} C_{a_1} \cdot C_{a_2} \cdot C_{a_3} \cdots C_{a_{2h}}$$

$$= \frac{1}{(2h-1)!} \sum_{i=0}^{h-1} \sum_{j=0}^{n} D(h,i+1) \cdot \binom{i+\frac{1}{2}}{j} \cdot (-4)^j \cdot \frac{(n-j+h+i)! \cdot C_{n-j+h+i}}{(n-j)!},$$

where $\binom{n+\frac{1}{2}}{i} = \left(n+\frac{1}{2}\right) \cdot \left(n-1+\frac{1}{2}\right) \cdots \left(n-i+1+\frac{1}{2}\right)/i!$, $D(k,i)$ are defined as $D(k,0) = 0$, $D(k,k) = 1$, $D(k+1,k) = D(k,k-1) - (8k-2)$, $D(k+1,1) = 24D(k,2) - 6D(k,1)$, and for all integers $1 \leq i \leq k-1$,

$$D(k+1,i) = D(k,i-1) - (8i-2) \cdot D(k,i) + 4i(4i+2) \cdot D(k,i+1).$$

To better illustrate the sequence $\{C(k,i)\}$ and $D(h,i)$, we compute them using mathematical software and list some values in the following Tables 1 and 2.

Table 1. Values of $C(k,i)$.

$C(k,i)$	$i=0$	$i=1$	$i=2$	$i=3$	$i=4$	$i=5$	$i=6$
$k=1$	-2	1					
$k=2$	12	-12	1				
$k=3$	-120	180	-30	1			
$k=4$	1680	-3360	840	-56	1		
$k=5$	$-30{,}240$	75,600	$-25{,}200$	2520	-90	1	
$k=6$	665,280	$-1{,}995{,}840$	831,600	$-110{,}880$	5940	-132	1

Table 2. Values of $D(k,i)$.

$D(k,i)$	$i=0$	$i=1$	$i=2$	$i=3$	$i=4$	$i=5$	$i=6$
$k=1$	0	1					
$k=2$	0	-6	1				
$k=3$	0	60	-20	1			
$k=4$	0	-840	420	-42	1		
$k=5$	0	15,120	$-10{,}080$	1512	-72	1	
$k=6$	0	$-332{,}640$	277,200	$-55{,}440$	3960	-110	1

Observing these two tables, we can easily find that if $2k-1 = p$ is a prime, then for all integers $0 \leq i < k$, we have the congruences $C(k,i) \equiv 0 \mod (2k-1)(2k)$ and $D(k,i) \equiv 0 \mod (2k-1)(2k-2)$. So we propose the following two conjectures:

Conjecture 1. *Let p be a prime. Then for any integer $0 \leq i < \frac{p+1}{2}$, we have the congruence*

$$C\left(\frac{p+1}{2}, i\right) \equiv 0 \mod p(p+1).$$

Conjecture 2. *Let p be a prime. Then for any integer $0 \leq i < \frac{p+1}{2}$, we have the congruence*

$$D\left(\frac{p+1}{2}, i\right) \equiv 0 \mod p(p-1).$$

For some special integers n and h, from Theorem 1 and Theorem 2 we can also deduce several interesting corollaries. In fact if we take $n = 0$ and $h = 1$ in the theorems respectively, then we have the following four corollaries:

Corollary 1. *For any positive integer h, we have the identity*

$$\sum_{i=0}^{h} C(h,i) \cdot (h+i)! \cdot C_{h+i} = (2h)!.$$

Corollary 2. *For any positive integer h, we have the identity*

$$\sum_{i=1}^{h} D(h,i) \cdot (h+i-1)! \cdot C_{h+i-1} = (2h-1)!.$$

Corollary 3. *For any integer $n \geq 0$, we have the identity*

$$\sum_{a+b+d=n} C_a \cdot C_b \cdot C_d = (n+1) \cdot \left[\frac{1}{2} \cdot (n+2) \cdot C_{n+2} - (2n+1) \cdot C_{n+1}\right].$$

Corollary 4. *For any integer $n \geq 0$, we have the identity*

$$\sum_{u+v+w+x+y=n} C_u \cdot C_v \cdot C_w \cdot C_x \cdot C_y = \frac{(n+1)(n+2)(4n^2+8n+3)}{6} \cdot C_{n+2}$$
$$- \frac{(n+3)(n+2)(n+1)(2n+3)}{6} \cdot C_{n+3} + \frac{(n+4)(n+3)(n+2)(n+1)}{24} \cdot C_{n+4}.$$

2. Several Simple Lemmas

To prove our theorems, we need following four simple lemmas. First we have:

Lemma 1. *Let function $f(x) = \frac{2}{1+\sqrt{1-4x}}$. Then for any positive integer h, we have the identity*

$$(2h)! \cdot f^{2h+1}(x) = \sum_{i=0}^{h} C(h,i) \cdot (1-4x)^i \cdot f^{(h+i)}(x),$$

where $f^{(i)}(x)$ denotes the i-order derivative of $f(x)$ for x, and $\{C(h,i)\}$ are defined as the same as in Theorem 1.

Proof. In fact, this identity and its generalization had appeared in D. S. Kim and T. Kim's important work [9] (see Theorem 3.1), but only in different forms. For the completeness of our results, here we give a different proof by mathematical induction. First from the properties of the derivative we have

$$f'(x) = \frac{4}{(1+\sqrt{1-4x})^2} \cdot \frac{1}{\sqrt{1-4x}} = \frac{f^2(x)}{\sqrt{1-4x}}$$

or identity

$$f^2(x) = (1-4x)^{\frac{1}{2}} \cdot f'(x). \tag{3}$$

From (3) and note that $C(1,0) = -2$ and $C(1,1) = 1$ we have

$$2f(x) \cdot f'(x) = -2(1-4x)^{-\frac{1}{2}} \cdot f'(x) + (1-4x)^{\frac{1}{2}} \cdot f''(x)$$

and

$$2!f^3(x) = -2f'(x) + (1-4x) \cdot f''(x) = \sum_{i=0}^{1} C(1,i) \cdot (1-4x)^i \cdot f^{(1+i)}(x).$$

That is, Lemma 1 is true for $h = 1$.
Assume that Lemma 1 is true for $h = k \geq 1$. That is,

$$(2k)! \cdot f^{2k+1}(x) = \sum_{i=0}^{k} C(k,i) \cdot (1-4x)^i \cdot f^{(k+i)}(x). \tag{4}$$

Then from (3), (4), the definition of $C(k,i)$, and the properties of the derivative we can deduce that

$$(2k+1)! \cdot f^{2k}(x) \cdot f'(x) = \sum_{i=0}^{k} C(k,i) \cdot (1-4x)^i \cdot f^{(k+i+1)}(x)$$
$$- \sum_{i=1}^{k} 4i \cdot C(k,i) \cdot (1-4x)^{i-1} \cdot f^{(k+i)}(x)$$

or

$$rrl(2k+1)! \cdot f^{2k+2}(x) = \sum_{i=0}^{k} C(k,i) \cdot (1-4x)^{i+\frac{1}{2}} \cdot f^{(k+i+1)}(x)$$
$$- \sum_{i=1}^{k} 4i \cdot C(k,i) \cdot (1-4x)^{i-\frac{1}{2}} \cdot f^{(k+i)}(x). \tag{5}$$

Applying (5) and the properties of the derivative we also have

$$(2k+2)! \cdot f^{2k+1}(x) \cdot f'(x) = \sum_{i=0}^{k} C(k,i) \cdot (1-4x)^{i+\frac{1}{2}} \cdot f^{(k+i+2)}(x)$$
$$- \sum_{i=0}^{k} (4i+2) \cdot C(k,i) \cdot (1-4x)^{i-\frac{1}{2}} \cdot f^{(k+i+1)}(x)$$
$$- \sum_{i=1}^{k} 4i \cdot C(k,i) \cdot (1-4x)^{i-\frac{1}{2}} \cdot f^{(k+i+1)}(x)$$
$$+ \sum_{i=1}^{k} (4i) \cdot (4i-2) \cdot C(k,i) \cdot (1-4x)^{i-\frac{3}{2}} \cdot f^{(k+i)}(x).$$

or note that identity (3) we have

$$(2k+2)! \cdot f^{2k+3}(x) = \sum_{i=0}^{k} C(k,i) \cdot (1-4x)^{i+1} \cdot f^{(k+i+2)}(x)$$

$$-\sum_{i=0}^{k}(4i+2) \cdot C(k,i) \cdot (1-4x)^{i} \cdot f^{(k+i+1)}(x)$$

$$-\sum_{i=1}^{k} 4i \cdot C(k,i) \cdot (1-4x)^{i} \cdot f^{(k+i+1)}(x)$$

$$+\sum_{i=1}^{k}(4i) \cdot (4i-2) \cdot C(k,i) \cdot (1-4x)^{i-1} \cdot f^{(k+i)}(x)$$

$$= C(k,k) \cdot (1-4x)^{k+1} \cdot f^{(2k+2)}(x) + \sum_{i=1}^{k} C(k,i-1) \cdot (1-4x)^{i} \cdot f^{(k+i+1)}(x)$$

$$-2C(k,0) \cdot f^{(k+1)}(x) - \sum_{i=1}^{k}(4i+2) \cdot C(k,i) \cdot (1-4x)^{i} \cdot f^{(k+i+1)}(x) \qquad (6)$$

$$-\sum_{i=1}^{k} 4i \cdot C(k,i) \cdot (1-4x)^{i} \cdot f^{(k+i+1)}(x) + 8 \cdot C(k,1) \cdot f^{(k+1)}(x)$$

$$+\sum_{i=1}^{k-1}(4i+4) \cdot (4i+2) \cdot C(k,i+1) \cdot (1-4x)^{i} \cdot f^{(k+i+1)}(x)$$

$$= (1-4x)^{k+1} \cdot f^{(2k+2)}(x) + (8 \cdot C(k,1) - 2 \cdot C(k,0)) \cdot f^{(k+1)}(x)$$
$$+ (C(k,k-1) - (8k+2) \cdot C(k,k)) \cdot (1-4x)^{k} \cdot f^{(2k+1)}(x)$$
$$+ \sum_{i=1}^{k-1}(C(k,i-1) - (8i+2) \cdot C(k,i) + (4i+4)(4i+2) \cdot C(k,i+1))$$
$$\times (1-4x)^{i} \cdot f^{(k+i+1)}(x)$$

$$= \sum_{i=0}^{k+1} C(k+1,i) \cdot (1-4x)^{i} \cdot f^{(k+i+1)}(x),$$

where we have used the identities $C(k+1,k) = C(k,k-1) - (8k+2) \cdot C(k,k)$, $C(k,k) = 1$, $C(k+1,0) = 8 \cdot C(k,1) - 2 \cdot C(k,0)$ and for all integers $1 \leq i \leq k-1$,

$$C(k+1,i) = C(k,i-1) - (8i+2) \cdot C(k,i) + (4i+4)(4i+2) \cdot C(k,i+1).$$

It is clear that (6) implies Lemma 1 is true for $h = k+1$.
This proves Lemma 1 by mathematical induction. □

Lemma 2. *For any positive integer h, we have the identity*

$$(2h-1)! \cdot f^{2h}(x) = \sum_{i=0}^{h-1} D(h,i+1) \cdot (1-4x)^{i+\frac{1}{2}} \cdot f^{(h+i)}(x),$$

where $D(h,i)$ are defined as the same as in Theorem 2.

Proof. It is clear that using the methods of proving Lemma 1 we can easily deduce Lemma 2. □

Lemma 3. *Let h be any positive integer. Then for any integer $k \geq 0$, we have the identity*

$$(1-4x)^{k} \cdot f^{(h+k)}(x) = \sum_{n=0}^{\infty} \left(\sum_{i=0}^{\min(n,k)} \frac{C_{n-i+h+k}}{(n-i)!} \binom{k}{i} \cdot (-4)^{i} \right) \cdot x^{n}.$$

Proof. From the binomial theorem we have

$$(1-4x)^k = \sum_{i=0}^{k} \binom{k}{i} \cdot (-4x)^i. \qquad (7)$$

On the other hand, from (1) we also have

$$f^{(h+k)}(x) = \sum_{n=0}^{\infty} \frac{(n+h+k)! \cdot C_{n+h+k}}{n!} \cdot x^n. \qquad (8)$$

Combining (7) and (8) we have

$$(1-4x)^k \cdot f^{(h+k)}(x)$$

$$= \left(\sum_{i=0}^{k} \binom{k}{i} \cdot (-4x)^i \right) \left(\sum_{n=0}^{\infty} \frac{(n+h+k)! \cdot C_{n+h+k}}{n!} \cdot x^n \right)$$

$$= \sum_{n=0}^{\infty} \sum_{i=0}^{k} \frac{(n+h+k)! \cdot C_{n+h+k}}{n!} \cdot \binom{k}{i} \cdot (-4)^i \cdot x^{n+i}$$

$$= \sum_{n=0}^{\infty} \left(\sum_{i=0}^{\min(n,k)} \frac{(n-i+h+k)! \cdot C_{n-i+h+k}}{(n-i)!} \binom{k}{i} \cdot (-4)^i \right) \cdot x^n.$$

This proves Lemma 3. □

Lemma 4. *Let h be any positive integer. Then for any integer $k \geq 0$, we have the identity*

$$(1-4x)^{k+\frac{1}{2}} \cdot f^{(h+k)}(x) = \sum_{n=0}^{\infty} \left(\sum_{i=0}^{n} \binom{k+\frac{1}{2}}{i} \cdot (-4)^i \cdot \frac{C_{n-i+h+k}}{(n-i)!} \right) \cdot x^n.$$

Proof. From the power series expansion of the function we know that

$$(1-4x)^{k+\frac{1}{2}} = \sum_{n=0}^{\infty} \binom{k+\frac{1}{2}}{n} \cdot (-4)^n \cdot x^n. \qquad (9)$$

Applying (8) and (9) we have

$$(1-4x)^{k+\frac{1}{2}} \cdot f^{(h+k)}(x)$$

$$= \left(\sum_{n=0}^{\infty} \binom{k+\frac{1}{2}}{n} \cdot (-4)^n \cdot x^n \right) \left(\sum_{n=0}^{\infty} \frac{(n+h+k)! \cdot C_{n+h+k}}{n!} \cdot x^n \right)$$

$$= \sum_{n=0}^{\infty} \left(\sum_{i=0}^{n} \binom{k+\frac{1}{2}}{i} \cdot (-4)^i \cdot \frac{(n-i+h+k)! \cdot C_{n-i+h+k}}{(n-i)!} \right) \cdot x^n.$$

This proves Lemma 4. □

3. Proofs of the Theorems

In this section, we shall complete the proofs of our theorems. First we prove Theorem 1. From (1) and the multiplicative properties of the power series we have

$$(2h)! \cdot f^{2h+1}(x) = (2h)! \sum_{n=0}^{\infty} \left(\sum_{a_1+a_2+\cdots+a_{2h+1}=n} C_{a_1} \cdot C_{a_2} \cdots C_{a_{2h+1}} \right) \cdot x^n. \qquad (10)$$

On the other hand, from Lemma 1 and Lemma 3 we also have

$$(2h)! \cdot f^{2h+1}(x) = \sum_{i=0}^{h} C(h,i) \cdot (1-4x)^i \cdot f^{(h+i)}(x)$$

$$= \sum_{n=0}^{\infty} \left(\sum_{i=0}^{h} C(h,i) \sum_{j=0}^{\min(n,i)} \frac{(n-j+h+i)! C_{n-j+h+i}}{(n-j)!} \binom{i}{j} (-4)^j \right) x^n. \tag{11}$$

Combining (10) and (11) we may immediately deduce the identity

$$\sum_{a_1+a_2+\cdots+a_{2h+1}=n} C_{a_1} \cdot C_{a_2} \cdot C_{a_3} \cdots C_{a_{2h+1}}$$

$$= \frac{1}{(2h)!} \sum_{i=0}^{h} C(h,i) \sum_{j=0}^{\min(n,i)} \frac{(n-j+h+i)! \cdot C_{n-j+h+i}}{(n-j)!} \cdot \binom{i}{j} \cdot (-4)^j.$$

This proves Theorem 1.

Now we prove Theorem 2. For any positive integer h, from (1) we have

$$f^{2h}(x) = \sum_{n=0}^{\infty} \left(\sum_{a_1+a_2+\cdots+a_{2h}=n} C_{a_1} \cdot C_{a_2} \cdots C_{a_{2h}} \right) \cdot x^n. \tag{12}$$

On the other hand, from Lemma 2 and Lemma 4 we also have

$$(2h-1)! \cdot f^{2h}(x) = \sum_{i=0}^{h-1} D(h,i+1) \cdot (1-4x)^{i+\frac{1}{2}} \cdot f^{(h+i)}(x)$$

$$= \sum_{i=0}^{h-1} D(h,i+1) \sum_{n=0}^{\infty} \left(\sum_{j=0}^{n} \binom{i+\frac{1}{2}}{j} (-4)^j \frac{(n-j+h+i)! \cdot C_{n-j+h+i}}{(n-j)!} \right) x^n \tag{13}$$

$$= \sum_{n=0}^{\infty} \sum_{i=0}^{h-1} \sum_{j=0}^{n} D(h,i+1) \binom{i+\frac{1}{2}}{j} (-4)^j \frac{(n-j+h+i)! C_{n-j+h+i}}{(n-j)!} x^n.$$

From (12), (13), and Lemma 2 we may immediately deduce the identity

$$\sum_{a_1+a_2+\cdots+a_{2h}=n} C_{a_1} \cdot C_{a_2} \cdot C_{a_3} \cdots C_{a_{2h}}$$

$$= \frac{1}{(2h-1)!} \sum_{i=0}^{h-1} \sum_{j=0}^{n} D(h,i+1) \cdot \binom{i+\frac{1}{2}}{j} \cdot (-4)^j \cdot \frac{(n-j+h+i)! \cdot C_{n-j+h+i}}{(n-j)!}.$$

This completes the proof of Theorem 2.

4. Conclusions

The main results of this paper are Theorem 1 and Theorem 2. They gave two special expressions for convolution (2). In addition, Corollary 1 gives a close relationship between $C(h,i)$ and C_{h+i}. Corollary 2 gives a close relationship between $D(h,i)$ and D_{h+i-1}. Corollary 3 and Corollary 4 give two exact representations for the special cases of Theorem 1 with $h=1$ and $h=2$.

About the new sequences $C(h,i)$ and $D(h,i)$, we proposed two interesting conjectures related to congruence mod p, where p is an odd prime. We believe that these conjectures are correct, but at the moment we cannot prove them. We also believe that these two conjectures will certainly attract the interest of many readers, thus further promoting the study of the properties of $C(h,i)$ and C_{h+i}.

Author Contributions: All authors have equally contributed to this work. All authors read and approved the final manuscript.

Funding: This work is supported by the National Natural Science Foundation (N. S. F.) (11771351) and (11826205) of China.

Acknowledgments: The authors would like to thank the Editor and referees for their very helpful and detailed comments, which have significantly improved the presentation of this paper.

Conflicts of Interest: The authors declare that there are no conflicts of interest regarding the publication of this paper.

References

1. Stanley, R.P. *Enumerative Combinatorics (Vol. 2)*; Cambridge Studieds in Advanced Mathematics; Cambridge University Press: Cambridge, UK, 1997; p. 49.
2. Chu, W.C. Further identities on Catalan numbers. *Discret. Math.* **2018**, *341*, 3159–3164. [CrossRef]
3. Joseph, A.; Lamprou, P. A new interpretation of the Catalan numbers arising in the theory of crystals. *J. Algebra* **2018**, *504*, 85–128. [CrossRef]
4. Allen, E.; Gheorghiciuc, I. A weighted interpretation for the super Catalan numbers. *J. Integer Seq.* **2014**, *9*, 17.
5. Tauraso, R. qq-Analogs of some congruences involving Catalan numbers. *Adv. Appl. Math.* **2012**, *48*, 603–614. [CrossRef]
6. Liu, J.C. Congruences on sums of super Catalan numbers. *Results Math.* **2018**, *73*, 73–140. [CrossRef]
7. Qi, F.; Shi, X.T.; Liu, F.F. An integral representation, complete monotonicity, and inequalities of the Catalan numbers. *Filomat* **2018**, *32*, 575–587. [CrossRef]
8. Qi, F.; Shi, X.T.; Mahmoud, M. The Catalan numbers: A generalization, an exponential representation, and some properties. *J. Comput. Anal. Appl.* **2017**, *23*, 937–944.
9. Kim, D.S.; Kim, T. A new approach to Catalan numbers using differential equations. *Russ. J. Math. Phys.* **2017**, *24*, 465–475. [CrossRef]
10. Kim, T.; Kim, D.S. Some identities of Catalan-Daehee polynomials arising from umbral calculus. *Appl. Comput. Math.* **2017**, *16*, 177–189.
11. Kim, T.; Kim, D.S. Differential equations associated with Catalan-Daehee numbers and their applications. *Revista de la Real Academia de Ciencias Exactas, Físicas y Naturales. Serie A. Matemáticas* **2017**, *111*, 1071–1081. [CrossRef]
12. Kim, D.S.; Kim, T. Triple symmetric identities for w-Catalan polynomials. *J. Korean Math. Soc.* **2017**, *54*, 1243–1264.
13. Kim, T.; Kwon, H.-I. Revisit symmetric identities for the λ-Catalan polynomials under the symmetry group of degree n. *Proc. Jangjeon Math. Soc.* **2016**, *19*, 711–716.
14. Kim, T. A note on Catalan numbers associated with p-adic integral on Zp. *Proc. Jangjeon Math. Soc.* **2016**, *19*, 493–501.
15. Basic, B. On quotients of values of Euler's function on the Catalan numbers. *J. Number Theory* **2016**, *169*, 160–173. [CrossRef]
16. Aker, K.; Gursoy, A.E. A new combinatorial identity for Catalan numbers. *Ars Comb.* **2017**, *135*, 391–398.
17. Qi, F.; Guo, B.N. Integral representations of the Catalan numbers and their applications. *Mathematics* **2017**, *5*, 40. [CrossRef]
18. Hein, N.; Huang, J. Modular Catalan numbers. *Eur. J. Combin.* **2017**, *61*, 197–218. [CrossRef]
19. Dilworth, S.J.; Mane, S.R. Applications of Fuss-Catalan numbers to success runs of Bernoulli trials. *J. Probab. Stat.* **2016**, *2016*, 2071582. [CrossRef]
20. Dolgy, D.V.; Jang, G.-W.; Kim, D.S.; Kim, T. Explicit expressions for Catalan-Daehee numbers. *Proc. Jangjeon Math. Soc.* **2017**, *20*, 1–9.
21. Zhang, Y.X.; Chen, Z.Y. A new identity involving the Chebyshev polynomials. *Mathematics* **2018**, *6*, 244. [CrossRef]

22. Zhao, J.H.; Chen, Z.Y. Some symmetric identities involving Fubini polynomials and Euler numbers. *Symmetry* **2018**, *10*, 359.
23. Zhang, W.P. Some identities involving the Fibonacci numbers and Lucas numbers. *Fibonacci Q.* **2004**, *42*, 149–154.

© 2019 by the authors. Licensee MDPI, Basel, Switzerland. This article is an open access article distributed under the terms and conditions of the Creative Commons Attribution (CC BY) license (http://creativecommons.org/licenses/by/4.0/).

Article

Representation by Chebyshev Polynomials for Sums of Finite Products of Chebyshev Polynomials

Taekyun Kim [1], Dae San Kim [2], Lee-Chae Jang [3,*] and Dmitry V. Dolgy [4]

[1] Department of Mathematics, Kwangwoon University, Seoul 139-701, Korea; tkkim@kw.ac.kr
[2] Department of Mathematics, Sogang University, Seoul 121-742, Korea; dskim@sogang.ac.kr
[3] Graduate School of Education, Konkuk University, Seoul 139-701, Korea
[4] Hanrimwon, Kwangwoon University, Seoul 139-701, Korea; d_dol@mail.ru
* Correspondence: lcjang@konkuk.ac.kr

Received: 29 November 2018; Accepted: 10 December 2018; Published: 11 December 2018

Abstract: In this paper, we consider sums of finite products of Chebyshev polynomials of the first, third, and fourth kinds, which are different from the previously-studied ones. We represent each of them as linear combinations of Chebyshev polynomials of all kinds whose coefficients involve some terminating hypergeometric functions $_2F_1$. The results may be viewed as a generalization of the linearization problem, which is concerned with determining the coefficients in the expansion of the product of two polynomials in terms of any given sequence of polynomials. These representations are obtained by explicit computations.

Keywords: Chebyshev polynomials of the first, second, third, and fourth kinds; sums of finite products; representation

1. Introduction and Preliminaries

We first fix some notations that will be used throughout this paper. For any nonnegative integer n, the falling factorial sequence $(x)_n$ and the rising factorial sequence $<x>_n$ are respectively given by:

$$(x)_n = x(x-1)\cdots(x-n+1), \quad (n \geq 1), \quad (x)_0 = 1, \tag{1}$$

$$<x>_n = x(x+1)\cdots(x+n-1), \quad (n \geq 1), \quad <x>_0 = 1. \tag{2}$$

Then, we easily see that the two factorial sequences are related by:

$$(-1)^n (x)_n = <-x>_n. \tag{3}$$

The Gauss hypergeometric function $_2F_1(a,b;c;x)$ is defined by:

$$_2F_1(a,b;c;x) = \sum_{n=0}^{\infty} \frac{<a>_n <b>_n}{<c>_n} \frac{x^n}{n!}, \quad (|x| < 1). \tag{4}$$

In this paper, we only need very basic facts about Chebyshev polynomials of the first, second, third, and fourth kinds, which we recall briefly in the following. The Chebyshev polynomials belong to the family of orthogonal polynomials. We let the interested reader refer to [1–4] for more details on these.

In terms of generating functions, the Chebyshev polynomials of the first, second, third, and fourth kinds are respectively given by:

$$F_1(t,x) = \frac{1-xt}{1-2xt+t^2} = \sum_{n=0}^{\infty} T_n(x)t^n, \tag{5}$$

$$F_2(t,x) = \frac{1}{1-2xt+t^2} = \sum_{n=0}^{\infty} U_n(x)t^n, \tag{6}$$

$$F_3(t,x) = \frac{1-t}{1-2xt+t^2} = \sum_{n=0}^{\infty} V_n(x)t^n, \tag{7}$$

$$F_4(t,x) = \frac{1+t}{1-2xt+t^2} = \sum_{n=0}^{\infty} W_n(x)t^n. \tag{8}$$

They are also explicitly given by the following expressions:

$$T_n(x) =_2 F_1\left(-n, n; \frac{1}{2}; \frac{1-x}{2}\right)$$
$$= \frac{n}{2} \sum_{l=0}^{[\frac{n}{2}]} (-1)^l \frac{1}{n-l} \binom{n-l}{l} (2x)^{n-2l}, \quad (n \geq 1), \tag{9}$$

$$U_n(x) = (n+1)_2 F_1\left(-n, n+2; \frac{3}{2}; \frac{1-x}{2}\right)$$
$$= \sum_{l=0}^{[\frac{n}{2}]} (-1)^l \binom{n-l}{l} (2x)^{n-2l}, \quad (n \geq 0), \tag{10}$$

$$V_n(x) =_2 F_1\left(-n, n+1; \frac{1}{2}; \frac{1-x}{2}\right)$$
$$= \sum_{l=0}^{n} \binom{n+l}{2l} 2^l (x-1)^l, \quad (n \geq 0), \tag{11}$$

$$W_n(x) = (2n+1)_2 F_1\left(-n, n+1; \frac{3}{2}; \frac{1-x}{2}\right)$$
$$= (2n+1) \sum_{l=0}^{n} \frac{2^l}{2l+1} \binom{n+l}{2l} (x-1)^l, \quad (n \geq 0), \tag{12}$$

The Chebyshev polynomials of all four kinds are also expressed by the Rodrigues formulas, which are given by:

$$T_n(x) = \frac{(-1)^n 2^n n!}{(2n)!} (1-x^2)^{\frac{1}{2}} \frac{d^n}{dx^n} (1-x^2)^{n-\frac{1}{2}}, \tag{13}$$

$$U_n(x) = \frac{(-1)^n 2^n (n+1)!}{(2n+1)!} (1-x^2)^{-\frac{1}{2}} \frac{d^n}{dx^n} (1-x^2)^{n+\frac{1}{2}}, \tag{14}$$

$$(1-x)^{-\frac{1}{2}}(1+x)^{\frac{1}{2}} V_n(x)$$
$$= \frac{(-1)^n 2^n n!}{(2n)!} \frac{d^n}{dx^n} (1-x)^{n-\frac{1}{2}} (1+x)^{n+\frac{1}{2}}, \tag{15}$$

$$(1-x)^{\frac{1}{2}}(1+x)^{-\frac{1}{2}} W_n(x)$$
$$= \frac{(-1)^n 2^n n!}{(2n)!} \frac{d^n}{dx^n} (1-x)^{n+\frac{1}{2}} (1-x)^{n-\frac{1}{2}}. \tag{16}$$

They satisfy orthogonalities with respect to various weight functions as given in the following:

$$\int_{-1}^{1} (1-x^2)^{-\frac{1}{2}} T_n(x) T_m(x) dx = \frac{\pi}{\varepsilon_n} \delta_{n,m}, \tag{17}$$

where:

$$\varepsilon_n = \begin{cases} 1, & \text{if } n = 0, \\ 2, & \text{if } n \geq 1, \end{cases} \quad (18)$$

$$\delta_{n,m} = \begin{cases} 0, & \text{if } n \neq m, \\ 1, & \text{if } n = m. \end{cases} \quad (19)$$

$$\int_{-1}^{1} (1-x^2)^{\frac{1}{2}} U_n(x) U_m(x) dx = \frac{\pi}{2} \delta_{n,m}, \quad (20)$$

$$\int_{-1}^{1} \left(\frac{1+x}{1-x}\right)^{\frac{1}{2}} V_n(x) V_m(x) dx = \pi \delta_{n,m}, \quad (21)$$

$$\int_{-1}^{1} \left(\frac{1-x}{1+x}\right)^{\frac{1}{2}} W_n(x) W_m(x) dx = \pi \delta_{n,m}. \quad (22)$$

For convenience, we let:

$$\alpha_{m,r}(x) = \sum_{i_1+\cdots+i_{r+1}=m} T_{i_1}(x) \cdots T_{i_{r+1}}(x), \quad (m, r \geq 0), \quad (23)$$

$$\beta_{m,r}(x) = \sum_{i_1+\cdots+i_{r+1}=m} V_{i_1}(x) \cdots V_{i_{r+1}}(x), \quad (m, r \geq 0), \quad (24)$$

$$\gamma_{m,r}(x) = \sum_{i_1+\cdots+i_{r+1}=m} W_{i_1}(x) \cdots W_{i_{r+1}}(x), \quad (m, r \geq 0), \quad (25)$$

Here, all the sums in (23)–(25) are over all nonnegative integers $i_1, \cdots, i_{r+1}$, with $i_1 + i_2 + \cdots + i_{r+1} = m$. Furthermore, note here that $\alpha_{m,r}(x), \beta_{m,r}(x), \gamma_{m,r}(x)$ all have degree m.

Further, let us put:

$$\sum_{l=0}^{m} \sum_{i_1+\cdots+i_{r+1}=m-l} \binom{r+l}{r} x^l T_{i_1}(x) \cdots T_{i_{r+1}}(x)$$

$$- \sum_{l=0}^{m-2} \sum_{i_1+\cdots+i_{r+1}=m-l-2} \binom{r+l}{r} x^l T_{i_1}(x) \cdots T_{i_{r+1}}(x), \quad (m \geq 2, r \geq 1), \quad (26)$$

$$\sum_{l=0}^{m} \sum_{i_1+\cdots+i_{r+1}=l} \binom{r-1+m-l}{r-1} V_{i_1}(x) \cdots V_{i_{r+1}}(x), \quad (m \geq 0, r \geq 1), \quad (27)$$

$$\sum_{l=0}^{m} \sum_{i_1+\cdots+i_{r+1}=l} (-1)^{m-l} \binom{r-1+m-l}{r-1} W_{i_1}(x) \cdots W_{i_{r+1}}(x), \quad (m \geq 0, r \geq 1). \quad (28)$$

We considered the expression (26) in [5] and (27) and (28) in [6] and were able to express each of them in terms of the Chebyshev polynomials of all four kinds. It is amusing to note that in such expressions, some terminating hypergeometric functions $_2F_1$ and $_3F_2$ appear respectively for (26)–(28). We came up with studying the sums in (26)–(28) by observing that they are respectively equal to $\frac{1}{2^{r-1}r!} T_{m+r}^{(r)}(x)$, $\frac{1}{2^r r!} V_{m+r}^{(r)}(x)$, and $\frac{1}{2^r r!} W_{m+r}^{(r)}(x)$. Actually, these easily follow by differentiating the generating functions in (5), (7), and (8).

In this paper, we consider the expressions $\alpha_{m,r}(x)$, $\beta_{m,r}(x)$, and $\gamma_{m,r}(x)$ in (23)–(25), which are sums of finite products of Chebyshev polynomials of the first, third, and fourth kinds, respectively. Then, we express each of them as linear combinations of $T_n(x), U_n(x), V_n(x)$, and $W_n(x)$. Here, we remark that $\alpha_{m,r}(x)$, $\beta_{m,r}(x)$, and $\gamma_{m,r}(x)$ are expressed in terms of $U_{m-j+r}^{(r)}(x)$, ($j = 0, 1, \cdots, m$) (see Lemmas 2 and 3) by making use of the generating function in (6). This is unlike the previous works for (26)–(28) (see [5,6]), where we showed they are respectively equal to $\frac{1}{2^{r-1}r!}T^{(r)}_{m+r}(x)$, $\frac{1}{2^r r!}V^{(r)}_{m+r}(x)$, and $\frac{1}{2^r r!}W^{(r)}_{m+r}(x)$ by exploiting the generating functions in (5), (7) and (8). Then, our results for $\alpha_{m,r}(x)$, $\beta_{m,r}(x)$, and $\gamma_{m,r}(x)$ will be found by making use of Lemmas 1 and 2, the general formulas in Propositions 1 and 2, and integration by parts. As we can notice here, generating functions play important roles in the present and the previous works in [5,6]. We would like to remark here that the technique of generating functions has been widely used not only in mathematics, but also in physics and biology. For this matter, we recommend the reader to refer to [7–9]. The next three theorems are our main results.

Theorem 1. *For any nonnegative integers m, r, the following identities hold true.*

$$\sum_{i_1+\cdots+i_{r+1}=m} T_{i_1}(x) \cdots T_{i_{r+1}}(x)$$

$$= \frac{1}{r!} \sum_{s=0}^{\left[\frac{m}{2}\right]} \sum_{l=0}^{s} \frac{\varepsilon_{m-2s}(-1)^l (m+r-1)!}{l!(m-s-1)!(s-1)!} {}_2F_1\left(2l-m, -r-1; l-m-r; \frac{1}{2}\right) T_{m-2s}(x) \quad (29)$$

$$= \frac{1}{r!} \sum_{s=0}^{\left[\frac{m}{2}\right]} \sum_{l=0}^{s} \frac{(-1)^l (m-2s+1)(m+r-1)!}{l!(m-s+1-l)!(s-1)!} {}_2F_1\left(2l-m, -r-1; l-m-r; \frac{1}{2}\right) U_{m-2s}(x) \quad (30)$$

$$= \frac{1}{r!} \sum_{s=0}^{m} \sum_{l=0}^{\left[\frac{s}{2}\right]} \frac{(-1)^l (m+r-1)!}{l! \left(m-\left[\frac{s}{2}\right]-1\right)! \left(\left[\frac{s}{2}\right]-l\right)!} {}_2F_1\left(2l-m, -r-1; l-m-r; \frac{1}{2}\right) V_{m-s}(x) \quad (31)$$

$$= \frac{1}{r!} \sum_{s=0}^{m} \sum_{l=0}^{\left[\frac{s}{2}\right]} \frac{(-1)^{s+l}(m+r-1)!}{l! \left(m-\left[\frac{s}{2}\right]-1\right)! \left(\left[\frac{s}{2}\right]-l\right)!} {}_2F_1\left(2l-m, -r-1; l-m-r; \frac{1}{2}\right) W_{m-s}(x). \quad (32)$$

Theorem 2. *For any nonnegative integers m, r, we have the following identities.*

$$\sum_{i_1+\cdots+i_{r+1}=m} V_{i_1}(x) \cdots V_{i_{r+1}}(x)$$

$$= \frac{1}{r!} \sum_{k=0}^{m} \sum_{l=0}^{\left[\frac{m-k}{2}\right]} \frac{(-1)^{m-k}\varepsilon_k(k+2s+r)!}{(s+k)!s!} \binom{r+1}{m-k-2s} {}_2F_1\left(-s, -s-k; -k-2s-r; 1\right) T_k(x) \quad (33)$$

$$= \frac{1}{r!} \sum_{k=0}^{m} \sum_{s=0}^{\left[\frac{m-k}{2}\right]} \frac{(-1)^{m-k}(k+1)(k+2s+r)!}{(s+k+1)!s!} \binom{r+1}{m-k-2s} {}_2F_1\left(-s, -s-k-1; -k-2s-r; 1\right) U_k(x) \quad (34)$$

$$= \frac{1}{r!} \sum_{k=0}^{m} \sum_{s=0}^{m-k} \frac{(-1)^{m-k-s}(k+r+s)!}{(k+\left[\frac{s+1}{2}\right])!\left[\frac{s}{2}\right]!} \binom{r+1}{m-k-s} {}_2F_1\left(-\left[\frac{s}{2}\right], -\left[\frac{s+1}{2}\right]-k; -k-s-r; 1\right) V_k(x) \quad (35)$$

$$= \frac{1}{r!} \sum_{k=0}^{m} \sum_{s=0}^{m-k} \frac{(-1)^{m-k}(k+r+s)!}{(k+\left[\frac{s+1}{2}\right])!\left[\frac{s}{2}\right]!} \binom{r+1}{m-k-s} {}_2F_1\left(-\left[\frac{s}{2}\right], -\left[\frac{s+1}{2}\right]-k; -k-s-r; 1\right) W_k(x) \quad (36)$$

Theorem 3. *For any nonnegative integers m, r, the following identities are valid.*

$$\sum_{i_1+\cdots+i_{r+1}=m} W_{i_1}(x) \cdots W_{i_{r+1}}(x)$$

$$= \frac{1}{r!} \sum_{k=0}^{m} \sum_{l=0}^{\left[\frac{m-k}{2}\right]} \frac{\varepsilon_k(k+2s+r)!}{(s+k)!s!} \binom{r+1}{m-k-2s} {}_2F_1\left(-s, -s-k; -k-2s-r; 1\right) T_k(x) \quad (37)$$

$$= \frac{1}{r!} \sum_{k=0}^{m} \sum_{s=0}^{\left[\frac{m-k}{2}\right]} \frac{(k+1)(k+2s+r)!}{(s+k+1)!s!} \binom{r+1}{m-k-2s} {}_2F_1\left(-s, -s-k-1; -k-2s-r; 1\right) U_k(x) \quad (38)$$

$$= \frac{1}{r!} \sum_{k=0}^{m} \sum_{s=0}^{m-k} \frac{(-1)^s (k+r+s)!}{\left(k + \left[\frac{s+1}{2}\right]\right)! \left[\frac{s}{2}\right]!} \binom{r+1}{m-k-s} {}_2F_1\left(-\left[\frac{s}{2}\right], -\left[\frac{s+1}{2}\right] - k; -k-s-r; 1\right) V_k(x) \quad (39)$$

$$= \frac{1}{r!} \sum_{k=0}^{m} \sum_{s=0}^{m-k} \frac{(k+r+s)!}{\left(k + \left[\frac{s+1}{2}\right]\right)! \left[\frac{s}{2}\right]!} \binom{r+1}{m-k-s} {}_2F_1\left(-\left[\frac{s}{2}\right], -\left[\frac{s+1}{2}\right] - k; -k-s-r; 1\right) W_k(x) \quad (40)$$

Before moving on to the next section, we would like to say a few words on the previous works that are associated with the results in the present paper. In terms of Bernoulli polynomials, quite a few sums of finite products of some special polynomials are expressed. They include Chebyshev polynomials of all four kinds, and Bernoulli, Euler, Genocchi, Legendre, Laguerre, Fibonacci, and Lucas polynomials (see [10–16]). All of these expressions in terms of Bernoulli polynomials have been derived from the Fourier series expansions of the functions closely related to each such polynomials. Further, as for Chebyshev polynomials of all four kinds and Legendre, Laguerre, Fibonacci, and Lucas polynomials, certain sums of finite products of such polynomials are also expressed in terms of all four kinds of Chebyshev polynomials in [5,6,17,18]. Finally, the reader may want to look at [19–21] for some applications of Chebyshev polynomials.

2. Proof of Theorem 1

In this section, we will prove Theorem 1. In order to do this, we first state Propositions 1 and 2 that are needed in proving Theorems 1–3. Here, we note that the facts (a), (b), (c), and (d) in Proposition 1 are stated respectively in the Equations (24) of [22], (36) of [22], (23) of [23], and (38) of [23]. All of them follow easily from the orthogonality relations in (17) and (20)–(22), Rodrigues' formulas in (13)–(16), and integration by parts.

Proposition 1. *For any polynomial $q(x) \in \mathbb{R}[x]$ of degree n, we have the following formulas.*

(a) $q(x) = \sum_{k=0}^{n} C_{k,1} T_k(x)$, where:

$$C_{k,1} = \frac{(-1)^k 2^k k! \varepsilon_k}{(2k)! \pi} \int_{-1}^{1} q(x) \frac{d^k}{dx^k} (1-x^2)^{k-\frac{1}{2}} dx,$$

(b) $q(x) = \sum_{k=0}^{n} C_{k,2} U_k(x)$, where:

$$C_{k,2} = \frac{(-1)^k 2^{k+1} (k+1)!}{(2k+1)! \pi} \int_{-1}^{1} q(x) \frac{d^k}{dx^k} (1-x^2)^{k+\frac{1}{2}} dx,$$

(c) $q(x) = \sum_{k=0}^{n} C_{k,3} V_k(x)$, where:

$$C_{k,3} = \frac{(-1)^k k! 2^k}{(2k)! \pi} \int_{-1}^{1} q(x) \frac{d^k}{dx^k} (1-x)^{k-\frac{1}{2}} (1+x)^{k+\frac{1}{2}} dx,$$

(d) $q(x) = \sum_{k=0}^{n} C_{k,4} W_k(x)$, where,

$$C_{k,4} = \frac{(-1)^k k! 2^k}{(2k)! \pi} \int_{-1}^{1} q(x) \frac{d^k}{dx^k} (1-x)^{k+\frac{1}{2}} (1+x)^{k+\frac{1}{2}} dx,$$

The next proposition is stated and proven in [17].

Proposition 2. *For any nonnegative integers m, k, we have the following formulas:*

(a)
$$\int_{-1}^{1}(1-x^2)^{k-\frac{1}{2}}x^m dx = \begin{cases} 0, & \text{if } m \equiv 1 \pmod{2}, \\ \frac{m!(2k)!\pi}{2^{m+2k}\left(\frac{m}{2}+k\right)!\left(\frac{m}{2}\right)!k!}, & \text{if } m \equiv 0 \pmod{2}. \end{cases}$$

(b)
$$\int_{-1}^{1}(1-x^2)^{k+\frac{1}{2}}x^m dx = \begin{cases} 0, & \text{if } m \equiv 1 \pmod{2}, \\ \frac{m!(2k+2)!\pi}{2^{m+2k+2}\left(\frac{m}{2}+k+1\right)!\left(\frac{m}{2}\right)!(k+1)!}, & \text{if } m \equiv 0 \pmod{2}. \end{cases}$$

(c)
$$\int_{-1}^{1}(1-x)^{k-\frac{1}{2}}(1+x)^{k+\frac{1}{2}}x^m dx = \begin{cases} \frac{(m+1)!(2k)!\pi}{2^{m+2k+1}\left(\frac{m+1}{2}+k\right)!\left(\frac{m+1}{2}\right)!k!}, & \text{if } m \equiv 1 \pmod{2}, \\ \frac{m!(2k)!\pi}{2^{m+2k}\left(\frac{m}{2}+k\right)!\left(\frac{m}{2}\right)!k!}, & \text{if } m \equiv 0 \pmod{2}. \end{cases}$$

(d)
$$\int_{-1}^{1}(1-x)^{k+\frac{1}{2}}(1+x)^{k-\frac{1}{2}}x^m dx = \begin{cases} -\frac{(m+1)!(2k)!\pi}{2^{m+2k+1}\left(\frac{m+1}{2}+k\right)!\left(\frac{m+1}{2}\right)!k!}, & \text{if } m \equiv 1 \pmod{2}, \\ \frac{m!(2k)!\pi}{2^{m+2k}\left(\frac{m}{2}+k\right)!\left(\frac{m}{2}\right)!k!}, & \text{if } m \equiv 0 \pmod{2}. \end{cases}$$

The following lemma was shown in [24] and can be derived by differentiating [23].

Lemma 1. *For any nonnegative integers n, r, the following identity holds:*

$$\sum_{i_1+\cdots+i_{r+1}=n} U_{i_1}(x)\cdots U_{i_{r+1}}(x) = \frac{1}{2^r r!}U^{(r)}_{n+k}(x), \tag{41}$$

where the sum is over all nonnegative integers $i_1, \cdots, i_{r+1}$, with $i_1+\cdots+i_{r+1}=n$.

Further, Equation (41) is equivalent to:

$$\left(\frac{1}{1-2xt+t^2}\right)^{r+1} = \frac{1}{2^r r!}\sum_{n=0}^{\infty}U^{(r)}_{n+r}(x)t^n. \tag{42}$$

In reference [24], the following lemma is stated for $m \geq r+1$. However, it holds for any nonnegative integer m, under the usual convention $\binom{r+1}{j} = 0$, for $j > r+1$. Therefore, we are going to give a proof for the next lemma.

Lemma 2. *Let m, r be any nonnegative integers. Then, the following identity holds.*

$$\sum_{i_1+\cdots+i_{r+1}=m} T_{i_1}(x)\cdots T_{i_{r+1}}(x)$$
$$= \frac{1}{2^r r!}\sum_{j=0}^{m}(-1)^j\binom{r+1}{j}x^j U^{(r)}_{m-j+r}(x), \tag{43}$$

where $\binom{r+1}{j} = 0$, for $j > r+1$.

Proof. By making use of (42), we have:

$$\sum_{m=0}^{\infty} \left(\sum_{i_1+\cdots+i_{r+1}=m} T_{i_1}(x) \cdots T_{i_{r+1}}(x) \right) t^m$$

$$= \left(\frac{1}{1-2xt+t^2} \right)^{r+1} (1-xt)^{r+1}$$

$$= \frac{1}{2^r r!} \sum_{n=0}^{\infty} U_{n+r}^{(r)}(x) t^n \sum_{j=0}^{r+1} \binom{r+1}{j} (-x)^j t^j \tag{44}$$

$$= \frac{1}{2^r r!} \sum_{m=0}^{\infty} \left(\sum_{j=0}^{\min\{m,r+1\}} (-1)^j \binom{r+1}{j} x^j U_{m-j+r}^{(r)}(x) \right) t^m$$

$$= \frac{1}{2^r r!} \sum_{m=0}^{\infty} \left(\sum_{j=0}^{m} (-1)^j \binom{r+1}{j} x^j U_{m-j+r}^{(r)}(x) \right) t^m.$$

Now, by comparing both sides of (44), we have the desired result. □

From (10), we see that the rth derivative of $U_n(x)$ is given by:

$$U_n^{(r)}(x) = \sum_{l=0}^{\left[\frac{n-r}{2}\right]} (-1)^l \binom{n-l}{l} (n-2l)_r 2^{n-2l} x^{n-2l-r}. \tag{45}$$

Especially, we have:

$$x^j U_{m-j+r}^{(r)}(x) = \sum_{l=0}^{\left[\frac{m-j}{2}\right]} (-1)^l \binom{m-j+r-l}{l} (m-j+r-2l)_r 2^{m-j+r-2l} x^{m-2l}. \tag{46}$$

In this section, we will show (29) and (31) of Theorem 1 and leave similar proofs for (30) and (32) as exercises to the reader. As in (23), let us put:

$$\alpha_{m,r}(x) = \sum_{i_1+\cdots+i_{r+1}=m} T_{i_1}(x) \cdots T_{i_{r+1}}(x),$$

and set:

$$\alpha_{m,r}(x) = \sum_{k=0}^{m} C_{k,1} T_k(x). \tag{47}$$

Then, we can now proceed as follows by using (a) of Proposition 1, (43) and (46), and integration by parts k times.

$$C_{k,1} = \frac{(-1)^k 2^k k! \varepsilon_k}{(2k)! \pi} \int_{-1}^{1} \alpha_{m,r}(x) \frac{d^k}{dx^k} (1-x^2)^{k-\frac{1}{2}} dx$$

$$= \frac{(-1)^k 2^k k! \varepsilon_k}{(2k)! \pi 2^r r!} \sum_{j=0}^{m} (-1)^j \binom{r+1}{j} \int_{-1}^{1} x^j U_{m-j+r}^{(r)}(x) \frac{d^k}{dx^k} (1-x^2)^{k-\frac{1}{2}} dx$$

$$= \frac{(-1)^k 2^k k! \varepsilon_k}{(2k)! \pi 2^r r!} \sum_{j=0}^{m} (-1)^j \binom{r+1}{j} \sum_{l=0}^{\left[\frac{m-j}{2}\right]} (-1)^l \binom{m-j+r-l}{l}$$

$$\times (m-j+r-2l)_r 2^{m-j+r-2l} \int_{-1}^{1} x^{m-2l} \frac{d^k}{dx^k} (1-x^2)^{k-\frac{1}{2}} dx.$$
(48)

$$= \frac{2^k k! \varepsilon_k}{(2k)! \pi 2^r r!} \sum_{j=0}^{m} (-1)^j \binom{r+1}{j} \sum_{l=0}^{\left[\frac{m-j}{2}\right]} (-1)^l \binom{m-j+r-l}{l}$$

$$\times (m-j+r-2l)_r 2^{m-j+r-2l} (m-2l)_k \int_{-1}^{1} x^{m-k-2l} (1-x^2)^{k-\frac{1}{2}} dx$$

$$= \frac{2^k k! \varepsilon_k}{(2k)! \pi 2^r r!} \sum_{l=0}^{\left[\frac{m-k}{2}\right]} \sum_{j=0}^{m-2l} (-1)^j \binom{r+1}{j} (-1)^l \binom{m-j+r-l}{l}$$

$$\times (m-j+r-2l)_r 2^{m-j+r-2l} (m-2l)_k \int_{-1}^{1} x^{m-k-2l} (1-x^2)^{k-\frac{1}{2}} dx.$$

Now, from (a) of Proposition 2 and after some simplifications, we see that:

$$\alpha_{m,r}(x) = \frac{1}{r!} \sum_{0 \le k \le m, k \equiv m \pmod{2}} \sum_{l=0}^{\left[\frac{m-k}{2}\right]} \sum_{j=0}^{m-2l} \varepsilon_k (-1)^j \binom{r+1}{j} 2^{-j}$$

$$\times \frac{(-1)^l (m-j+r-l)!(m-2l)!}{l!(m-j-2l)! \left(\frac{m+k}{2}-l\right)! \left(\frac{m-k}{2}-l\right)!} T_k(x)$$

$$= \frac{1}{r!} \sum_{s=0}^{\left[\frac{m}{2}\right]} \sum_{l=0}^{s} \frac{\varepsilon_{m-2s}(-1)^l (m-2l)!}{l!(m-s-l)!(s-l)!}$$
(49)

$$\times \sum_{j=0}^{m-2l} \frac{2^{-j}(-1)^j (m+r-l-j)!(r+1)_j}{j!(m-2l-j)!} T_{m-2s}(x)$$

$$= \frac{1}{r!} \sum_{s=0}^{\left[\frac{m}{2}\right]} \sum_{l=0}^{s} \frac{\varepsilon_{m-2s}(-1)^l (m+r-l)!}{l!(m-s-l)!(s-l)!}$$

$$\times \sum_{j=0}^{m-2l} \frac{2^{-j}(-1)^j (m-2l)_j (r+1)_j}{j!(m+r-l)_j} T_{m-2s}(x)$$

$$= \frac{1}{r!} \sum_{s=0}^{\left[\frac{m}{2}\right]} \sum_{l=0}^{s} \frac{\varepsilon_{m-2s}(-1)^l (m+r-l)!}{l!(m-s-l)!(s-l)!}$$

$$\times \sum_{j=0}^{m-2l} \frac{2^{-j} <2l-m>_j <-r-1>_j}{j! <l-m-r>_j} T_{m-2s}(x)$$

$$= \frac{1}{r!} \sum_{s=0}^{\left[\frac{m}{2}\right]} \sum_{l=0}^{s} \frac{\varepsilon_{m-2s}(-1)^l (m+r-l)!}{l!(m-s-l)!(s-l)!}$$

$$\times {}_2F_1 \left(2l-m, -r-1; l-m-r; \frac{1}{2}\right) T_{m-2s}(x),$$

where we note that we made the change of variables $m - k = 2s$.

This completes the proof for (29). Next, we let:

$$\alpha_{m,r}(x) = \sum_{k=0}^{m} C_{k,3} V_k(x). \tag{50}$$

Then, we can obtain the following by making use of (c) of Proposition 1, (43) and (46), and integration by parts k times.

$$C_{k,3} = \frac{k! 2^k}{(2k)! \pi 2^r r!} \sum_{l=0}^{\left[\frac{m-k}{2}\right]} \sum_{j=0}^{m-2l} (-1)^j \binom{r+1}{j} (-1)^l \binom{m-j+r-l}{l} (m-j+r-2l)_r 2^{m-j+r-2l}$$
$$\times (m-2l)_k \int_{-1}^{1} x^{m-2l-k} (1-x)^{k-\frac{1}{2}} (1+x)^{k+\frac{1}{2}} dx. \tag{51}$$

where we note from (c) of Proposition 2 that:

$$\int_{-1}^{1} x^{m-2l-k} (1-x)^{k-\frac{1}{2}} (1+x)^{k+\frac{1}{2}} dx$$
$$= \begin{cases} \frac{(m-2l-k+1)!(2k)!\pi}{2^{m+k-2l+1} \left(\frac{m+k+1}{2}-l\right)! \left(\frac{m-k+1}{2}-l\right)! k!}, & \text{if } k \not\equiv m \pmod{2}, \\ \frac{(m-2l-k)!(2k)!\pi}{2^{m+k-2l} \left(\frac{m+k}{2}-l\right)! \left(\frac{m-k}{2}-l\right)! k!}, & \text{if } k \equiv m \pmod{2}. \end{cases} \tag{52}$$

From (50)–(52), and after some simplifications, we get:

$$\alpha_{m,r}(x) = \sum_1 + \sum_2, \tag{53}$$

where:

$$\sum_1 = \frac{1}{r!} \sum_{0 \leq k \leq m, k \not\equiv m \pmod 2} \sum_{l=0}^{\left[\frac{m-k}{2}\right]} \sum_{j=0}^{m-2l} (-1)^j \binom{r+1}{j} 2^{-j-1}$$
$$\times \frac{(-1)^l (m-j+r-l)!(m-2l)!(m-2l-k+1)}{l!(m-j-2l)! \left(\frac{m+k+1}{2}-1\right)! \left(\frac{m-k+1}{2}-1\right)!} V_k(x),$$

$$\sum_2 = \frac{1}{r!} \sum_{0 \leq k \leq m, k \equiv m \pmod 2} \sum_{l=0}^{\left[\frac{m-k}{2}\right]} \sum_{j=0}^{m-2l} (-1)^j \binom{r+1}{j} 2^{-j} \tag{54}$$
$$\times \frac{(-1)^l (m-j+r-l)!(m-2l)!}{l!(m-j-2l)! \left(\frac{m+k}{2}-1\right)! \left(\frac{m-k}{2}-1\right)!} V_k(x).$$

Proceeding analogously to the case of (29), we observe from (54) that:

$$\sum_1 = \frac{1}{r!} \sum_{s=0}^{\left[\frac{m-1}{2}\right]} \sum_{l=0}^{s} \frac{(-1)^l (m+r-l)!}{l!(m-s-l)!(s-l)!}$$
$$\times \sum_{j=0}^{m-2l} \frac{2^{-j} (-1)^j (r+1)_j (m-2l)_j}{j!(m+r-l)_j} V_{m-2s-1}(x)$$
$$= \frac{1}{r!} \sum_{s=0}^{\left[\frac{m-1}{2}\right]} \sum_{l=0}^{s} \frac{(-1)^l (m+r-l)!}{l!(m-s-l)!(s-l)!} \tag{55}$$
$$\times {}_2F_1\left(2l-m, -r-1; l-m-r; \frac{1}{2}\right) V_{m-2s-1}(x),$$

$$\Sigma_2 = \frac{1}{r!} \sum_{s=0}^{\left[\frac{m}{2}\right]} \sum_{l=0}^{s} \frac{(-1)^l (m+r-l)!}{l!(m-s-l)!(s-l)!}$$
$$\times \sum_{j=0}^{m-2l} \frac{2^{-j}(-1)^j (r+1)_j (m-2l)_j}{j!(m+r-l)_j} V_{m-2s}(x)$$
(56)
$$= \frac{1}{r!} \sum_{s=0}^{\left[\frac{m}{2}\right]} \sum_{l=0}^{s} \frac{(-1)^l (m+r-l)!}{l!(m-s-l)!(s-l)!}$$
$$\times {}_2F_1\left(2l-m, -r-1; l-m-r; \frac{1}{2}\right) V_{m-2s}(x).$$

We now obtain the result in (31) from (53), (55) and (56).

3. Proofs of Theorems 2 and 3

In this section, we will show (34) and (36) for Theorem 2, leaving (33) and (35) as exercises to the reader, and note that Theorem 3 follows from (33)–(36) by simple observation. The next lemma can be shown analogously to Lemma 1.

Lemma 3. *For any nonnegative integers* m, r, *the following identities are valid.*

$$\sum_{i_1+\cdots+i_{r+1}=m} V_{i_1}(x) \cdots V_{i_{r+1}}(x)$$
$$= \frac{1}{2^r r!} \sum_{j=0}^{m} (-1)^j \binom{r+1}{j} U_{m-j+r}^{(r)}(x),$$
(57)

$$\sum_{i_1+\cdots+i_{r+1}=m} W_{i_1}(x) \cdots W_{i_{r+1}}(x)$$
$$= \frac{1}{2^r r!} \sum_{j=0}^{m} \binom{r+1}{j} U_{m-j+r}^{(r)}(x),$$
(58)

where $\binom{r+1}{j} = 0$, *for* $j > r+1$.

As in (24), let us set:
$$\beta_{m,r}(x) = \sum_{i_1+\cdots+i_{r+1}=m} V_{i_1}(x) \cdots V_{i_{r+1}}(x),$$

and put:
$$\beta_{m,r}(x) = \sum_{k=0}^{m} C_{k,2} U_k(x).$$
(59)

First, we note:

$$U_{m-j+r}^{(r+k)}(x) = \sum_{l=0}^{\left[\frac{m-j-k}{2}\right]} (-1)^l \binom{m-j+r-l}{l} (m-j+r-2l)_{r+k} 2^{m-j+r-2l} x^{m-j-k-2l}.$$
(60)

Then, we have the following by exploiting (b) of Proposition 1, (57) and (60), and integration by parts k times.

$$C_{k,2} = \frac{(-1)^k 2^{k+1}(k+1)!}{(2k+1)!\pi} \int_{-1}^{1} \beta_{m,r}(x) \frac{d^k}{dx^k}(1-x^2)^{k+\frac{1}{2}} dx$$

$$= \frac{(-1)^k 2^{k+1}(k+1)!}{(2k+1)!\pi 2^r r!} \sum_{j=0}^{m}(-1)^j \binom{r+1}{j} \int_{-1}^{1} U_{m-j+r}^{(r)}(x) \frac{d^k}{dx^k}(1-x^2)^{k+\frac{1}{2}} dx$$

$$= \frac{2^{k+1}(k+1)!}{(2k+1)!\pi 2^r r!} \sum_{j=0}^{m-k}(-1)^j \binom{r+1}{j} \int_{-1}^{1} U_{m-j+r}^{(r+k)}(x)(1-x^2)^{k+\frac{1}{2}} dx \quad (61)$$

$$= \frac{2^{k+1-r}(k+1)!}{(2k+1)!\pi r!} \sum_{j=0}^{m-k}(-1)^j \binom{r+1}{j} \sum_{l=0}^{\left[\frac{m-j-k}{2}\right]}(-1)^l \binom{m-j+r-l}{l}$$

$$\times (m-j+r-2l)_{r+k} 2^{m-j+r-2l} \int_{-1}^{1} x^{m-j-k-2l}(1-x^2)^{k+\frac{1}{2}} dx$$

where we note from (b) of Proposition 2 that:

$$\int_{-1}^{1} x^{m-j-k-2l}(1-x^2)^{k+\frac{1}{2}} dx$$
$$= \begin{cases} 0, & \text{if } j \not\equiv m-k \pmod{2}, \\ \frac{(m-j-k-2l)!(2k+2)!\pi}{2^{m-j+k-2l+2}\left(\frac{m-j+k}{2}+1-l\right)!\left(\frac{m-j-k}{2}-l\right)!(k+1)!}, & \text{if } j \equiv m-k \pmod{2}. \end{cases} \quad (62)$$

From (59), (61) and (62), and after some simplifications, we obtain:

$$\beta_{m,r}(x) = \frac{1}{r!} \sum_{k=0}^{m} \sum_{\substack{0 \le j \le m-k, j \equiv m-k \pmod{2}}} \sum_{l=0}^{\left[\frac{m-k-j}{2}\right]} (-1)^j \binom{r+1}{j}(k+1)$$

$$\times \frac{(-1)^l(m-j+r-l)!}{l!\left(\frac{m-j+k}{2}+1-l\right)!\left(\frac{m-j-k}{2}-l\right)!} U_k(x)$$

$$= \frac{1}{r!} \sum_{k=0}^{m} \sum_{s=0}^{\left[\frac{m-k}{2}\right]} \frac{(-1)^{m-k}(k+1)(k+2s+r)!}{(s+k+1)!s!} \binom{r+1}{m-k-2s} \quad (63)$$

$$\times \sum_{l=0}^{s} \frac{(-1)^l(s+k+1)_l(s)_l}{l!(k+2s+r)_l} U_k(x)$$

$$= \frac{1}{r!} \sum_{k=0}^{m} \sum_{s=0}^{\left[\frac{m-k}{2}\right]} \frac{(-1)^{m-k}(k+1)(k+2s+r)!}{(s+k+1)!s!} \binom{r+1}{m-k-2s}$$

$$\times \sum_{l=0}^{s} \frac{<-s>_l<-s-k-1>_l}{l!<-k-2s-r>_l} U_k(x)$$

$$= \frac{1}{r!} \sum_{k=0}^{m} \sum_{s=0}^{\left[\frac{m-k}{2}\right]} \frac{(-1)^{m-k}(k+1)(k+2s+r)!}{(s+k+1)!s!} \binom{r+1}{m-k-2s}$$

$$\times {}_2F_1(-s, -s-k-1; -k-2s-r; 1) U_k(x).$$

This completes the proof for (34). Next, we let:

$$\beta_{m,r}(x) = \sum_{k=0}^{m} C_{k,4} W_k(x). \quad (64)$$

Then, from (d) of Proposition 1, (57) and (60), and integration by parts k times, we have:

$$C_{k,4} = \frac{k! 2^{k-r}}{(2k)! \pi r!} \sum_{j=0}^{m-k} (-1)^j \binom{r+1}{j} \sum_{l=0}^{\left[\frac{m-j-k}{2}\right]} (-1)^l \binom{m-j+r-l}{l} \tag{65}$$

$$\times (m-j+r-2l)_{r+k} 2^{m-j+r-2l} \int_{-1}^{1} x^{m-j-k-2l}(1-x)^{k+\frac{1}{2}}(1-x)^{k-\frac{1}{2}} dx$$

From (d) of Proposition 2, we observe that:

$$\int_{-1}^{1} x^{m-j-k-2l}(1-x)^{k+\frac{1}{2}}(1-x)^{k-\frac{1}{2}} dx$$

$$= \begin{cases} -\dfrac{(m-j-k-2l+1)!(2k)!\pi}{2^{m-j+k-2l+1}\left(\frac{m-j+k+1}{2}-1\right)!\left(\frac{m-j-k+1}{2}-l\right)!k!}, & \text{if } j \not\equiv m-k \pmod{2}, \\ \dfrac{(m-j-k-2l)!(2k)!\pi}{2^{m-j+k-2l}\left(\frac{m-j+k}{2}-l\right)!\left(\frac{m-j-k}{2}-l\right)!k!}, & \text{if } j \equiv m-k \pmod{2}. \end{cases} \tag{66}$$

By (64)–(66), and after some simplifications, we get:

$$\beta_{m,r}(x) = -\frac{1}{2r!} \sum_{k=0}^{m} \sum_{\substack{0 \le j \le m-k \\ j \not\equiv m-k \pmod 2}} (-1)^j \binom{r+1}{j} \sum_{l=0}^{\left[\frac{m-j-k}{2}\right]}$$

$$\times \frac{(-1)^l (m-j+r-1)!(m-j-k-2l+1)}{l! \left(\frac{m-j+k+1}{2}-1\right)! \left(\frac{m-j-k+1}{2}-1\right)!} W_k(x)$$

$$+ \frac{1}{r!} \sum_{k=0}^{m} \sum_{\substack{0 \le j \le m-k \\ j \equiv m-k \pmod 2}} (-1)^j \binom{r+1}{j} \sum_{l=0}^{\left[\frac{m-j-k}{2}\right]}$$

$$\times \frac{(-1)^l (m-j+r-1)!}{l! \left(\frac{m-j+k}{2}-1\right)! \left(\frac{m-j-k}{2}-1\right)!} W_k(x) \tag{67}$$

$$= \frac{1}{r!} \sum_{k=0}^{m} \sum_{s=0}^{\left[\frac{m-k-1}{2}\right]} (-1)^{m-k} \binom{r+1}{m-k-2s-1} \frac{(k+2s+r+1)!}{(s+k+1)!s!}$$

$$\times \sum_{l=0}^{s} \frac{(-1)^l (s+k+1)_l (s)_l}{l!(k+2s+r+1)_l} W_k(x)$$

$$+ \frac{1}{r!} \sum_{k=0}^{m} \sum_{s=0}^{\left[\frac{m-k}{2}\right]} (-1)^{m-k} \binom{r+1}{m-k-2s} \frac{(k+2s+r)!}{(s+k)!s!}$$

$$\times \sum_{l=0}^{s} \frac{(-1)^l (s+k)_l (s)_l}{l!(k+2s+r)_l} W_k(x).$$

Further modification of (67) gives us:

$$\begin{aligned}
\beta_{m,r}(x) &= \frac{1}{r!} \sum_{k=0}^{m} \sum_{s=0}^{\left[\frac{m-k-1}{2}\right]} (-1)^{m-k} \frac{(k+2s+r+1)!}{(s+k+1)!s!} \binom{r+1}{m-k-2s-1} \\
&\quad \times {}_2F_1(-s,-s-k-1;-k-2s-r-1;1) W_k(x) \\
&\quad + \frac{1}{r!} \sum_{k=0}^{m} \sum_{l=0}^{\left[\frac{m-k}{2}\right]} (-1)^{m-k} \frac{(k+2s+r)!}{(s+k)!s!} \binom{r+1}{m-k-2s} \\
&\quad \times {}_2F_1(-s,-s-k;-k-2s-r;1) W_k(x) \\
&= \frac{1}{r!} \sum_{k=0}^{m} \sum_{s=0}^{m-k} \frac{(-1)^{m-k}(k+r+s)!}{\left(k+\left[\frac{s+1}{2}\right]\right)! \left[\frac{s}{2}\right]!} \binom{r+1}{m-k-s} \\
&\quad \times {}_2F_1\left(-\left[\frac{s}{2}\right], -\left[\frac{s+1}{2}\right]-k; -k-s-r; 1\right) W_k(x).
\end{aligned} \qquad (68)$$

This finishes up the proof for (36).

Remark 1. *We note from (57) and (58) that the only difference between $\beta_{m,r}(x)$ and $\gamma_{m,r}(x)$ (see (24) and (25)) is the alternating sign $(-1)^j$ in their sums, which corresponds to $(-1)^{m-k}$ in (33)–(36). This remark gives the results in (37)–(40) of Theorem 3.*

4. Conclusions

Our paper can be viewed as a generalization of the linearization problem, which is concerned with determining the coefficients in the expansion $a_n(x)b_m(x) = \sum_{k=0}^{n+m} c_k(nm) p_k(x)$ of the product $a_n(x)b_m(x)$ of two polynomials $a_n(x)$ and $b_m(x)$ in terms of an arbitrary polynomial sequence $\{p_k(x)\}_{k \geq 0}$. Our pursuit of this line of research can also be justified from another fact; namely, the famous Faber–Pandharipande–Zagier and Miki identities follow by expressing the sum $\sum_{k=1}^{m-1} \frac{1}{k(m-k)} B_k(x) B_{m-k}(x)$ as a linear combination of Bernoulli polynomials. For some details on this, we let the reader refer to the Introduction of [15]. Here, we considered sums of finite products of the Chebyshev polynomials of the first, third, and fourth kinds and represented each of those sums of finite products as linear combinations of $T_n(x)$, $U_n(x)$, $V_n(x)$, and $W_n(x)$, which involve some terminating hypergeometric function ${}_2F_1$. Here, we remark that $\alpha_{m,r}(x)$, $\beta_{m,r}(x)$, and $\gamma_{m,r}(x)$ are expressed in terms of $U_{m-j+r}^{(r)}(x)$, $(j = 0, 1, \cdots, m)$ (see Lemmas 2 and 3) by making use of the generating function in (6). This is unlike the previous works for (26)–(28) (see [5,6]), where we showed they are respectively equal to $\frac{1}{2^{r-1}r!} T_{m+r}^{(r)}(x)$, $\frac{1}{2^r r!} V_{m+r}^{(r)}(x)$, and $\frac{1}{2^r r!} W_{m+r}^{(r)}(x)$ by exploiting the generating functions in (5), (7) and (8). Then, our results for $\alpha_{m,r}(x)$, $\beta_{m,r}(x)$, and $\gamma_{m,r}(x)$ were found by making use of Lemmas 1 and 2, the general formulas in Propositions 1 and 2, and integration by parts. It is certainly possible to represent such sums of finite products by other orthogonal polynomials, which is one of our ongoing projects. More generally, along the same line as the present paper, we are planning to consider some sums of finite products of many special polynomials and want to find their applications.

Author Contributions: T.K. and D.S.K. conceived of the framework and structured the whole paper; T.K. wrote the paper; L.-C.J. and D.V.D. checked the results of the paper; D.S.K. and T.K. completed the revision of the article.

Funding: This research received no external funding.

Conflicts of Interest: The authors declare no conflict of interest.

References

1. Andrews, G.E.; Askey, R.; Roy, R. *Special Functions*; Encyclopedia of Mathematics and Its Applications 71; Cambridge University Press: Cambridge, UK, 1999.
2. Beals, R.; Wong, R. *Special Functions and Orthogonal Polynomials*; Cambridge Studies in Advanced Mathematics 153; Cambridge University Press: Cambridge, UK, 2016.

3. Kim, T.; Kim, D.S.; Jang, G.-W.; Jang, L.C. Fourier series of functions involving higher-order ordered Bell polynomials. *Open Math.* **2017**, *15*, 1606–1617. [CrossRef]
4. Mason, J.C.; Handscomb, D.C. *Chebyshev Polynomials*; Chapman&Hall/CRC: Boca Raton, FC, USA, 2003.
5. Kim, T.; Kim, D.S.; Dolgy, D.V.; Kwon, J. Representing sums of finite products of chebyshev polynomials of the first kind and lucas polynomials by chebyshev polynomials. *Math. Comput. Sci.* **2018**. [CrossRef]
6. Kim, T.; Kim, D.S.; Dolgy, D.V.; Ryoo, C.S. Representing sums of finite products of Chebyshev polynomials of the third and fourth kinds by Chebyshev polynomials. *Symmetry* **2018**, *10*, 258. [CrossRef]
7. Shang, Y. Unveiling robustness and heterogeneity through percolation triggered by random-link breakdown. *Phys. Rev. E* **2014**, *90*, 032820. [CrossRef] [PubMed]
8. Shang, Y. Effect of link oriented self-healing on resilience of networks. *J. Stat. Mech. Theory Exp.* **2016**, *2016*, 083403. [CrossRef]
9. Shang, Y. Modeling epidemic spread with awareness and heterogeneous transmission rates in networks. *J. Biol. Phys.* **2013**, *39*, 489–500. [PubMed]
10. Agarwal, R.P.; Kim, D.S.; Kim, T.; Kwon, J. Sums of finite products of Bernoulli functions. *Adv. Differ. Equ.* **2017**, *2017*, 237. [CrossRef]
11. Kim, T.; Kim, D.S.; Dolgy, D.V.; Kwon, J. Sums of finite products of Chebyshev polynomials of the third and fourth kinds. *Adv. Differ. Equ.* **2018**, *2018*, 283. [CrossRef]
12. Kim, T.; Kim, D.S.; Dolgy, D.V.; Park, J.-W. Sums of finite products of Chebyshev polynomials of the second kind and of Fibonacci polynomials. *J. Inequal. Appl.* **2018**, *2018*, 148. [CrossRef] [PubMed]
13. Kim, T.; Kim, D.S.; Dolgy, D.V.; Park, J.-W. Sums of finite products of Legendre and Laguerre polynomials. *Adv. Differ. Equ.* **2018**, *2018*, 277. [CrossRef]
14. Kim, T.; Kim, D.S.; Jang, G.-W.; Kwon, J. Sums of finite products of Euler functions. In *Advances in Real and Complex Analysis with Applications*; Trends in Mathematics; Birkhäuser: Basel, Switzerland, 2017; pp. 243–260.
15. Kim, T.; Kim, D.S.; Jang, L.C.; Jang, G.-W. Sums of finite products of Genocchi functions. *Adv. Differ. Equ.* **2017**, *2017*, 268. [CrossRef]
16. Kim, T.; Kim, D.S.; Jang, L.C.; Jang, G.-W. Fourier series for functions related to Chebyshev polynomials of the first kind and Lucas polynomials. *Mathematics* **2018**, *6*, 276. [CrossRef]
17. Kim, T.; Dolgy, D.V.; Kim, D.S. Representing sums of finite products of Chebyshev polynomials of the second kind and Fibonacci polynomials in terms of Chebyshev polynomials. *Adv. Stud. Contemp. Math.* **2018**, *28*, 321–335.
18. Kim, T.; Kim, D.S.; Kwon, J.; Jang, G.-W. Sums of finite products of Legendre and Laguerre polynomials by Chebyshev polynomials. *Adv. Stud. Contemp. Math.* **2018**, *28*, 551–565.
19. Doha, E.H.; Abd-Elhameed, W.M.; Alsuyuti, M.M. On using third and fourth kinds Chebyshev polynomials for solving the integrated forms of high odd-order linear boundary value problems. *J. Egypt. Math. Soc.* **2015**, *23*, 397–405. [CrossRef]
20. Eslahchi, M.R.; Dehghan, M.; Amani, S. The third and fourth kinds of Chebyshev polynomials and best uniform approximation. *Math. Comput. Model.* **2012**, *55*, 1746–1762. [CrossRef]
21. Mason, J.C. Chebyshev polynomials of the second, third and fourth kinds in approximation, indefinite integration, and integral transforms. *J. Comput. Appl. Math.* **1993**, *49*, 169–178. [CrossRef]
22. Kim, D.S.; Kim, T.; Lee, S.-H. Some identities for Bernoulli polynomials involving Chebyshev polynomials. *J. Comput. Anal. Appl.* **2014**, *16*, 172–180.
23. Kim, D.S.; Dolgy, D.V.; Kim, T.; Rim, S.-H. Identities involving Bernoulli and Euler polynomials arising from Chebyshev polynomials. *Proc. Jangjeon Math. Soc.* **2012**, *15*, 361–370.
24. Zhang, W. Some identities involving the Fibonacci numbers and Lucas numbers. *Fibonacci Q.* **2004**, *42*, 149–154.

© 2018 by the authors. Licensee MDPI, Basel, Switzerland. This article is an open access article distributed under the terms and conditions of the Creative Commons Attribution (CC BY) license (http://creativecommons.org/licenses/by/4.0/).

Article

Symmetry Identities of Changhee Polynomials of Type Two

Joohee Jeong [1,†,‡], **Dong-Jin Kang** [2,‡] **and Seog-Hoon Rim** [1,*,‡]

1. Department of Mathematics Education, Kyungpook National University, Daegu 41566, Korea; jhjeong@knu.ac.kr
2. Department of Computer Engineering, Information Technology Services, Kyungpook National University, Daegu 41566, Korea; djkang@knu.ac.kr
* Correspondence: shrim@knu.ac.kr; Tel.: +82-53-950-5890
† Current address: 80 Daehakro, Bukgu, Daegu 41566, Korea.
‡ These authors contributed equally to this work.

Received: 31 October 2018 ; Accepted: 6 December 2018 ; Published: 11 December 2018

Abstract: In this paper, we consider Changhee polynomials of type two, which are motivated from the recent work of D. Kim and T. Kim. We investigate some symmetry identities for the Changhee polynomials of type two which are derived from the properties of symmetry for the fermionic p-adic integral on $\mathbb{Z}_p$.

Keywords: Changhee polynomials; Changhee polynomials of type two; fermionic p-adic integral on $\mathbb{Z}_p$

1. Introduction

Let p be a fixed odd prime number. Throughout this paper, $\mathbb{Z}_p$, $\mathbb{Q}_p$ and $\mathbb{C}_p$ will denote the ring of p-adic integers, the field of p-adic rational numbers and the completion of the algebraic closure of $\mathbb{Q}_p$.

The p-adic norm $|\cdot|_p$ is normalized as $|p|_p = \frac{1}{p}$.

Let $f(x)$ be a continulus funciton on $\mathbb{Z}_p$. Then the fermionic p-adic integral on $\mathbb{Z}_p$ is defined by Kim in [1] as

$$\int_{\mathbb{Z}_p} f(x) d\mu_{-1}(x) = \lim_{N \to \infty} \sum_{x=0}^{p^N-1} f(x) \mu_{-1}(x) = \lim_{x \to \infty} \sum_{x=0}^{p^N-1} f(x)(-1)^x. \tag{1}$$

For $n \in \mathbb{N}$, by (1), we get

$$\int_{\mathbb{Z}_p} f(x+n) d\mu_{-1}(x) + (-1)^{n-1} \int_{\mathbb{Z}_p} f(x) d\mu_{-1}(x)$$
$$= 2 \sum_{\ell=0}^{n-1} f(\ell)(-1)^{n-1-\ell} \tag{2}$$

as shown in [2–5]. In particular, if we take $n = 1$, then we have

$$\int_{\mathbb{Z}_p} f(x+1) d\mu_{-1}(x) + \int_{\mathbb{Z}_p} f(x) d\mu_{-1}(x) = 2f(0), \tag{3}$$

which is noted in [6,7].

In the previous paper [8], D. Kim and T. Kim introduced the Changhee polynomials $\widetilde{Ch}_n(x)$ of type two by the generating function

$$\sum_{n=0}^{\infty} \widetilde{Ch}_n(x) \frac{t^n}{n!} = \frac{2}{(1+t) + (1+t)^{-1}} (1+t)^x. \tag{4}$$

By exploiting the method of fermionic p-adic integral on $\mathbb{Z}_p$, the Changhee polynomials of type two can be represented by the fermionic p-adic integrals of $\mathbb{Z}_p$: for $t \in \mathbb{C}_p$ with $|t|_p < p^{-\frac{1}{p-1}}$,

$$\int_{\mathbb{Z}_p} (1+t)^{2y+1+2x} d\mu_{-1}(y) = \frac{2}{(1+t)^2 + 1}(1+t)^{2x+1}$$
$$= \sum_{n=0}^{\infty} \widetilde{Ch}_n(x) \frac{t^n}{n!} \tag{5}$$

When $x = 0$, $\widetilde{Ch}_n = \widetilde{Ch}_n(0)$ are called the Changhee numbers of type two.

In this paper, we will introduce further generalization of Changhee polynomials of type two, by using again fermionic p-adic integration on $\mathbb{Z}_p$.

We investigate some symmetry identities for the w-Changhee polynomials of type two which are derived from the properties of symmetry for the fermionic p-adic integral on $\mathbb{Z}_p$. Many authors investigated symmetric properties of special polynomials and numbers. See [9–12] and their references.

We introduce w-Changhee polynomials of type two in Section 3.

2. Changhee Polynomials and Numbers of Type Two

In this section, we use the techniques presented in the articles of C. Cesarano, C. Fornaro [13] and C. Cesarno [14], in particular the similarity of Chebyshev polynomials.

By using the generating functions of Changhee numbers and polynomials of type two, we have the following result.

Proposition 1. *For $n \in \mathbb{N}$ and $1 \leq k \leq n$, we have*

$$\widetilde{Ch}_n(x) = \sum_{m=0}^{n} \binom{n}{m} (2x)_m \widetilde{Ch}_{n-m}, \tag{6}$$

where $(x)_n = x(x-1)\cdots(x-n+1)$, $(n \geq 1)$, $(x)_0 = 1$.

Proof of Proposition 1.

$$\sum_{n=0}^{\infty} \widetilde{Ch}_n(x) \frac{t^n}{n!} = \frac{2}{(1+t) + (1+t)^{-1}} (1+t)^{2x}$$
$$= \sum_{m=0}^{\infty} \widetilde{Ch}_m \frac{t^m}{m!} \sum_{\ell=0}^{\infty} (2x)_\ell \frac{t^\ell}{\ell!}$$
$$= \sum_{n=0}^{\infty} \left(\sum_{m=0}^{n} \binom{n}{m} \widetilde{Ch}_m (2x)_{n-m} \right) \frac{t^n}{n!}$$

□

The Stirling number $S_1(\ell, n)$ of the first kind is defined in [2–5,15] by the generating function

$$(\log(1+t))^n = n! \sum_{\ell=n}^{\infty} S_1(\ell, n),$$

and the Stirling number $S_2(m, n)$ of the second kind is given in [4] by the generating function

$$(e^t - 1)^n = n! \sum_{m=n}^{\infty} S_2(m, n) \frac{t^m}{m!}.$$

As is well known, the Euler polynomials $E_n(x)$ are defined in [16–18] by the generating function

$$\frac{2}{e^t+1}e^{xt} = \sum_{n=0}^{\infty} E_n(x)\frac{t^n}{n!}. \tag{7}$$

When $x=0$, $E_n = E_n(0)$, $(n \geq 0)$, are called the n-th Euler numbers, whereas the Euler numbers E_n^* of the second kind are given by the generating function

$$\operatorname{sech}(t) = \frac{2}{e^t+e^{-t}} = \sum_{n=0}^{\infty} E_n^* \frac{t^n}{n!} \tag{8}$$

as noted in [16,19].

Before we proceed, we study some relevant relations between the Changhee numbers of type two and the Euler numbers of the second kind.

Proposition 2. *For $n \in \mathbb{N}$ and $0 \leq k \leq n$, we have*

$$\widetilde{Ch}_n = \sum_{k=0}^{n} E_k^* S_1(n,k). \tag{9}$$

Proof of Proposition 2. From the generating functions of Changhee numbers of type two shown in (8), we have

$$\sum_{n=0}^{\infty} \widetilde{Ch}_n \frac{t^n}{n!} = \frac{2}{(1+t)+(1+t)^{-1}} = \frac{2}{e^{\log(1+t)}+e^{-\log(1+t)}}$$

$$= \operatorname{sech}(\log(1+t)) \tag{10}$$

$$= \sum_{n=0}^{\infty} E_n^* \frac{(\log(1+t))^n}{n!} = \sum_{n=0}^{\infty}\left(\sum_{k=0}^{n} E_k^* S_1(n,k)\right)\frac{t^n}{n!}.$$

Thus we have the result. □

The result above helps us to derive some values of Changhee numbers of type two $\widetilde{Ch}_n$'s as follows: from $E_0^* = 1$, $E_1^* = 0$, $E_2^* = -1$, $E_3^* = 0$, $E_4^* = 5$, $E_5^* = 0$ and $S_1(n,n) = 0$ for $n \geq 0$, $S_1(n,0) = 0$ for $n \geq 1$, $S_1(2,1) = 1$, $S_1(3,1) = 2$, $S_1(4,1) = 6$, $S_1(5,1) = 24$, $S_1(3,2) = 3$, $S_1(4,2) = 11$, $S_1(5,2) = 50$, $S_1(4,3) = 6$, $S_1(5,3) = 35$, $S_1(5,4) = 10$,

$$\widetilde{Ch}_0 = E_0^* S_1(0,0) = 1,$$
$$\widetilde{Ch}_1 = E_0^* S_1(1,0) + E_1^* S_1(1,1) = 0+0 = 0,$$
$$\widetilde{Ch}_2 = E_0^* S_1(2,0) + E_1^* S_1(2,1) + E_2^* S_1(2,2) = 0+0-1 = -1.$$
$$\widetilde{Ch}_3 = E_0^* S_1(3,0) + E_1^* S_1(3,1) + E_2^* S_1(3,2) + E_3^* S_1(3,3)$$
$$= 0+0-3+0 = -3,$$
$$\widetilde{Ch}_4 = E_0^* S_1(4,0) + E_1^* S_1(4,1) + E_2^* S_1(4,2) + E_3^* S_1(4,3) + E_4^* S_1(4,4)$$
$$= 0+0-11+0+5 = -6,$$
$$\widetilde{Ch}_5 = E_0^* S_1(5,0) + E_1^* S_1(5,1) + E_2^* S_1(5,2) + E_3^* S_1(5,3) + E_4^* S_1(5,4)$$
$$+ E_5^* S_1(5,5) = 0+0-50+0+50+0 = 0.$$

For the inversion formulas for Proposition 2, we have the following.

Proposition 3. *For $n \in \mathbb{N}$ and $0 \le k \le n$, we have*

$$E_n^* = \sum_{k=0}^{n} \widetilde{Ch}_k S_2(n,k).$$

Proof of Proposition 3. From (6) and (8), we get the following, by replacing t by $e^t - 1$:

$$\frac{2}{(1+t)^2+1}(1+t) = \sum_{n=0}^{\infty} \widetilde{Ch}_n \frac{t^n}{n!}$$

$$\frac{2}{e^{2t}+1}e^t = \sum_{k=0}^{\infty} \widetilde{Ch}_k \frac{1}{k!}(e^t-1)^k$$

$$= \sum_{n=0}^{\infty} \left(\sum_{k=0}^{n} \widetilde{Ch}_k S_2(n,k) \right) \frac{t^n}{n!} \qquad (11)$$

$$= \frac{2}{e^t+e^{-t}} = \sum_{n=0}^{\infty} E_n^* \frac{t^n}{n!}.$$

Now (11) gives us the desired result $E_n^* = \sum_{k=0}^{n} \widetilde{Ch}_k S_2(n,k)$. □

Also by using the fermionic p-adic integration on $\mathbb{Z}_p$, we can represent Changhee numbers of type two as follows.

Proposition 4 (Witt's formula for Changhee numbers of type two).

For $n \in \mathbb{N}$, we have

$$\widetilde{Ch}_n = \int_{\mathbb{Z}_p} (2x+1)_n d\mu_{-1}(x). \qquad (12)$$

Proof of Proposition 4. First, we observe

$$\int_{\mathbb{Z}_p} (1+t)^{2x+1} d\mu_{-1}(x) = \int_{\mathbb{Z}_p} \sum_{n=0}^{\infty} (2x+1)_n \frac{t^n}{n!} d\mu_{-1}(x)$$

$$= \sum_{n=0}^{\infty} \int_{\mathbb{Z}_p} (2x+1)_n d\mu_{-1} \frac{t^n}{n!}, \qquad (13)$$

On the other hand, by the definition of fermionic p-adic integration on $\mathbb{Z}_p$,

$$\int_{\mathbb{Z}_p} (1+t)^{2x+1} d\mu_{-1}(x) = \frac{2}{(1+t)^2+1}(1+t) = \sum_{n=0}^{\infty} \widetilde{Ch}_n \frac{t^n}{n!}. \qquad (14)$$

Thus, by comparing the coefficients of both sides of (13) and (14), we have the desired result. □

3. Symmetry of w-Changhee Polynomials of Type Two

Motivated from D. Kim and T. Kim [20], for $w \in \mathbb{N}$, we define w-Changhee polynomials of type two by the following generating function

$$\frac{2}{(1+t)^{2w}+1}(1+t)^{2wx+1} = \sum_{n=0}^{\infty} \widetilde{Ch}_{n,w}(x) \frac{t^n}{n!}. \qquad (15)$$

When $x = 0$, $\widetilde{Ch}_{n,w} = \widetilde{Ch}_{n,w}(0)$ are called the w-Changhee numbers of type two. When $w = 1$, $\widetilde{Ch}_{n,1}(x) = \widetilde{Ch}_n(x)$ are just the Changhee polynomials of type two in (4). For the case of $w = \frac{1}{2}$, the $\frac{1}{2}$-Changhee polynomials of type two are related to the well-known Changhee polynomials of type two, i.e., $\widetilde{Ch}_{n,\frac{1}{2}}(x) = \widetilde{Ch}_n(x+1)$.

The generating function of w-Changhee polynomials of type two can be related with Changhee polynomials of type two or Changhee numbers of type two as follows.

Proposition 5. *For $n, w, \ell \in \mathbb{N}$ and $1 \leq \ell \leq n$, we have*

$$(1)\ \widetilde{Ch}_{n,w}(x) = \sum_{\ell=0}^{n} \widetilde{Ch}_{\ell}(2wx),\ \text{and}$$

$$(2)\ \widetilde{Ch}_{n,w}(x) = \sum_{\ell=0}^{n} \binom{n}{\ell} (2wx)_{\ell} \widetilde{Ch}_{n-\ell}.$$

Proof of Proposition 5. (1) is immediate from the definition. For (2), we have

$$\sum_{n=0}^{\infty} \widetilde{Ch}_{n,w}(x) \frac{t^n}{n!} = \left(\sum_{\ell=0}^{\infty} \widetilde{Ch}_{\ell} \frac{t^{\ell}}{\ell!} \right) (1+t)^{2wx}$$

$$= \left(\sum_{\ell=0}^{\infty} \widetilde{Ch}_{\ell} \frac{t^{\ell}}{\ell!} \right) \left(\sum_{m=0}^{\infty} (2wx)_m \frac{t^m}{m!} \right)$$

$$= \sum_{n=0}^{\infty} \left(\sum_{\ell=0}^{n} \binom{n}{\ell} (2wx)_{\ell} \widetilde{Ch}_{n-\ell} \right) \frac{t^n}{n!}.$$

□

From (3), we can easily derive the following:

$$2 \sum_{\ell=0}^{n} (-1)^{\ell} (1+t)^{2\ell} = \frac{2\{1 + (-1)^{n+1}(1+t)^{2(n+1)}\}}{(1+t)^2 + 1} \tag{16}$$

The left hand side of (16) can be written as

$$2 \sum_{\ell=0}^{n} (-1)^{\ell} (1+t)^{2\ell} = \sum_{n=0}^{\infty} \left(\sum_{\ell=0}^{n-1} (-1)^{\ell} (2\ell)_n \right) \frac{t^n}{n!} \tag{17}$$

We use the notation of λ-falling factorial in [12,21] for $\lambda \in \mathbb{R}$,

$$(\ell \mid \lambda)_n = \begin{cases} \ell(\ell - \lambda) \cdots (\ell - \lambda(n-1)), & (\text{if } n \geq 1) \\ 1, & (\text{if } n = 0). \end{cases}$$

Then the right hand side of (17) can be written as

$$2 \sum_{\ell=0}^{n-1} (-1)^{\ell} (1+t)^{2\ell} = \sum_{n=0}^{\infty} T_m(n; (\ell \mid \tfrac{1}{2})) \frac{t^n}{n!}. \tag{18}$$

where we denote, for $\lambda \in \mathbb{R}$,

$$T_m(n; (\ell \mid \lambda)) = \sum_{\ell=0}^{n} (-1)^{\ell} (\ell \mid \lambda)_m.$$

For $n \in \mathbb{N}, n \equiv 1 \pmod{2}, m \geq 0$ we have

$$\sum_{m=0}^{\infty} 2 \left(\sum_{\ell=0}^{n} (-1)^{\ell} (-2\ell)_m \right) \frac{t^m}{m!} = \frac{2(1 + (1+t)^{2(n+1)})}{(1+t)^2 + 1}. \tag{19}$$

On the other hand, by (4) and (18), we have

$$\sum_{m=0}^{\infty} \left(\widetilde{Ch}_m + \widetilde{Ch}_m(n+1) \right) \frac{t^m}{m!} = \frac{2(1+t)}{(1+t)^2+1} + \frac{2(1+t)^{2(n+1)}(1+t)}{(1+t)^2+1}$$

$$= 2\sum_{\ell=0}^{n}(-1)^{\ell}(1+t)^{2\ell+1} \tag{20}$$

$$= 2T_m(n;(\ell+\tfrac{1}{2}\,|\,\tfrac{1}{2})).$$

Now we consider a quotient of fermionic p-adic integrals on $\mathbb{Z}_p$,

$$\frac{2\int_{\mathbb{Z}_p}(1+t)^{2w_2x_2}d\mu_{-1}(x_2)}{\int_{\mathbb{Z}_p}(1+t)^{2w_1w_2x_1}d\mu_{-1}(x_1)} = \sum_{\ell=0}^{w_1-1}(-1)(1+t)^{2w_2\ell}$$

$$= \sum_{m=0}^{\infty}\sum_{\ell=0}^{w_1-1}(-1)^{\ell}(2w_2\ell)_m \tag{21}$$

$$= \sum_{m=0}^{\infty}\sum_{\ell=0}^{w_1-1}(2w_2)^m(-1)^{\ell}\left(\ell\,\big|\,\tfrac{1}{2w_2}\right)_m$$

$$= \sum_{m=0}^{\infty}(2w_2)^m T_m(w_1-1\,|\,(\ell\,|\,\tfrac{1}{2w_2})),$$

where $T_m(n\,|\,(\ell\,|\,\lambda)) = \sum_{\ell=0}^{n}(-1)^{\ell}(\ell\,|\,\lambda)_m$ for $\lambda \in \mathbb{R}$.

For the symmetry of w-Changhee polynomials of type two, we consider the following quotient form of fermionic p-adic integration on $\mathbb{Z}_p$.

$$T(w_1,w_2) = \frac{2\int_{\mathbb{Z}_p}\int_{\mathbb{Z}_p}(1+t)^{2w_1x_1+2w_2x_2+2}d\mu_{-1}(x_1)\,d\mu_{-1}(x_2)}{\int_{\mathbb{Z}_p}(1+t)^{2w_1w_2x_1+1}d\mu_{-1}(x_1)}(1+t)^{2w_1w_2x}$$

$$= \int_{\mathbb{Z}_p}(1+t)^{2w_1x_1+1}d\mu_{-1}(x_1)(1+t)^{2w_1w_2x}$$

$$\times \frac{\int_{\mathbb{Z}_p}(1+t)^{2w_2x_2}d\mu_{-1}(x_2)}{\int_{\mathbb{Z}_p}(1+t)^{2w_1w_2x_1}d\mu_{-1}(x_1)} \tag{22}$$

$$= \left(\sum_{\ell=0}^{\infty}\widetilde{Ch}_{\ell,w_1}(w_2x)\frac{t^{\ell}}{\ell!}\right)\left(\sum_{k=0}^{\infty}(2w_2)^k T_k(w_1-1\,|\,(k\,|\,\tfrac{1}{2w_2}))\right)$$

$$= \sum_{n=0}^{\infty}\left(\sum_{k=0}^{n}\binom{n}{k}\widetilde{Ch}_{n-k,w_1}(w_2x)(2w_k)^k T_k(w_1-1\,|\,(k\,|\,\tfrac{1}{2w_2}))\right)\frac{t^n}{n!}.$$

Similarly we have the following identity for $T(w_1,w_2)$ because $T(w_1,w_2)$ is symmetric on w_1 and w_2.

$$T(w_1,w_2) = \sum_{n=0}^{\infty}\left(\sum_{k=0}^{n}\binom{n}{k}\widetilde{Ch}_{n-k,w_2}(w_1x)(2w_1)^k T_k(w_2-1\,|\,(k\,|\,\tfrac{1}{2w_1}))\right)\frac{t^n}{n!}. \tag{23}$$

Thus, by (22) and (23), we have the following theorem.

Theorem 1. *For $w_1, w_2 \in \mathbb{N}$ with $w_1 \equiv 1 \pmod{2}$, $w_2 \equiv 1 \pmod{2}$ and $n \geq 0$, we have*

$$\sum_{k=0}^{n}\binom{n}{k}\widetilde{Ch}_{n-k,w_2}(w_1x)(2w_1)^k T_k(w_2-1\,|\,(k\,|\,\tfrac{1}{2w_1}))$$

$$= \sum_{k=0}^{n}\binom{n}{k}\widetilde{Ch}_{n-k,w_1}(w_2x)(2w_2)^k T_k(w_1-1\,|\,(k\,|\,\tfrac{1}{2w_2})).$$

If we take $w_2 = 1$ in Theorem 1, we have the following

Corollary 1. *For $w_1 \in \mathbb{N}$ with $w_1 \equiv 1 \pmod 2$ and $n \geq 0$, we have*

$$\widetilde{Ch}_n(w_1 x) = \sum_{k=0}^{n} \binom{n}{k} \widetilde{Ch}_{n-k,w_1}(x) 2^k T_k(w_1 - 1 \mid (k \mid \tfrac{1}{2})).$$

From (22), we rewrite $T(w_1, w_2)$ as follows:

$$\begin{aligned}
T(w_1, w_2) &= \int_{\mathbb{Z}_p} (1+t)^{2w_1 x_1} d\mu_{-1}(x_1)(1+t)^{2w_1 w_2 x} \\
&\quad \times \frac{2 \int_{\mathbb{Z}_p} (1+t)^{2w_2 x_2} d\mu_{-1}(x_2)}{\int_{\mathbb{Z}_p} (1+t)^{2w_1 w_2 x_1} d\mu_{-1}(x_1)} \\
&= \int_{\mathbb{Z}_p} (1+t)^{2w_1 x_1} d\mu_{-1}(x_1)(1+t)^{2w_1 w_2 x} \\
&\quad \times 2 \sum_{\ell=0}^{w_1-1} (1+t)^{2w_2 \ell} (-1)^{\ell} \\
&= 2 \sum_{\ell=0}^{w_1-1} (-1)^{\ell} \int_{\mathbb{Z}_p} (1+t)^{2w_1 x_1 + 2w_1 w_2 x + 2w_2 \ell} d\mu_{-1}(x_1) \\
&= 2 \sum_{\ell=0}^{w_1-1} (-1)^{\ell} \int_{\mathbb{Z}_p} (1+t)^{2w_1 x_1 + 2w_1 w_2 x + \frac{w_2}{w_1}\ell} d\mu_{-1}(x_1) \\
&= 2 \sum_{\ell=0}^{w_1-1} (-1)^{\ell} \sum_{k=0}^{\infty} \widetilde{Ch}_{k,w_1}\left(w_2 x + \tfrac{w_2}{w_1}\ell\right) \frac{t^k}{k!} \\
&= \sum_{n=0}^{\infty} \left(2 \sum_{\ell=0}^{w_1-1} (-1)^{\ell} \widetilde{Ch}_{n,w_1}\left(\tfrac{w_2}{w_1}\ell + w_2 x\right)\right) \frac{t^n}{n!}
\end{aligned} \qquad (24)$$

Similarly, by the symmetry of $T(w_1, w_2)$, we have the following identity

$$T(w_1, w_2) = \sum_{n=0}^{\infty} \left(2 \sum_{\ell=0}^{w_2-1} (-1)^{\ell} \widetilde{Ch}_{n,w_2}\left(\tfrac{w_1}{w_2}\ell + w_1 x\right)\right) \frac{t^n}{n!}. \qquad (25)$$

Now from (24) and (25), we have the following theorem.

Theorem 2. *For $w_1, w_2 \in \mathbb{N}$ with $w_1 \equiv 1 \pmod 2$, $w_2 \equiv 1 \pmod 2$ and $n \geq 0$, we have*

$$\sum_{\ell=0}^{w_1-1} (-1)^{\ell} \widetilde{Ch}_{n,w_1}\left(\tfrac{w_2}{w_1}\ell + w_2 x\right) = \sum_{\ell=0}^{w_2-1} (-1)^{\ell} \widetilde{Ch}_{n,w_2}\left(\tfrac{w_1}{w_2}\ell + w_1 x\right).$$

When we take $w_2 = 1$, we have

$$\widetilde{Ch}_n(w_1 \ell + w_1 x) = \sum_{\ell=0}^{w_1-1} (-1)^{\ell} \widetilde{Ch}_{n,w_1}\left(\tfrac{\ell}{w_1} + x\right).$$

4. Conclusions

The Changhee polynomials of type two are considered by D. Kim and T. Kim (see [8]) and various properties on their polynomials and numbers are investigated.

In this paper, we investigate some symmetry identities for the Changhee polynomials of type two which are derived from the properties of symmetry for the fermionic p-adic integrals on $\mathbb{Z}_p$. The techniques presented in the articles by Cesarano and Fornaro [13,14], paticularly the Chebyshev polynomials, are used.

Especially we introduce w-Changhee polynomials of type two and investigate interesting symmetry identities.

For the cases of $w = 1$, $w = \frac{1}{2}$ and $w = \frac{1}{4}$, the symmetry of the w-Changhee polynomials of type two are related to the works of Changhee polynomials of type two, those of well-known Changhee polynomials (see [4,22]), and those of the Catalan polynomials (see [20]) respectively.

Recently, many works are done on some identities of special polynomials in the view point of degenerate sense (see [15,20,21]). Our result could be developed in that direction also: i.e., on the symmetry of the degenerate w-Changhee polynomials of type two.

Finally, we remark that our results on symmetry of two variables could be extended to the three variables case.

Author Contributions: All authors contributed equally to this work. All authors read and approved the final manuscript.

Funding: This research received no external funding.

Acknowledgments: The authors would like to thank the referees for their valuable comments which improved the original manuscript in its present form.

Conflicts of Interest: The authors declare no conflict of interest.

References

1. Kim, T. q-Volkenborn integration. *Russ. J. Math. Phys.* **2002**, *9*, 288–299.
2. Kim, D.S. Identities associated with generalized twisted Euler polynomials twisted by ramified roots of unity. *Adv. Stud. Contemp. Math. (Kyungshang)* **2012**, *22*, 363–377.
3. Kim, D.S.; Kim, T.; Kim, Y.H.; Lee, S.H. Some arithmetic properties of Bernoulli and Euler numbers. *Adv. Stud. Contemp. Math. (Kyungshang)* **2012**, *22*, 467–480.
4. Kim, D.S.; Kim, T.; Seo, J.J. A note on Changhee polynomials and numbers. *Adv. Stud. Theor. Phys.* **2013**, *7*, 993–1003. [CrossRef]
5. Kim, T.; Rim, S.-H. New Changhee q-Euler numbers and polynomials associated with p-adic q-integerals. *Comput. Math. Appl.* **2007**, *54*, 484–489. [CrossRef]
6. Kim, T. Non Archmedean q-integrals associated with multiple Changhee q-Bernoulli polynomials. *Russ. Math. Phys.* **2003**, *10*, 91–98.
7. Kim, T. p-adic q-integrals associated with the Changhee-Barnes' q-Bernoulli polynomials. *Integr. Transf. Spec. Funct.* **2004**, *15*, 415–420. [CrossRef]
8. Kim, D.S.; Kim, T. A note on type 2 Changhee and Daehee polynomials. *arXiv* **2018**, arXiv:1809.05217.
9. Kim, D.S.; Lee, N.; Na, H.; Park, K.H. Abundant symmetry for higher-order Bernoulli polynomials (I). *Adv. Stud. Contemp. Math.* **2013**, *23*, 461–482.
10. Kim, D.S.; Lee, N.; Na, J.; Park, K.H. Abundant symmetry for higher-order Bernoulli polynomials (II). *Proc. Jangjeon Math. Soc.* **2013**, *16*, 359–378.
11. Kim, T. Symmetry p-adic invariant integral on $\mathbb{Z}_p$ for Bernoulli and Euler polynomials. *J. Differ. Equ. Appl.* **2008**, *14*, 1267–1277. [CrossRef]
12. Kim, T.; Dolgy, D. On the identities of symmetry for degenerate Bernoulli polynomials of order r. *Adv. Stud. Contemp. Math.* **2015**, *25*, 457–462.
13. Cesarano, C.; Fornaro, C. A note on two-variable Chebyshev polynomials. *Georgian Math. J.* **2017**, *24*, 339–349. [CrossRef]
14. Cesarno, C. Generalized Chebyshev polynomials. *Hacet. J. Math. Stat.* **2014**, *43*, 731–740.
15. Kim, T.; Kim, D.S. Identities for degenerate Bernoulli polynomials and Korobov polynomials of the first kind. *Sci. China Math.* **2018**. [CrossRef]
16. Kim, D.S.; Kim, T. Some p-adic integrals on $\mathbb{Z}_p$ associated with trigonometric functions. *Russ. J. Math. Phys.* **2018**, *25*, 300–308. [CrossRef]
17. Simsek, Y. Identities on Changhee numbers and Apostol-type Daehee polynomials. *Adv. Stud. Contemp. Math. (Kyungshang)* **2017**, *27*, 199–212.
18. Zhang, W.P. Number of solutions to a congruence equations mod p. *J. Northwest Univ. Natl. Sci.* **2016**, *46*, 313–316. (In Chinese)
19. Knuth, D.E.; Buckholtz, T.J. Computation of Tangent, Euler, and Bernoulli Numbers. *Math. Comput.* **1967**, *21*, 663–688. [CrossRef]

20. Kim, D.S.; Kim, T. Triple symmetric identities for w-Caralan polynomials. *J. Korean Math. Soc.* **2017**, *54*, 1243–1264.
21. Kim, T.; Kim, D.S. Identities of symmetry for degenerate Euler polynomials and alternating generalized falling factorial sums. *Iran. J. Sci. Technol. Trans. A Sci.* **2017**, *41*, 939–949. [CrossRef]
22. Kim, T. Symmetry of power sum polynomials and multivariate fermionic p-adic invariant integral on $\mathbb{Z}_p$. *Russ. J. Math. Phys.* **2009**, *16*, 93–96. [CrossRef]

© 2018 by the authors. Licensee MDPI, Basel, Switzerland. This article is an open access article distributed under the terms and conditions of the Creative Commons Attribution (CC BY) license (http://creativecommons.org/licenses/by/4.0/).

Article

Symmetric Identities of Hermite-Bernoulli Polynomials and Hermite-Bernoulli Numbers Attached to a Dirichlet Character χ

Serkan Araci [1,*], Waseem Ahmad Khan [2] and Kottakkaran Sooppy Nisar [3]

[1] Department of Economics, Faculty of Economics, Administrative and Social Sciences, Hasan Kalyoncu University, TR-27410 Gaziantep, Turkey
[2] Department of Mathematics, Faculty of Science, Integral University, Lucknow-226026, India; waseem08_khan@rediffmail.com
[3] Department of Mathematics, College of Arts and Science-Wadi Aldawaser, Prince Sattam bin Abdulaziz University, 11991 Riyadh Region, Kingdom of Saudi Arabia; n.sooppy@psau.edu.sa or ksnisar1@gmail.com
* Correspondence: mtsrkn@hotmail.com

Received: 14 November 2018; Accepted: 27 November 2018; Published: 29 November 2018

Abstract: We aim to introduce arbitrary complex order Hermite-Bernoulli polynomials and Hermite-Bernoulli numbers attached to a Dirichlet character χ and investigate certain symmetric identities involving the polynomials, by mainly using the theory of p-adic integral on $\mathbb{Z}_p$. The results presented here, being very general, are shown to reduce to yield symmetric identities for many relatively simple polynomials and numbers and some corresponding known symmetric identities.

Keywords: q-Volkenborn integral on $\mathbb{Z}_p$; Bernoulli numbers and polynomials; generalized Bernoulli polynomials and numbers of arbitrary complex order; generalized Bernoulli polynomials and numbers attached to a Dirichlet character χ

1. Introduction and Preliminaries

For a fixed prime number p, throughout this paper, let $\mathbb{Z}_p$, $\mathbb{Q}_p$, and $\mathbb{C}_p$ be the ring of p-adic integers, the field of p-adic rational numbers, and the completion of algebraic closure of $\mathbb{Q}_p$, respectively. In addition, let $\mathbb{C}$, $\mathbb{Z}$, and $\mathbb{N}$ be the field of complex numbers, the ring of rational integers and the set of positive integers, respectively, and let $\mathbb{N}_0 := \mathbb{N} \cup \{0\}$. Let $\mathrm{UD}(\mathbb{Z}_p)$ be the space of all uniformly differentiable functions on $\mathbb{Z}_p$. The notation $[z]_q$ is defined by

$$[z]_q := \frac{1-q^z}{1-q} \quad (z \in \mathbb{C}; q \in \mathbb{C} \setminus \{1\}; q^z \neq 1).$$

Let v_p be the normalized exponential valuation on $\mathbb{C}_p$ with $|p|_p = p^{v_p(p)} = p^{-1}$. For $f \in \mathrm{UD}(\mathbb{Z}_p)$ and $q \in \mathbb{C}_p$ with $|1-q|_p < 1$, q-Volkenborn integral on $\mathbb{Z}_p$ is defined by Kim [1]

$$I_q(f) = \int_{\mathbb{Z}_p} f(x)\, d\mu_q(x) = \lim_{N\to\infty} \frac{1}{[p^N]_q} \sum_{x=0}^{p^N-1} f(x)\, q^x. \tag{1}$$

For recent works including q-Volkenborn integration see References [1–10].

The ordinary p-adic invariant integral on $\mathbb{Z}_p$ is given by [7,8]

$$I_1(f) = \lim_{q\to 1} I_q(f) = \int_{\mathbb{Z}_p} f(x)\, dx. \tag{2}$$

It follows from Equation (2) that
$$I_1(f_1) = I_1(f) + f'(0), \qquad (3)$$
where $f_n(x) := f(x+n)$ $(n \in \mathbb{N})$ and $f'(0)$ is the usual derivative. From Equation (3), one has
$$\int_{\mathbb{Z}_p} e^{xt} dx = \frac{t}{e^t - 1} = \sum_{n=0}^{\infty} B_n \frac{t^n}{n!}, \qquad (4)$$
where B_n are the nth Bernoulli numbers (see References [11–14]; see also Reference [15] (Section 1.7)). From Equation (2) and (3), one gets
$$\frac{n \int_{\mathbb{Z}_p} e^{xt} dx}{\int_{\mathbb{Z}_p} e^{nxt} dx} = \frac{1}{t}\left(\int_{\mathbb{Z}_p} e^{(x+n)t} dx - \int_{\mathbb{Z}_p} e^{xt} dx\right)$$
$$= \sum_{j=0}^{n-1} e^{jt} = \sum_{k=0}^{\infty} \left(\sum_{j=0}^{n-1} j^k\right) \frac{t^k}{k!} = \sum_{k=0}^{\infty} S_k(n-1) \frac{t^k}{k!}, \qquad (5)$$
where
$$S_k(n) = 1^k + \cdots + n^k \quad (k \in \mathbb{N},\ n \in \mathbb{N}_0). \qquad (6)$$

From Equation (4), the generalized Bernoulli polynomials $B_n^{(\alpha)}(x)$ are defined by the following p-adic integral (see Reference [15] (Section 1.7))
$$\underbrace{\int_{\mathbb{Z}_p} \cdots \int_{\mathbb{Z}_p}}_{\alpha \text{ times}} e^{(x+y_1+y_2+\cdots+y_\alpha)t} dy_1 dy_2 \cdots dy_\alpha = \left(\frac{t}{e^t - 1}\right)^\alpha e^{xt} = \sum_{n=0}^{\infty} B_n^{(\alpha)}(x) \frac{t^n}{n!} \qquad (7)$$
in which $B_n^{(1)}(x) := B_n(x)$ are classical Bernoulli numbers (see, e.g., [1–10]).

Let $d, p \in \mathbb{N}$ be fixed with $(d, p) = 1$. For $N \in \mathbb{N}$, we set
$$X = X_d = \varprojlim_{N} \left(\mathbb{Z}/dp^N\mathbb{Z}\right);$$
$$a + dp^N\mathbb{Z}_p = \left\{x \in X \mid x \equiv a \pmod{dp^N}\right\}$$
$$\left(a \in \mathbb{Z} \text{ with } 0 \le a < dp^N\right); \qquad (8)$$
$$X^* = \bigcup_{\substack{0 < a < dp \\ (a,p)=1}} (a + dp\mathbb{Z}_p), \quad X_1 = \mathbb{Z}_p.$$

Let χ be a Dirichlet character with conductor $d \in \mathbb{N}$. The generalized Bernoulli polynomials attached to χ are defined by means of the generating function (see, e.g., [16])
$$\int_X \chi(y) e^{(x+y)t} dy = \frac{t \sum_{j=0}^{d-1} \chi(j) e^{jt}}{e^{dt} - 1} e^{xt} = \sum_{n=0}^{\infty} B_{n,\chi}(x) \frac{t^n}{n!}. \qquad (9)$$

Here $B_{n,\chi} := B_{n,\chi}(0)$ are the generalized Bernoulli numbers attached to χ. From Equation (9), we have (see, e.g., [16])
$$\int_X \chi(x) x^n dx = B_{n,\chi} \quad \text{and} \quad \int_X \chi(y)(x+y)^n\, dy = B_{n,\chi}(x). \qquad (10)$$

Define the p-adic functional $T_k(\chi, n)$ by (see, e.g., [16])

$$T_k(\chi, n) = \sum_{\ell=0}^{n} \chi(\ell) \ell^k \quad (k \in \mathbb{N}). \tag{11}$$

Then one has (see, e.g., [16])

$$B_{k,\chi}(nd) - B_{k,\chi} = kT_{k-1}(\chi, nd-1) \quad (k, n, d \in \mathbb{N}). \tag{12}$$

Kim et al. [16] (Equation (2.14)) presented the following interesting identity

$$\frac{dn \int_X \chi(x) e^{xt} dx}{\int_X e^{dnxt} dx} = \sum_{\ell=0}^{nd-1} \chi(\ell) e^{\ell t} = \sum_{k=0}^{\infty} T_k(\chi, nd-1) \frac{t^k}{k!} \quad (n \in \mathbb{N}). \tag{13}$$

Very recently, Khan [17] (Equation (2.1)) (see also Reference [11]) introduced and investigated λ-Hermite-Bernoulli polynomials of the second kind ${}_HB_n(x, y|\lambda)$ defined by the following generating function

$$\int_{\mathbb{Z}_p} (1 + \lambda t)^{\frac{x+u}{\lambda}} (1 + \lambda t^2)^{\frac{y}{\lambda}} d\mu_0(u)$$

$$= \frac{\log(1 + \lambda t)^{\frac{1}{\lambda}}}{(1 + \lambda t)^{\frac{1}{\lambda}} - 1} (1 + \lambda t)^{\frac{x}{\lambda}} (1 + \lambda t^2)^{\frac{y}{\lambda}} = \sum_{m=0}^{\infty} {}_HB_m(x, y|\lambda) \frac{t^m}{m!} \tag{14}$$

$$\left(\lambda, t \in \mathbb{C}_p \text{ with } \lambda \neq 0, \ |\lambda t| < p^{-\frac{1}{p-1}}\right).$$

Hermite-Bernoulli polynomials ${}_HB_k^{(\alpha)}(x, y)$ of order α are defined by the following generating function

$$\left(\frac{t}{e^t - 1}\right)^{\alpha} e^{xt + yt^2} = \sum_{k=0}^{\infty} {}_HB_k^{(\alpha)}(x, y) \frac{t^k}{k!} \quad (\alpha, x, y \in \mathbb{C}; |t| < 2\pi) \tag{15}$$

where ${}_HB_k^{(1)}(x, y) := {}_HB_k(x, y)$ are Hermite-Bernoulli polynomials, cf. [18,19]. For more information related to systematic works of some special functions and polynomials, see References [20–29].

We aim to introduce arbitrary complex order Hermite-Bernoulli polynomials attached to a Dirichlet character χ and investigate certain symmetric identities involving the polynomials (15) and (31), by mainly using the theory of p-adic integral on $\mathbb{Z}_p$. The results presented here, being very general, are shown to reduce to yield symmetric identities for many relatively simple polynomials and numbers and some corresponding known symmetric identities.

2. Symmetry Identities of Hermite-Bernoulli Polynomials of Arbitrary Complex Number Order

Here, by mainly using Kim's method in References [30,31], we establish certain symmetry identities of Hermite-Bernoulli polynomials of arbitrary complex number order.

Theorem 1. *Let $\alpha, x, y, z \in \mathbb{C}, \eta_1, \eta_2 \in \mathbb{N}$, and $n \in \mathbb{N}_0$. Then,*

$$\sum_{m=0}^{n} \sum_{\ell=0}^{m} \binom{n}{m} \binom{m}{\ell} {}_HB_{n-m}^{(\alpha)}(\eta_2 x, \eta_2^2 z) S_{m-\ell}(\eta_1 - 1) B_{\ell}^{(\alpha-1)}(\eta_1 y) \eta_1^{n-m-1} \eta_2^m$$

$$= \sum_{m=0}^{n} \sum_{\ell=0}^{m} \binom{n}{m} \binom{m}{\ell} {}_HB_{n-m}^{(\alpha)}(\eta_1 x, \eta_1^2 z) S_{m-\ell}(\eta_2 - 1) B_{\ell}^{(\alpha-1)}(\eta_2 y) \eta_2^{n-m-1} \eta_1^m \tag{16}$$

and

$$\sum_{m=0}^{n} \sum_{j=0}^{\eta_1-1} \binom{n}{m} \eta_1^{m-1} \eta_2^{n-m} B_{n-m}^{(\alpha-1)} (\eta_1 y) \, _H B_m^{(\alpha)} \left(\eta_2 x + \frac{\eta_2}{\eta_1} j, \eta_2^2 z \right) \qquad (17)$$

$$= \sum_{m=0}^{n} \sum_{j=0}^{\eta_1-1} \binom{n}{m} \eta_2^{m-1} \eta_1^{n-m} B_{n-m}^{(\alpha-1)} (\eta_2 y) \, _H B_m^{(\alpha)} \left(\eta_1 x + \frac{\eta_1}{\eta_2} j, \eta_1^2 z \right).$$

Proof. Let

$$F(\alpha; \eta_1, \eta_2)(t) := \frac{e^{\eta_1 \eta_2 t} - 1}{\eta_1 \eta_2 t} \left(\frac{\eta_1 t}{e^{\eta_1 t} - 1} \right)^{\alpha} e^{\eta_1 \eta_2 x t + \eta_1^2 \eta_2^2 z t^2} \left(\frac{\eta_2 t}{e^{\eta_2 t} - 1} \right)^{\alpha} e^{\eta_1 \eta_2 y t} \qquad (18)$$

$(\alpha, x, y, z \in \mathbb{C}; t \in \mathbb{C} \setminus \{0\}; \eta_1, \eta_2 \in \mathbb{N}; 1^{\alpha} := 1)$.

Since $\lim_{t \to 0} \eta t / (e^{\eta t} - 1) = 1 = \lim_{t \to 0} (e^{\eta t} - 1) / (\eta t)$ $(\eta \in \mathbb{N})$, $F(\alpha; \eta_1, \eta_2)(t)$ may be assumed to be analytic in $|t| < 2\pi/(\eta_1 \eta_2)$. Obviously $F(\alpha; \eta_1, \eta_2)(t)$ is symmetric with respect to the parameters η_1 and η_2.

Using Equation (4), we have

$$F(\alpha; \eta_1, \eta_2)(t) := \left(\frac{\eta_1 t}{e^{\eta_1 t} - 1} \right)^{\alpha} e^{\eta_1 \eta_2 x t + \eta_1^2 \eta_2^2 z t^2} \frac{\int_{\mathbb{Z}_p} e^{\eta_2 t u} du}{\int_{\mathbb{Z}_p} e^{\eta_1 \eta_2 t u} du} \left(\frac{\eta_2 t}{e^{\eta_2 t} - 1} \right)^{\alpha - 1} e^{\eta_1 \eta_2 y t}. \qquad (19)$$

Using Equations (5) and (15), we find

$$F(\alpha; \eta_1, \eta_2)(t) = \sum_{n=0}^{\infty} {}_H B_n^{(\alpha)}(\eta_2 x, \eta_2^2 z) \frac{(\eta_1 t)^n}{n!} \cdot \frac{1}{\eta_1} \sum_{m=0}^{\infty} S_m(\eta_1 - 1) \frac{(\eta_2 t)^m}{m!} \\ \cdot \sum_{\ell=0}^{\infty} B_\ell^{(\alpha-1)}(\eta_1 y) \frac{(\eta_2 t)^\ell}{\ell!}. \qquad (20)$$

Employing a formal manipulation of double series (see, e.g., [32] (Equation (1.1)))

$$\sum_{n=0}^{\infty} \sum_{k=0}^{\infty} A_{k,n} = \sum_{n=0}^{\infty} \sum_{k=0}^{[n/p]} A_{k,n-pk} \quad (p \in \mathbb{N}) \qquad (21)$$

with $p = 1$ in the last two series in Equation (20), and again, the resulting series and the first series in Equation (20), we obtain

$$F(\alpha; \eta_1, \eta_2)(t) = \sum_{n=0}^{\infty} \sum_{m=0}^{n} \sum_{\ell=0}^{m} \frac{{}_H B_{n-m}^{(\alpha)}(\eta_2 x, \eta_2^2 z) S_{m-\ell}(\eta_1 - 1) B_\ell^{(\alpha-1)}(\eta_1 y)}{(n-m)! (m-\ell)! \ell!} \\ \times \eta_1^{n-m-1} \eta_2^m t^n. \qquad (22)$$

Noting the symmetry of $F(\alpha; \eta_1, \eta_2)(t)$ with respect to the parameters η_1 and η_2, we also get

$$F(\alpha; \eta_1, \eta_2)(t) = \sum_{n=0}^{\infty} \sum_{m=0}^{n} \sum_{\ell=0}^{m} \frac{{}_H B_{n-m}^{(\alpha)}(\eta_1 x, \eta_1^2 z) S_{m-\ell}(\eta_2 - 1) B_\ell^{(\alpha-1)}(\eta_2 y)}{(n-m)! (m-\ell)! \ell!} \\ \times \eta_2^{n-m-1} \eta_1^m t^n. \qquad (23)$$

Equating the coefficients of t^n in the right sides of Equations (22) and (23), we obtain the first equality of Equation (16).

For (17), we write

$$F(\alpha;\eta_1,\eta_2)(t) = \frac{1}{\eta_1}\left(\frac{\eta_1 t}{e^{\eta_1 t}-1}\right)^\alpha e^{\eta_1\eta_2 xt+\eta_1^2\eta_2^2 zt^2}\frac{e^{\eta_1\eta_2 t}-1}{e^{\eta_2 t}-1}\left(\frac{\eta_2 t}{e^{\eta_2 t}-1}\right)^{\alpha-1}e^{\eta_1\eta_2 yt}. \quad (24)$$

Noting

$$\frac{e^{\eta_1\eta_2 t}-1}{e^{\eta_2 t}-1} = \sum_{j=0}^{\eta_1-1} e^{\eta_2 j t} = \sum_{j=0}^{\eta_1-1} e^{\eta_1 \frac{\eta_2}{\eta_1} j t},$$

we have

$$F(\alpha;\eta_1,\eta_2)(t) = \frac{1}{\eta_1}\sum_{j=0}^{\eta_1-1}\left(\frac{\eta_1 t}{e^{\eta_1 t}-1}\right)^\alpha e^{\eta_1\left(\eta_2 x+\frac{\eta_2}{\eta_1}j\right)t+\eta_1^2\eta_2^2 zt^2}\left(\frac{\eta_2 t}{e^{\eta_2 t}-1}\right)^{\alpha-1}e^{\eta_1\eta_2 yt}. \quad (25)$$

Using Equation (15), we obtain

$$\begin{aligned}F(\alpha;\eta_1,\eta_2)(t) &= \frac{1}{\eta_1}\sum_{n=0}^{\infty} B_n^{(\alpha-1)}(\eta_1 y)\frac{(\eta_2 t)^n}{n!}\\ &\times \sum_{m=0}^{\infty}\sum_{j=0}^{\eta_1-1} {}_H B_m^{(\alpha)}\left(\eta_2 x+\frac{\eta_2}{\eta_1}j, \eta_2^2 z\right)\frac{(\eta_1 t)^m}{m!}.\end{aligned} \quad (26)$$

Applying Equation (21) with $p=1$ to the right side of Equation (26), we get

$$\begin{aligned}F(\alpha;\eta_1,\eta_2)(t) &= \sum_{n=0}^{\infty}\sum_{m=0}^{n}\sum_{j=0}^{\eta_1-1} B_{n-m}^{(\alpha-1)}(\eta_1 y)\\ &\times {}_H B_m^{(\alpha)}\left(\eta_2 x+\frac{\eta_2}{\eta_1}j, \eta_2^2 z\right)\frac{\eta_1^{m-1}\eta_2^{n-m}}{m!(n-m)!}t^n.\end{aligned} \quad (27)$$

In view of symmetry of $F(\alpha;\eta_1,\eta_2)(t)$ with respect to the parameters η_1 and η_2, we also obtain

$$\begin{aligned}F(\alpha;\eta_1,\eta_2)(t) &= \sum_{n=0}^{\infty}\sum_{m=0}^{n}\sum_{j=0}^{\eta_2-1} B_{n-m}^{(\alpha-1)}(\eta_2 y)\\ &\times {}_H B_m^{(\alpha)}\left(\eta_1 x+\frac{\eta_1}{\eta_2}j, \eta_1^2 z\right)\frac{\eta_2^{m-1}\eta_1^{n-m}}{m!(n-m)!}t^n.\end{aligned} \quad (28)$$

Equating the coefficients of t^n in the right sides of Equation (27) and Equation (28), we have Equation (17). □

Corollary 1. *By substituting $\alpha = 1$ in Theorem 1, we have*

$$\begin{aligned}&\sum_{m=0}^{n}\sum_{\ell=0}^{m}\binom{n}{m}\binom{m}{\ell} {}_H B_{n-m}(\eta_2 x, \eta_2^2 z) S_{m-\ell}(\eta_1-1)(\eta_1 y)^\ell \eta_1^{n-m-1}\eta_2^m\\ &= \sum_{m=0}^{n}\sum_{\ell=0}^{m}\binom{n}{m}\binom{m}{\ell} B_{n-m}(\eta_1 x, \eta_1^2 z) S_{m-\ell}(\eta_2-1)(\eta_2 y)^\ell \eta_2^{n-m-1}\eta_1^m\end{aligned}$$

and

$$\sum_{m=0}^{n} \sum_{j=0}^{\eta_1-1} \binom{n}{m} \eta_1^{m-1} \eta_2^{n-m} (\eta_1 y)^{n-m} {}_H B_m \left(\eta_2 x + \frac{\eta_2}{\eta_1} j, \eta_2^2 z \right)$$

$$= \sum_{m=0}^{n} \sum_{j=0}^{\eta_1-1} \binom{n}{m} \eta_2^{m-1} \eta_1^{n-m} (\eta_2 y)^{n-m} {}_H B_m \left(\eta_1 x + \frac{\eta_1}{\eta_2} j, \eta_1^2 z \right). \tag{29}$$

Corollary 2. *Taking $\alpha = 1$ and $z = 0$ in Theorem 1, we have*

$$\sum_{m=0}^{n} \sum_{\ell=0}^{m} \binom{n}{m} \binom{m}{\ell} B_{n-m}(\eta_2 x) S_{m-\ell}(\eta_1 - 1) (\eta_1 y)^{\ell} \eta_1^{n-m-1} \eta_2^{m}$$

$$= \sum_{m=0}^{n} \sum_{\ell=0}^{m} \binom{n}{m} \binom{m}{\ell} B_{n-m}(\eta_1 x) S_{m-\ell}(\eta_2 - 1) (\eta_2 y)^{\ell} \eta_2^{n-m-1} \eta_1^{m}$$

and

$$\sum_{m=0}^{n} \sum_{j=0}^{\eta_1-1} \binom{n}{m} \eta_1^{m-1} \eta_2^{n-m} (\eta_1 y)^{n-m} B_m \left(\eta_2 x + \frac{\eta_2}{\eta_1} j \right)$$

$$= \sum_{m=0}^{n} \sum_{j=0}^{\eta_1-1} \binom{n}{m} \eta_2^{m-1} \eta_1^{n-m} (\eta_2 y)^{n-m} B_m \left(\eta_1 x + \frac{\eta_1}{\eta_2} j \right). \tag{30}$$

3. Symmetry Identities of Arbitrary Order Hermite-Bernoulli Polynomials Attached to a Dirichlet Character χ

We begin by introducing generalized Hermite-Bernoulli polynomials attached to a Dirichlet character χ of order $\alpha \in \mathbb{C}$ defined by means of the following generating function:

$$\left(\frac{t \sum_{j=0}^{d-1} \chi(j) e^{jt}}{e^{dt} - 1} \right)^{\alpha} e^{xt + yt^2} = \sum_{n=0}^{\infty} {}_H B_{n,\chi}^{(\alpha)}(x,y) \frac{t^n}{n!} \tag{31}$$

$$(\alpha, x, y \in \mathbb{C}),$$

where χ is a Dirichlet character with conductor d.

Here, $B_{n,\chi}^{(\alpha)}(x) := {}_H B_{n,\chi}^{(\alpha)}(x,0)$, $B_{n,\chi}^{(\alpha)} := {}_H B_{n,\chi}^{(\alpha)}(0,0)$, and $B_{n,\chi} := {}_H B_{n,\chi}^{(1)}(0,0)$ are called the generalized Hermite-Bernoulli polynomials and numbers attached to χ of order α and Hermite-Bernoulli numbers attached to χ, respectively.

Remark 1. *Taking $y = 0$ in Equation (31) gives ${}_H B_{n,\chi}^{(\alpha)}(x,0) := {}_H B_{n,\chi}^{(\alpha)}(x)$, cf. [33].*

Remark 2. *Equation (15) is obtained when $\chi := 1$ in Equation (31).*

Remark 3. *The Hermite-Bernoulli polynomials ${}_H B_n(x,y)$ are obtained when $\chi := 1$ and $\alpha = 1$ in Equation (31).*

Remark 4. *The generalized Bernoulli polynomials $B_n^{(\alpha)}(x)$ is obtained when $\chi := 1$ and $y = 0$ in Equation (31).*

Remark 5. *The classical Bernoulli polynomials attached to χ is obtained when $\alpha = 1$ and $y = 0$ in Equation (31).*

Theorem 2. Let $\alpha, x, y, z \in \mathbb{C}$, $\eta_1, \eta_2 \in \mathbb{N}$, and $n \in \mathbb{N}_0$. Then,

$$\sum_{m=0}^{n}\sum_{\ell=0}^{m}\binom{n}{m}\binom{m}{\ell}\eta_1^{n-m-1}\eta_2^m \,_H B_{n-m,\chi}^{(\alpha)}\left(\eta_2 x, \eta_2^2 z\right) B_{m-\ell,\chi}^{(\alpha-1)}(\eta_1 y)\, T_\ell(\chi, d\eta_1 - 1)$$
$$= \sum_{m=0}^{n}\sum_{\ell=0}^{m}\binom{n}{m}\binom{m}{\ell}\eta_2^{n-m-1}\eta_1^m \,_H B_{n-m,\chi}^{(\alpha)}\left(\eta_1 x, \eta_1^2 z\right) B_{m-\ell,\chi}^{(\alpha-1)}(\eta_2 y)\, T_\ell(\chi, d\eta_2 - 1) \tag{32}$$

and

$$\sum_{m=0}^{n}\sum_{\ell=0}^{d\eta_1-1}\chi(\ell)\binom{n}{m}\eta_1^{n-m-1}\eta_2^m \,_H B_{n-m,\chi}^{(\alpha)}\left(\eta_2 x + \frac{\ell\eta_2}{\eta_1}, \eta_2^2 z\right) B_{m,\chi}^{(\alpha-1)}(\eta_1 y)$$
$$= \sum_{m=0}^{n}\sum_{\ell=0}^{d\eta_2-1}\chi(\ell)\binom{n}{m}\eta_2^{n-m-1}\eta_1^m \,_H B_{n-m,\chi}^{(\alpha)}\left(\eta_1 x + \frac{\ell\eta_1}{\eta_2}, \eta_1^2 z\right) B_{m,\chi}^{(\alpha-1)}(\eta_2 y), \tag{33}$$

where χ is a Dirichlet character with conductor d.

Proof. Let

$$G(\alpha; \eta_1, \eta_2)(t) := \frac{d}{\int_X e^{d\eta_1 \eta_2 u t} du}\left(\frac{\eta_1 t \sum_{j=0}^{d-1}\chi(j)e^{j\eta_1 t}}{e^{d\eta_1 t}-1}\right)^\alpha e^{\eta_1 \eta_2 x t + \eta_1^2 \eta_2^2 z t^2} \tag{34}$$
$$\times \left(\frac{\eta_2 t \sum_{j=0}^{d-1}\chi(j)e^{j\eta_2 t}}{e^{d\eta_2 t}-1}\right)^\alpha e^{\eta_1 \eta_2 y t}$$

$(\alpha, x, y, z \in \mathbb{C}; t \in \mathbb{C}\setminus\{0\}; \eta_1, \eta_2 \in \mathbb{N}; 1^\alpha := 1)$.

Obviously $G(\alpha; \eta_1, \eta_2)(t)$ is symmetric with respect to the parameters η_1 and η_2. As in the function $F(\alpha; \eta_1, \eta_2)(t)$ in Equation (18), $G(\alpha; \eta_1, \eta_2)(t)$ can be considered to be analytic in a neighborhood of $t=0$. Using Equation (9), we have

$$G(\alpha; \eta_1, \eta_2)(t) = \frac{d\int_X \chi(u)e^{\eta_2 u t}du}{\int_X e^{d\eta_1 \eta_2 u t}du}\left(\frac{\eta_1 t \sum_{j=0}^{d-1}\chi(j)e^{j\eta_1 t}}{e^{d\eta_1 t}-1}\right)^\alpha e^{\eta_1 \eta_2 x t + \eta_1^2 \eta_2^2 z t^2} \tag{35}$$
$$\times \left(\frac{\eta_2 t \sum_{j=0}^{d-1}\chi(j)e^{j\eta_2 t}}{e^{d\eta_2 t}-1}\right)^{\alpha-1} e^{\eta_1 \eta_2 y t}.$$

Applying Equations (13) and (31) to Equation (35), we obtain

$$G(\alpha; \eta_1, \eta_2)(t) := \frac{1}{\eta_1}\sum_{n=0}^{\infty}{}_H B_{n,\chi}^{(\alpha)}\left(\eta_2 x, \eta_2^2 z\right)\frac{(\eta_1 t)^n}{n!}\sum_{m=0}^{\infty}B_{m,\chi}^{(\alpha-1)}(\eta_1 y)\frac{(\eta_2 t)^m}{m!} \tag{36}$$
$$\times \sum_{\ell=0}^{\infty}T_\ell(\chi, d\eta_1 - 1)\frac{(\eta_2 t)^\ell}{\ell!}.$$

Similarly as in the proof of Theorem 1, we find

$$G(\alpha; \eta_1, \eta_2)(t) = \sum_{n=0}^{\infty}\sum_{m=0}^{n}\sum_{\ell=0}^{m}\frac{\eta_1^{n-m-1}\eta_2^m}{(n-m)!(m-\ell)!\ell!} \tag{37}$$
$$\times {}_H B_{n-m,\chi}^{(\alpha)}\left(\eta_2 x, \eta_2^2 z\right) B_{m-\ell,\chi}^{(\alpha-1)}(\eta_1 y)\, T_\ell(\chi, d\eta_1 - 1)\, t^n.$$

In view of the symmetry of $G(\alpha; \eta_1, \eta_2)(t)$ with respect to the parameters η_1 and η_2, we also get

$$G(\alpha; \eta_1, \eta_2)(t) = \sum_{n=0}^{\infty} \sum_{m=0}^{n} \sum_{\ell=0}^{m} \frac{\eta_2^{n-m-1} \eta_1^m}{(n-m)!(m-\ell)!\ell!} \tag{38}$$
$$\times {}_H B_{n-m,\chi}^{(\alpha)}\left(\eta_1 x, \eta_1^2 z\right) B_{m-\ell,\chi}^{(\alpha-1)}(\eta_2 y) T_\ell(\chi, d\eta_2 - 1) t^n.$$

Equating the coefficients of t^n of the right sides of Equations (37) and (38), we obtain Equation (32).

From Equation (13), we have

$$\frac{d \int_X \chi(u) e^{\eta_2 u t} du}{\int_X e^{d\eta_1 \eta_2 u t} du} = \frac{1}{\eta_1} \sum_{\ell=0}^{d\eta_1 - 1} \chi(\ell) e^{\ell \eta_2 t}. \tag{39}$$

Using Equation (39) in Equation (35), we get

$$G(\alpha; \eta_1, \eta_2)(t) = \frac{1}{\eta_1} \sum_{\ell=0}^{d\eta_1-1} \chi(\ell) \left(\frac{\eta_1 t \sum_{j=0}^{d-1} \chi(j) e^{j\eta_1 t}}{e^{d\eta_1 t} - 1} \right)^{\alpha} e^{\left(\eta_2 x + \frac{\ell \eta_2}{\eta_1}\right)\eta_1 t + \eta_1^2 \eta_2^2 z t^2} \tag{40}$$
$$\times \left(\frac{\eta_2 t \sum_{j=0}^{d-1} \chi(j) e^{j\eta_2 t}}{e^{d\eta_2 t} - 1} \right)^{\alpha-1} e^{\eta_1 \eta_2 y t}.$$

Using Equation (31), similarly as above, we obtain

$$G(\alpha; \eta_1, \eta_2)(t) = \sum_{n=0}^{\infty} \sum_{m=0}^{n} \sum_{\ell=0}^{d\eta_1-1} \chi(\ell) {}_H B_{n-m,\chi}^{(\alpha)}\left(\eta_2 x + \frac{\ell \eta_2}{\eta_1}, \eta_2^2 z\right) \tag{41}$$
$$\times B_{m,\chi}^{(\alpha-1)}(\eta_1 y) \frac{\eta_1^{n-m-1} \eta_2^m}{(n-m)!m!} t^n.$$

Since $G(\alpha; \eta_1, \eta_2)(t)$ is symmetric with respect to the parameters η_1 and η_2, we also have

$$G(\alpha; \eta_1, \eta_2)(t) = \sum_{n=0}^{\infty} \sum_{m=0}^{n} \sum_{\ell=0}^{d\eta_2-1} \chi(\ell) {}_H B_{n-m,\chi}^{(\alpha)}\left(\eta_1 x + \frac{\ell \eta_1}{\eta_2}, \eta_1^2 z\right) \tag{42}$$
$$\times B_{m,\chi}^{(\alpha-1)}(\eta_2 y) \frac{\eta_2^{n-m-1} \eta_1^m}{(n-m)!m!} t^n.$$

Equating the coefficients of t^n of the right sides in Equation (41) and Equation (42), we get Equation (33). □

4. Conclusions

The results in Theorems 1 and 2, being very general, can reduce to yield many symmetry identities associated with relatively simple polynomials and numbers using Remarks 1–5. Setting $z = 0$ and $\alpha \in \mathbb{N}$ in the results in Theorem 1 and Theorem 2 yields the corresponding known identities in References [33,34], respectively.

Author Contributions: All authors contributed equally.

Funding: Dr. S. Araci was supported by the Research Fund of Hasan Kalyoncu University in 2018.

Conflicts of Interest: The authors declare no conflict of interest.

References

1. Kim, T. *q*-Volkenborn integration. *Russ. J. Math. Phys.* **2002**, *9*, 288–299.
2. Cenkci, M. The *p*-adic generalized twisted *h*, *q*-Euler-*l*-function and its applications. *Adv. Stud. Contem. Math.* **2007**, *15*, 37–47.
3. Cenkci, M.; Simsek, Y.; Kurt, V. Multiple two-variable *p*-adic *q*-*L*-function and its behavior at $s = 0$. *Russ. J. Math. Phys.* **2008**, *15*, 447–459. [CrossRef]
4. Kim, T. On a *q*-analogue of the *p*-adic log gamma functions and related integrals. *J. Numb. Theor.* **1999**, *76*, 320–329. [CrossRef]
5. Kim, T. A note on *q*-Volkenborn integration. *Proc. Jangeon Math. Soc.* **2005**, *8*, 13–17.
6. Kim, T. *q*-Euler numbers and polynomials associated with *p*-adic *q*-integrals. *J. Nonlinear Math. Phys.* **2007**, *14*, 15–27. [CrossRef]
7. Kim, T. A note on *p*-adic *q*-integral on $\mathbb{Z}_p$ associated with *q*-Euler numbers. *Adv. Stud. Contem. Math.* **2007**, *15*, 133–137.
8. Kim, T. On *p*-adic *q*-*l*-functions and sums of powers. *J. Math. Anal. Appl.* **2007**, *329*, 1472–1481. [CrossRef]
9. Kim, T.; Choi, J.Y.; Sug, J.Y. Extended *q*-Euler numbers and polynomials associated with fermionic *p*-adic *q*-integral on $\mathbb{Z}_p$. *Russ. J. Math. Phy.* **2007**, *14*, 160–163. [CrossRef]
10. Simsek, Y. On *p*-adic twisted *q*-*L*-functions related to generalized twisted Bernoulli numbers. *Russ. J. Math. Phy.* **2006**, *13*, 340–348. [CrossRef]
11. Haroon, H.; Khan, W.A. Degenerate Bernoulli numbers and polynomials associated with degenerate Hermite polynomials. *Commun. Korean Math. Soc.* **2017**, in press.
12. Khan, N.; Usman, T.; Choi, J. A new generalization of Apostol-type Laguerre-Genocchi polynomials. *C. R. Acad. Sci. Paris Ser. I* **2017**, *355*, 607–617. [CrossRef]
13. Pathan, M.A.; Khan, W.A. Some implicit summation formulas and symmetric identities for the generalized Hermite-Bernoulli polynomials. *Mediterr. J. Math.* **2015**, *12*, 679–695. [CrossRef]
14. Pathan, M.A.; Khan, W.A. A new class of generalized polynomials associated with Hermite and Euler polynomials. *Mediterr. J. Math.* **2016**, *13*, 913–928. [CrossRef]
15. Srivastava, H.M.; Choi, J. *Zeta and q-Zeta Functions and Associated Series and Integrals*; Elsevier Science Publishers: Amsterdam, The Netherlands; London, UK; New York, NY, USA, 2012.
16. Kim, T.; Rim, S.H.; Lee, B. Some identities of symmetry for the generalized Bernoulli numbers and polynomials. *Abs. Appl. Anal.* **2009**, *2009*, 848943. [CrossRef]
17. Khan, W.A. Degenerate Hermite-Bernoulli numbers and polynomials of the second kind. *Prespacetime J.* **2016**, *7*, 1297–1305.
18. Cesarano, C. Operational Methods and New Identities for Hermite Polynomials. *Math. Model. Nat. Phenom.* **2017**, *12*, 44–50. [CrossRef]
19. Dattoli, G.; Lorenzutta, S.; Cesarano, C. Finite sums and generalized forms of Bernoulli polynomials. *Rend. Mat.* **1999**, *19* , 385–391.
20. Bell, E.T. Exponential polynomials. *Ann. Math.* **1934**, *35*, 258–277. [CrossRef]
21. Andrews, L.C. *Special Functions for Engineers and Applied Mathematicians*; Macmillan Publishing Company: New York, NY, USA, 1985.
22. Jang, L.C.; Kim, S.D.; Park, D.W.; Ro, Y.S. A note on Euler number and polynomials. *J. Inequ. Appl.* **2006**, *2006*, 34602. [CrossRef]
23. Kim, T. On the *q*-extension of Euler and Genocchi numbers, *J. Math. Anal. Appl.* **2007**, *326*, 1458–1465. [CrossRef]
24. Kim, T. *q*-Bernoulli numbers and polynomials associated with Gaussian binomial coefficients. *Russ. J. Math. Phys.* **2008**, *15*, 51–57. [CrossRef]
25. Kim, T. On the multiple *q*-Genocchi and Euler numbers. *Russ. J. Math. Phy.* **2008**, *15*, 481–486. [CrossRef]
26. Kim, T. New approach to *q*-Euler, Genocchi numbers and their interpolation functions. *Adv. Stud. Contem. Math.* **2009**, *18*, 105–112.
27. Kim, T. Sums of products of *q*-Euler numbers. *J. Comput. Anal. Appl.* **2010**, *12*, 185–190.
28. Kim, Y.H.; Kim, W.; Jang, L.C. On the *q*-extension of Apostol-Euler numbers and polynomials. *Abs. Appl. Anal.* **2008**, *2008*, 296159.

29. Simsek, Y. Complete sum of products of (h,q)-extension of the Euler polynomials and numbers. *J. Differ. Eqn. Appl.* **2010**, *16*, 1331–1348. [CrossRef]
30. Kim, T.; Kim, D.S. An identity of symmetry for the degenerate Frobenius-Euler polynomials. *Math. Slovaca* **2018**, *68*, 239–243. [CrossRef]
31. Kim, T. Symmetry p-adic invariant integral on $\mathbb{Z}_p$ for Bernoulli and Euler polynomials. *J. Differ. Equ. Appl.* **2008**, *14*, 1267–1277. [CrossRef]
32. Choi, J. Notes on formal manipulations of double series. *Commun. Korean Math. Soc.* **2003**, *18*, 781–789. [CrossRef]
33. Kim, T.; Jang, L.C.; Kim, Y.H.; Hwang, K.W. On the identities of symmetry for the generalized Bernoulli polynomials attached to χ of higher order. *J. Inequ. Appl.* **2009**, *2009*, 640152. [CrossRef]
34. Kim, T.; Hwang, K.W.; Kim, Y.H. Symmetry properties of higher order Bernoulli polynomials. *Adv. Differ. Equ.* **2009**, *2009*, 318639. [CrossRef]

© 2018 by the authors. Licensee MDPI, Basel, Switzerland. This article is an open access article distributed under the terms and conditions of the Creative Commons Attribution (CC BY) license (http://creativecommons.org/licenses/by/4.0/).

Article

A New Class of Hermite-Apostol Type Frobenius-Euler Polynomials and Its Applications

Serkan Araci [1,*], Mumtaz Riyasat [2], Shahid Ahmad Wani [2] and Subuhi Khan [2]

[1] Department of Economics, Faculty of Economics, Administrative and Social Sciences, Hasan Kalyoncu University, Gaziantep TR-27410, Turkey
[2] Department of Mathematics, Faculty of Science, Aligarh Muslim University, Aligarh 202 002, India; mumtazrst@gmail.com (M.R.); shahidwani177@gmail.com (S.A.W.); subuhi2006@gmail.com (S.K.)
* Correspondence: serkan.araci@hku.edu.tr; Tel.: +90-5366750331

Received: 6 November 2018; Accepted: 15 November 2018; Published: 19 November 2018

Abstract: The article is written with the objectives to introduce a multi-variable hybrid class, namely the Hermite–Apostol-type Frobenius–Euler polynomials, and to characterize their properties via different generating function techniques. Several explicit relations involving Hurwitz–Lerch Zeta functions and some summation formulae related to these polynomials are derived. Further, we establish certain symmetry identities involving generalized power sums and Hurwitz–Lerch Zeta functions. An operational view for these polynomials is presented, and corresponding applications are given. The illustrative special cases are also mentioned along with their generating equations.

Keywords: Apostol-type Frobenius–Euler polynomials; three-variable Hermite polynomials; symmetric identities; explicit relations; operational connection

MSC: 11B68; 05A10; 11B65

1. Introduction and Preliminaries

The multi-variable forms of the special polynomials of mathematical physics help in deriving several useful identities and in introducing new families of special polynomials. We know that the generalized Hermite polynomials are important to deal with quantum mechanical and optical beam transport problems [1] (also see [2,3]). The generating equation for the three-variable Hermite polynomials (3VHP) $H_n(x,y,z)$ [4] is given by:

$$e^{xt+yt^2+zt^3} = \sum_{n=0}^{\infty} H_n(x,y,z)\frac{t^n}{n!}, \qquad (1)$$

which for $z = 0$ reduce to the two-variable Hermite–Kampé de Fériet polynomials (2VHKdFP) $H_n(x,y)$ [5] and for $z = 0$, $x = 2x$ and $y = -1$ become the classical Hermite polynomials $H_n(x)$ [6].

For $u \in \mathbb{C}$, $u \neq 1$, the generating equation for the Apostol-type Frobenius–Euler polynomials (ATFEP) $\mathfrak{F}_n^{(\alpha)}(x;u;\lambda)$, of order $\alpha \in \mathbb{C}$, is given by [7]:

$$\left(\frac{1-u}{\lambda e^t - u}\right)^\alpha e^{xt} = \sum_{n=0}^{\infty} \mathfrak{F}_n^{(\alpha)}(x;u;\lambda)\frac{t^n}{n!}, \qquad (2)$$

which for $x = 0$ gives the Apostol-type Frobenius–Euler numbers (ATFEN) $\mathfrak{F}_n^{(\alpha)}(u;\lambda)$, of order α such that:

$$\left(\frac{1-u}{\lambda e^t - u}\right)^\alpha = \sum_{n=0}^{\infty} \mathfrak{F}_n^{(\alpha)}(u;\lambda)\frac{t^n}{n!}. \qquad (3)$$

For $u = -1$, the ATFEP reduce to the Apostol–Euler polynomials $\mathcal{E}_n^{(\alpha)}(x;\lambda)$ [8], which for $\lambda = 1$, become the Euler polynomials $E_n^{(\alpha)}(x)$ [9]. Furthermore, the ATFEP for $\lambda = 1$ becomes the Frobenius–Euler polynomials $\mathfrak{F}_n^{(\alpha)}(x;u)$ [10].

The generating equations for the special polynomials are important from different view points and help in finding connection formulas, recursive relations and difference equations and in solving enumeration problems in combinatorics and encoding their solutions.

We intended to introduce a new hybrid class, namely the class of three-variable Hermite–Apostol-type Frobenius–Euler polynomials (3VHATFEP).

Upon replacing the powers x^n by the polynomials $H_n(x,y,z)$ for $(n = 0,1,2,\ldots)$ in Equation (2) and upon the use of Equation (1), we have:

For $u, \lambda \in \mathbb{C}, u \neq 1$, the three-variable Hermite–Apostol-type Frobenius–Euler polynomials ${}_H\mathcal{F}_n^{(\alpha)}(x,y,z;u;\lambda)$, of order $\alpha \in \mathbb{C}$, are defined by the following generating function:

$$\left(\frac{1-u}{\lambda e^t - u}\right)^\alpha e^{xt+yt^2+zt^3} = \sum_{n=0}^{\infty} {}_H\mathfrak{F}_n^{(\alpha)}(x,y,z;u;\lambda)\frac{t^n}{n!}, \quad (4)$$

which for $\lambda = 1$ becomes the three-variable Hermite–Frobenius–Euler polynomials ${}_H\mathfrak{F}_n^{(\alpha)}(x,y,z;u)$, of order α, which again for $\alpha = 1$, give the three-variable Hermite-Frobenius–Euler polynomials ${}_H\mathfrak{F}_n(x,y,z;u)$.

Again, the 3VHATFEP for $u = -1$ give the three-variable Hermite–Apostol–Euler polynomials ${}_H\mathcal{E}_n^{(\alpha)}(x,y,z;\lambda)$ of order α, which for $\lambda = 1$ reduce to the three-variable Hermite–Euler polynomials ${}_HE_n^{(\alpha)}(x,y,z)$.

The 3VHATFEP are also defined as the discrete Apostol-type Frobenius–Euler convolution of the 3VHP given by:

$${}_H\mathfrak{F}_n^{(\alpha)}(x,y,z;u;\lambda) = n! \sum_{k=0}^{n} \sum_{r=0}^{[k/3]} \frac{\mathfrak{F}_{n-k}^{(\alpha)}(u;\lambda)z^r H_{k-3r}(x,y)}{(n-k)!r!(k-3r)!}, \quad (5)$$

where $H_n(x,y)$ are the 2VHKdFP.

Next, we deduce certain special cases related to the 3VHATFEP family. Some of these cases are known in the literature. These polynomials are given in Table 1 below.

In this article, the 3VHATFEP are introduced, and certain properties including the explicit relations, summation formulae and symmetric identities for these polynomials are proven using different generating function methods. Some applications for the aforementioned hybrid class of polynomials are given.

Table 1. Special polynomials related to the ${}_H\mathfrak{F}_n^{(\alpha)}(x,y,z;u;\lambda)$ family.

S.No.	Cases	Name of Polynomial	Generating Function
I.	$z=0$	2-variable Hermite–Apostol-type Frobenius–Euler polynomials of order α	$\left(\dfrac{1-u}{\lambda e^t - u}\right)^\alpha e^{xt+yt^2} = \sum_{n=0}^\infty {}_H\mathfrak{F}_n^{(\alpha)}(x,y;u;\lambda)\dfrac{t^n}{n!}$
	$z=0, \lambda=1$	2-variable Hermite-Frobenius–Euler polynomials of order α	$\left(\dfrac{1-u}{e^t - u}\right)^\alpha e^{xt+yt^2} = \sum_{n=0}^\infty {}_H\mathfrak{F}_n^{(\alpha)}(x,y;u)\dfrac{t^n}{n!}$
	$z=0, \lambda=\alpha=1$	2-variable Hermite–Frobenius–Euler polynomials	$\left(\dfrac{1-u}{e^t - u}\right) e^{xt+yt^2} = \sum_{n=0}^\infty {}_H\mathfrak{F}_n(x,y;u)\dfrac{t^n}{n!}$
II.	$x=2x,\ y=-1; z=0$	Hermite–Apostol-type Frobenius–Euler polynomials of order α	$\left(\dfrac{1-u}{\lambda e^t - u}\right)^\alpha e^{2xt-t^2} = \sum_{n=0}^\infty {}_H\mathfrak{F}_n^{(\alpha)}(x;u;\lambda)\dfrac{t^n}{n!}$
	$x=2x,\ y=-1,\ z=0;\lambda=1$	Hermite–Frobenius–Euler polynomials of order α	$\left(\dfrac{1-u}{e^t - u}\right)^\alpha e^{2xt-t^2} = \sum_{n=0}^\infty {}_H\mathfrak{F}_n^{(\alpha)}(x;u)\dfrac{t^n}{n!}$
	$x=2x,\ y=-1,\ z=0;\alpha=\lambda=1$	Hermite–Frobenius–Euler polynomials	$\left(\dfrac{1-u}{e^t - u}\right) e^{2xt-t^2} = \sum_{n=0}^\infty {}_H\mathfrak{F}_n(x;u)\dfrac{t^n}{n!}$
III.	$u=-1$	3-variable Hermite–Apostol-Euler polynomials of order α [11]	$\left(\dfrac{2}{\lambda e^t + 1}\right)^\alpha e^{xt+yt^2+zt^3} = \sum_{n=0}^\infty {}_H\mathfrak{E}_n^{(\alpha)}(x,y,z;\lambda)\dfrac{t^n}{n!}$
	$u=-1,\ \lambda=1$	3-variable Hermite–Euler polynomials [11]	$\left(\dfrac{2}{e^t + 1}\right)^\alpha e^{xt+yt^2+zt^3} = \sum_{n=0}^\infty {}_H E_n^{(\alpha)}(x,y,z)\dfrac{t^n}{n!}$
	$u=-1,\ \lambda=\alpha=1$	3-variable Hermite–Euler polynomials [11]	$\left(\dfrac{2}{e^t + 1}\right) e^{xt+yt^2+zt^3} = \sum_{n=0}^\infty {}_H E_n(x,y,z)\dfrac{t^n}{n!}$
IV.	$u=-1,\ z=0$	2-variable Hermite–Apostol-Euler polynomials of order α [11]	$\left(\dfrac{2}{\lambda e^t + 1}\right)^\alpha e^{xt+yt^2} = \sum_{n=0}^\infty {}_H\mathfrak{E}_n^{(\alpha)}(x,y;\lambda)\dfrac{t^n}{n!}$
	$u=-1,\ \lambda=1;\ z=0$	2-variable Hermite–Euler polynomials	$\left(\dfrac{2}{e^t + 1}\right)^\alpha e^{xt+yt^2} = \sum_{n=0}^\infty {}_H E_n^{(\alpha)}(x,y)\dfrac{t^n}{n!}$
	$u=-1,\ \lambda=\alpha=1;\ z=0$	2-variable Hermite–Euler polynomials [11]	$\left(\dfrac{2}{e^t + 1}\right) e^{xt+yt^2} = \sum_{n=0}^\infty {}_H E_n(x,y)\dfrac{t^n}{n!}$
V.	$u=-1,\ x=2x,$ $y=-1;\ z=0$	Hermite–Apostol–Euler polynomials of order α [12]	$\left(\dfrac{2}{\lambda e^t + 1}\right)^\alpha e^{2xt-t^2} = \sum_{n=0}^\infty {}_H\mathfrak{E}_n^{(\alpha)}(x;\lambda)\dfrac{t^n}{n!}$
	$u=-1,\ \lambda=1;$ $x=2x,\ y=-1;\ z=0$	Hermite–Euler polynomials of order α [12]	$\left(\dfrac{2}{e^t + 1}\right)^\alpha e^{2xt-t^2} = \sum_{n=0}^\infty {}_H E_n^{(\alpha)}(x)\dfrac{t^n}{n!}$
	$u=-1,\ \lambda=\alpha=1;$ $x=2x,\ y=-1;\ z=0$	Hermite–Euler polynomials [12]	$\left(\dfrac{2}{e^t + 1}\right) e^{2xt-t^2} = \sum_{n=0}^\infty {}_H E_n(x)\dfrac{t^n}{n!}$

2. Relations

To derive some relations for the 3VHATFEP, the following results are proven:

Theorem 1. *Let $\alpha, \beta \in \mathbb{Z}$, then we have the following relation for the 3VHATFEP of order α:*

$$_H\mathfrak{F}_n^{(\alpha\pm\beta)}(x,y,z;u;\lambda) = \sum_{k=0}^{n} \binom{n}{k} \mathfrak{F}_k^{(\alpha)}(u;\lambda) {}_H\mathfrak{F}_{n-k}^{(\pm\beta)}(x,y,z;u;\lambda). \tag{6}$$

Proof. We write the generating Function (4) in the following form:

$$\sum_{n=0}^{\infty} {}_H\mathfrak{F}_n^{(\alpha\pm\beta)}(x,y,z;u;\lambda)\frac{t^n}{n!} = \left(\frac{1-u}{\lambda e^t - u}\right)^{(\alpha\pm\beta)} e^{xt+yt^2+zt^3}, \tag{7}$$

for which, upon using Equations (3) and (4) and then after simplification, we get Equation (6). □

Corollary 1. *For $\alpha, \beta \in \mathbb{Z}$, the following relation for the 3VHAEP of order α holds true:*

$$_H\mathfrak{E}_n^{(\alpha\pm\beta)}(x,y,z;\lambda) = \sum_{k=0}^{n} \binom{n}{k} \mathfrak{E}_k^{(\alpha)}(\lambda) {}_H\mathfrak{E}_{n-k}^{(\pm\beta)}(x,y,z;\lambda), \tag{8}$$

$\mathfrak{E}_k^{(\alpha)}(\lambda)$ *means Apostol–Euler numbers of order α.*

Theorem 2. *The following recurrence relation for the 3VHATFEP holds true:*

$$_H\mathfrak{F}_{n+1}(x,y,z;u;\lambda) = x \, {}_H\mathfrak{F}_n(x,y,z;u;\lambda) + 2yn \, {}_H\mathfrak{F}_{n-1}(x,y,z;u;\lambda) + 3zn(n-1) \\ {}_H\mathfrak{F}_{n-2}(x,y,z;u;\lambda) - \frac{\lambda}{1-u} \sum_{k=0}^{n} \binom{n}{k} {}_H\mathfrak{F}_{n-k}(x,y,z;u;\lambda) {}_H\mathfrak{F}_k(1,0,0;u;\lambda). \tag{9}$$

Proof. Taking $\alpha = 1$ and then taking the derivative with respect to t in Equation (4), we find:

$$\sum_{n=0}^{\infty} {}_H\mathfrak{F}_{n+1}(x,y,z;u;\lambda)\frac{t^n}{n!} = \left(\frac{1-u}{\lambda e^t - u}\right) e^{xt+yt^2+zt^3}(x+2yt+3zt^2) - \frac{(1-u)\lambda e^t}{(\lambda e^t - u)^2} e^{xt+yt^2+zt^3}, \tag{10}$$

from which, upon using Equation (4) (for $\alpha = 1$) and after simplifying the resultant equation, it follows that:

$$\sum_{n=0}^{\infty} {}_H\mathfrak{F}_{n+1}(x,y,z;u;\lambda)\frac{t^n}{n!} = x \sum_{n=0}^{\infty} {}_H\mathfrak{F}_n(x,y,z;u;\lambda)\frac{t^n}{n!} + 2y \sum_{n=0}^{\infty} {}_H\mathfrak{F}_n(x,y,z;u;\lambda)\frac{t^{n+1}}{n!} + 3z \\ \sum_{n=0}^{\infty} {}_H\mathfrak{F}_n(x,y,z;u;\lambda)\frac{t^{n+2}}{n!} - \frac{\lambda}{1-u} \sum_{n=0}^{\infty} {}_H\mathfrak{F}_n(x,y,z;u;\lambda)\frac{t^n}{n!} \sum_{k=0}^{\infty} {}_H\mathfrak{F}_k(1,0,0;u;\lambda)\frac{t^k}{k!}. \tag{11}$$

Replacing $n \to n-1, n-2$ and $n-k$ consecutively in the second, third and last term of the above equation on the r.h.s., it follows that:

$$\sum_{n=0}^{\infty} {}_H\mathfrak{F}_{n+1}(x,y,z;u;\lambda)\frac{t^n}{n!} = x \sum_{n=0}^{\infty} {}_H\mathfrak{F}_n(x,y,z;u;\lambda)\frac{t^n}{n!} + 2y \sum_{n=0}^{\infty} {}_H\mathfrak{F}_{n-1}(x,y,z;u;\lambda)\frac{t^n}{(n-1)!} + 3z \\ \sum_{n=0}^{\infty} {}_H\mathfrak{F}_{n-2}(x,y,z;u;\lambda)\frac{t^n}{(n-2)!} - \frac{\lambda}{1-u} \sum_{n=0}^{\infty} \sum_{k=0}^{\infty} {}_H\mathfrak{F}_{n-k}(x,y,z;u;\lambda) {}_H\mathfrak{F}_k(1,0,0;u;\lambda)\frac{t^n}{k!(n-k)!},$$

which, upon comparing the coefficients of like powers of $t^n/n!$ on both sides, gives the recurrence Relation (9). □

Corollary 2. *The following recurrence relation for the 3VHAEP holds true:*

$$_H\mathfrak{E}_{n+1}(x,y,z;\lambda) = x\,_H\mathfrak{E}_n(x,y,z;\lambda) + 2yn\,_H\mathfrak{E}_{n-1}(x,y,z;\lambda) + 3zn(n-1)\,_H\mathfrak{E}_{n-2}(x,y,z;\lambda)$$
$$-\frac{\lambda}{2}\sum_{k=0}^{n}\binom{n}{k}\,_H\mathfrak{E}_{n-k}(x,y,z;\lambda)\,_H\mathfrak{E}_k(1,0,0;\lambda). \tag{12}$$

Theorem 3. *For $\gamma > 0$, the following relation for the 3VHATFEP of order α holds true:*

$$(1-u)^{\gamma}\,_H\mathfrak{F}_n^{(\alpha-\gamma)}(x,y,z;u;\lambda) = \sum_{k=0}^{n}\binom{n}{k}\,_H\mathfrak{F}_{n-k}^{(\alpha)}(x,y,z;u;\lambda)\sum_{p=0}^{\gamma}\binom{\gamma}{p}\lambda^p p^k(-u)^{\gamma-p}. \tag{13}$$

Proof. We write the generating Function (4) in the following form:

$$\sum_{n=0}^{\infty}{}_H\mathfrak{F}_n^{(\alpha-\gamma)}(x,y,z;u;\lambda)\frac{t^n}{n!} = \left(\frac{1-u}{\lambda e^t - u}\right)^{\alpha} e^{xt+yt^2+zt^3}(\lambda e^t - u)^{\gamma}(1-u)^{-\gamma}, \tag{14}$$

which, upon simplifying and again using Equation (4), gives:

$$\sum_{n=0}^{\infty}{}_H\mathfrak{F}_n^{(\alpha-\gamma)}(x,y,z;u;\lambda)\frac{t^n}{n!} = (1-u)^{-\gamma}\sum_{n=0}^{\infty}{}_H\mathfrak{F}_n^{(\alpha)}(x,y,z;u;\lambda)\frac{t^n}{n!}\sum_{k=0}^{\infty}\sum_{p=0}^{\gamma}\binom{\gamma}{p}\lambda^p p^k(-u)^{\gamma-p}\frac{t^k}{k!}. \tag{15}$$

Now, simplifying and then comparing the coefficients of the same powers of t in the resultant equation yield Assertion (13). □

Corollary 3. *For $\gamma > 0$, the following relation for the 3VHAEP of order α holds true:*

$$2^{\gamma}\,_H\mathfrak{E}_n^{(\alpha-\gamma)}(x,y,z;\lambda) = \sum_{k=0}^{n}\binom{n}{k}\,_H\mathfrak{E}_{n-k}^{(\alpha)}(x,y,z;\lambda)\sum_{p=0}^{\gamma}\binom{\gamma}{p}\lambda^p p^k. \tag{16}$$

Theorem 4. *For $u, \alpha \in \mathbb{C}$, $u \neq 1$, there is the following relationship between the 3VHATFEP of order α and the generalized Hurwitz–Lerch Zeta function (GHLZF) $\Phi_\mu(z,s,a)$:*

$$_H\mathfrak{F}_n^{(\alpha)}(x,y,z;u;\lambda) = \left(\frac{u-1}{u}\right)^{\alpha}\sum_{l=0}^{n}\binom{n}{l}\Phi_\alpha\left(\frac{\lambda}{u}, l-n, x\right)H_l(0,y,z). \tag{17}$$

Proof. We write the generating Function (4) in the following form:

$$\sum_{n=0}^{\infty}{}_H\mathfrak{F}_n^{(\alpha)}(x,y,z;u;\lambda)\frac{t^n}{n!} = (1-u)^{\alpha}(\lambda e^t - u)^{-\alpha}e^{xt+yt^2+zt^3}, \tag{18}$$

which, upon simplification, becomes:

$$\sum_{n=0}^{\infty}{}_H\mathfrak{F}_n^{(\alpha)}(x,y,z;u;\lambda)\frac{t^n}{n!} = (1-u)^{\alpha}(-u)^{-\alpha}\sum_{n=0}^{\infty}\sum_{k=0}^{\infty}\frac{(\alpha)_k}{k!}\left(\frac{\lambda}{u}\right)^k\frac{(k+x)^n t^n}{n!}e^{yt^2+zt^3}. \tag{19}$$

Using Equation (1) and the following formula for the GHLZF $\Phi_\mu(z,s,a)$ [13]:

$$\Phi_\mu(z,s,a) = \sum_{n=0}^{\infty}\frac{(\mu)_n}{n!}\frac{z^n}{(n+a)^s}, \tag{20}$$

and after simplifying the resultant equation yield Relation (17). □

Corollary 4. *There is the following relationship between the 3VHAEP of order α and generalized Hurwitz–Lerch Zeta function $\Phi_\mu(z,s,a)$:*

$$_H\mathfrak{E}_n^{(\alpha)}(x,y,z;\lambda) = 2^\alpha \sum_{l=0}^{n} \binom{n}{l} \Phi_\alpha(-\lambda, 1-n, x) H_l(0,y,z). \tag{21}$$

Theorem 5. *Let α and γ be nonnegative integers. There is the following relationship between the numbers $S(n,k,\lambda)$ and the 3VHATFEP of order α:*

$$\alpha! \sum_{l=0}^{n} \binom{n}{l} {}_H\mathfrak{F}_{n-l}^{(\alpha)}(x,y,z;u;\lambda) S\left(l,\alpha,\frac{\lambda}{u}\right) = \left(\frac{1-u}{u}\right)^\alpha H_n(x,y,z), \tag{22}$$

$$_H\mathfrak{F}_n^{(\alpha-\gamma)}(x,y,z;u;\lambda) = \gamma! \left(\frac{u}{1-u}\right)^\gamma \sum_{l=0}^{n} \binom{n}{l} {}_H\mathfrak{F}_{n-l}^{(\alpha)}(x,y,z;u;\lambda) S\left(l,\gamma,\frac{\lambda}{u}\right). \tag{23}$$

Proof. The generating Equation (4) can be formulated as:

$$\sum_{n=0}^{\infty} {}_H\mathcal{F}_n^{(\alpha)}(x,y,z;u;\lambda)\frac{t^n}{n!} = (1-u)^\alpha \frac{1}{(\lambda e^t - u)^\alpha} e^{xt+yt^2+zt^3}, \tag{24}$$

which, upon rearranging the terms using Equation (1) and the following expansion:

$$\frac{(\lambda e^t - 1)^k}{k!} = \sum_{n=0}^{\infty} S(n,k,\lambda)\frac{t^n}{n!}. \tag{25}$$

becomes:

$$\alpha! \sum_{n=0}^{\infty} {}_H\mathfrak{F}_n^{(\alpha)}(x,y,z;u;\lambda)\frac{t^n}{n!} \sum_{l=0}^{\infty} S\left(l,\alpha,\frac{\lambda}{u}\right)\frac{t^l}{l!} = \left(\frac{1-u}{u}\right)^\alpha \sum_{n=0}^{\infty} H_n(x,y,z)\frac{t^n}{n!}. \tag{26}$$

which, upon rearranging the summation and then simplifying the resultant equation, yields Relation (22).

Again, we consider the following arrangement of the generating Function (4):

$$\sum_{n=0}^{\infty} {}_H\mathfrak{F}_n^{(\alpha-\gamma)}(x,y,z;u;\lambda)\frac{t^n}{n!} = \left(\frac{1-u}{\lambda e^t - u}\right)^\alpha e^{xt+yt^2+zt^3} \left(\frac{u}{1-u}\right)^\gamma \gamma! \frac{(\frac{\lambda}{u} e^t - 1)^\gamma}{\gamma!}, \tag{27}$$

which, upon the use of Equations (4) and (25), applying the Cauchy product rule and then canceling the same powers of t in resultant the equation, yields Relation (23). □

Corollary 5. *There is the following relationship between the numbers $S(n,k,\lambda)$ and the 3VHAEP of order α:*

$$\alpha! \sum_{l=0}^{n} \binom{n}{l} {}_H\mathfrak{E}_{n-l}^{(\alpha)}(x,y,z;\lambda) S\left(l,\alpha,-\lambda\right) = (-2)^\alpha H_n(x,y,z).$$

$$_H\mathfrak{E}_n^{(\alpha-\gamma)}(x,y,z;\lambda) = \gamma! \left(\frac{-1}{2}\right)^\gamma \sum_{l=0}^{n} \binom{n}{l} {}_H\mathfrak{E}_{n-l}^{(\alpha)}(x,y,z;\lambda) S\left(l,\gamma,-\lambda\right). \tag{28}$$

In the next section, we derive some summation formulae for the 3VHATFEP.

3. Summation Formulae

In order to prove the summation formulae for the 3VHATFEP ${}_H\mathfrak{F}_n^{(\alpha)}(x,y,z;u;\lambda)$, we have the following theorems:

Theorem 6. *The following implicit summation formula for the 3VHATFEP of order α holds true:*

$$_H\mathfrak{F}_n^{(\alpha)}(x+w,y,z;u;\lambda) = \sum_{k=0}^{n} \binom{n}{k} {}_H\mathfrak{F}_k^{(\alpha)}(x,y,z;u;\lambda) w^{n-k}. \tag{29}$$

Proof. Substituting $x \to x+w$ in (4), then making use of Equation (4) and with the series expansion of e^{wt} in the resultant equation, we have:

$$\sum_{n=0}^{\infty} {}_H\mathfrak{F}_n^{(\alpha)}(x+w,y,z;u;\lambda) \frac{t^n}{n!} = \sum_{n=0}^{\infty}\sum_{k=0}^{\infty} {}_H\mathfrak{F}_k^{(\alpha)}(x,y,z;u;\lambda) w^n \frac{t^{n+k}}{n!k!}, \tag{30}$$

which, upon simplification, gives Assertion (29). □

Corollary 6. *For $w = 1$ in Equation (29), we have:*

$$_H\mathfrak{F}_n^{(\alpha)}(x+1,y,z;u;\lambda) = \sum_{k=0}^{n} \binom{n}{k} {}_H\mathfrak{F}_k^{(\alpha)}(x,y,z;u;\lambda). \tag{31}$$

Theorem 7. *The following implicit summation formula for the 3VHATFEP of order α holds true:*

$$_H\mathfrak{F}_n^{(\alpha)}(x+v,y+w,z+r;u;\lambda) = \sum_{k=0}^{n} \binom{n}{k} {}_H\mathfrak{F}_{n-k}^{(\alpha)}(x,y,z;u;\lambda)\, H_k(v,w,r). \tag{32}$$

Proof. Replacing $x \to x+v$, $y \to y+w$ and $z \to z+r$ in the generating Function (4) and by the help of Equations (1) and (4), we find:

$$\sum_{n=0}^{\infty} {}_H\mathfrak{F}_n^{(\alpha)}(x+v,y+w,z+r;u;\lambda) \frac{t^n}{n!} = \sum_{n=0}^{\infty}\sum_{k=0}^{\infty} {}_H\mathfrak{F}_n^{(\alpha)}(x,y,z;\lambda;u) H_k(v,w,r) \frac{t^{n+k}}{n!k!}, \tag{33}$$

which, after simplification, gives Formula (32). □

Corollary 7. *For $r = 0$ in Equation (32), we have:*

$$_H\mathfrak{F}_n^{(\alpha)}(x+v,y+w,z;u;\lambda) = \sum_{k=0}^{n} \binom{n}{k} {}_H\mathfrak{F}_{n-k}^{(\alpha)}(x,y,z;u;\lambda)\, H_k(v,w). \tag{34}$$

Theorem 8. *The following implicit summation formula for the 3VHATFEP of order α holds true:*

$$_H\mathfrak{F}_{n+k}^{(\alpha)}(p,y,z;u;\lambda) = \sum_{l,m=0}^{n,k} \binom{n}{l}\binom{k}{m}(p-x)^{l+m}\, {}_H\mathfrak{F}_{n+k-l-m}^{(\alpha)}(x,y,z;u;\lambda). \tag{35}$$

Proof. Reestablishing t by $t+v$ and after using the following rule:

$$\sum_{N=0}^{\infty} f(N) \frac{(x+y)^N}{N!} = \sum_{l,m=0}^{\infty} f(l+m) \frac{x^l\, y^m}{l!\, m!} \tag{36}$$

in Equation (4) and then simplifying the resultant equation, it follows that:

$$e^{-x(t+v)} \sum_{n,k=0}^{\infty} {}_H\mathfrak{F}_{n+k}^{(\alpha)}(x,y,z;\lambda;u) \frac{t^n v^k}{n!\, k!} = \left(\frac{1-u}{\lambda e^{t+v}-u}\right)^{\alpha} e^{y(t+v)^2 + z(t+v)^3}. \tag{37}$$

123

Replacing x by p in the above equation, equating the resultant equation to the above equation and then expanding the exponential function give:

$$\sum_{n,k=0}^{\infty} {}_H\mathfrak{F}_{n+k}^{(\alpha)}(p,y,z;u;\lambda)\frac{t^n v^k}{n!\, k!} = \sum_{N=0}^{\infty} (p-x)^N \frac{(t+v)^N}{N!} \sum_{n,k=0}^{\infty} {}_H\mathfrak{F}_{n+k}^{(\alpha)}(x,y,z;u;\lambda)\frac{t^n v^k}{n!\, k!}. \tag{38}$$

Now, using Formula (36) in the above equation and then replacing $n \to n-l$ and $k \to k-m$ in the resultant equation, it follows that:

$$\sum_{n,k=0}^{\infty} {}_H\mathfrak{F}_{n+k}^{(\alpha)}(p,y,z;u;\lambda)\frac{t^n v^k}{n!\, k!} = \sum_{n,k=0}^{\infty} \sum_{l,m=0}^{n,k} \frac{(p-x)^{l+m}}{l!\, m!} {}_H\mathfrak{F}_{n+k-l-m}^{(\alpha)}(x,y,z;u;\lambda)\frac{t^n v^k}{(n-l)!\,(k-m)!}, \tag{39}$$

which gives Formula (35). □

Corollary 8. *For $n=0$ in Equation (35), we have:*

$$ {}_H\mathfrak{F}_k^{(\alpha)}(p,y,z;u;\lambda) = \sum_{m=0}^{k} \binom{k}{m}(p-x)^m {}_H\mathfrak{F}_{k-m}^{(\alpha)}(x,y,z;u;\lambda). \tag{40}$$

Corollary 9. *Replacing p by $p+x$ and taking $z=0$ in Equation (35), we have:*

$$ {}_H\mathfrak{F}_{n+k}^{(\alpha)}(p+x,y;u;\lambda) = \sum_{l,m=0}^{n,k} \binom{n}{l}\binom{k}{m} p^{l+m} {}_H\mathfrak{F}_{n+k-l-m}^{(\alpha)}(x,y;u;\lambda). \tag{41}$$

Corollary 10. *Replacing p by $p+x$ and taking $y=0\ z=0$ in Equation (35), we have:*

$$ {}_H\mathfrak{F}_{n+k}^{(\alpha)}(p+x;u;\lambda) = \sum_{l,m=0}^{n,k} \binom{n}{l}\binom{k}{m} p^{l+m} {}_H\mathfrak{F}_{n+k-l-m}^{(\alpha)}(x;u;\lambda). \tag{42}$$

Corollary 11. *For $p=0$ in Equation (35), we have:*

$$ {}_H\mathfrak{F}_{n+k}^{(\alpha)}(y,z;u;\lambda) = \sum_{l,m=0}^{n,k} \binom{n}{l}\binom{k}{m}(-x)^{l+m} {}_H\mathfrak{F}_{n+k-l-m}^{(\alpha)}(x,y,z;u;\lambda). \tag{43}$$

Theorem 9. *The following relation for the 3VHATFEP of order α holds true:*

$$ {}_H\mathfrak{F}_n^{(\alpha)}(x,y,z;u;\lambda) = \sum_{k=0}^{[\frac{n}{3}]} \frac{n!}{(n-3k)!\, k!} {}_H\mathfrak{F}_{n-3k}^{(\alpha)}(x,y;u;\lambda) z^k. \tag{44}$$

Proof. Using the equation from Table 1(I), the expansion of e^{zt^3} in Equation (4) and then simplifying the resulting equation give:

$$\sum_{n=0}^{\infty} {}_H\mathfrak{F}_n^{(\alpha)}(x,y,z;u;\lambda)\frac{t^n}{n!} = \sum_{n=0}^{\infty} \left(\sum_{k=0}^{[\frac{n}{3}]} \frac{n!}{(n-3k)!\, k!} {}_H\mathfrak{F}_{n-3k}^{(\alpha)}(x,y;u;\lambda) z^k \right) \frac{t^n}{n!}. \tag{45}$$

After comparing the coefficients of same powers of $t^n/n!$ in the above equation, we are led to Relation (44). □

Theorem 10. *The following relation for the 3VHATFEP of order α holds true:*

$$_H\mathfrak{F}_n^{(\alpha)}(x,y,z;u;\lambda) = \sum_{k=0}^{n}\sum_{s=0}^{[\frac{k}{3}]} \frac{n!}{(n-k)!(k-3s)!s!} \mathfrak{F}_{n-k}^{(\alpha)}(u;\lambda) H_{k-3s}(x,y)z^s. \qquad (46)$$

Proof. Using Equations (3) and (1) (for $z = 0$), the expansion of e^{zt^3} in Equation (4) and after rearranging the terms, it follows that:

$$\sum_{n=0}^{\infty} {}_H\mathfrak{F}_n^{(\alpha)}(x,y,z;u;\lambda)\frac{t^n}{n!} = \sum_{n=0}^{\infty}\sum_{k=0}^{n}\binom{n}{k}\mathfrak{F}_{n-k}^{(\alpha)}(u;\lambda)\left(\sum_{s=0}^{[\frac{k}{3}]}\frac{k!}{(k-3s)!s!}H_{k-3s}(x,y)z^s\right)\frac{t^n}{n!}. \qquad (47)$$

Upon canceling the coefficients of like powers of t in Equation (47), we get Assertion (46). □

Theorem 11. *The following relation for the 3VHATFEP of order α holds true:*

$$_H\mathfrak{F}_n^{(\alpha)}(x,y,z;u;\lambda) = \sum_{s=0}^{[\frac{n}{3}]}\sum_{k=0}^{[\frac{n-3s}{2}]} \frac{n!}{s!(n-3s-2k)!k!} \mathfrak{F}_{n-3s-2k}^{(\alpha)}(x;u;\lambda) y^k z^s. \qquad (48)$$

Proof. With the use of Equation (2), the expansions of e^{yt^2} and e^{zt^3} in Equation (4) and upon simplifying the resulting equation, we obtain:

$$\sum_{n=0}^{\infty} {}_H\mathfrak{F}_n^{(\alpha)}(x,y,z;u;\lambda)\frac{t^n}{n!} = \sum_{n=0}^{\infty}\left(\sum_{s=0}^{[\frac{n}{3}]}\sum_{k=0}^{[\frac{n-3s}{2}]} \frac{n!}{s!(n-3s-2k)!k!}\mathfrak{F}_{n-3s-2k}^{(\alpha)}(x;u;\lambda)y^k z^s\right)\frac{t^n}{n!} \qquad (49)$$

Finally, upon equating the coefficients of the same powers of t in the above equation, Relation (48) is proven. □

In the next section, we establish some symmetric identities for the 3VHATFEP.

4. Symmetric Identities

The identities for the generalized special functions are useful in electromagnetic processes, combinatorics, numerical analysis, etc. Several types of identities and relations related to Apostol-type polynomials and related polynomials are considered in [14–27]. This provides the motivation to explore symmetry identities for the 3VHATFEP. We recall the following:

For any $\gamma \in \mathbb{R}$ or $\mathbb{C}$, the generalized sum of integer powers $S_k(p;\gamma)$ is given by:

$$\frac{\gamma^{p+1}e^{(p+1)t} - 1}{\gamma e^t - 1} = \sum_{k=0}^{\infty} S_k(p;\gamma)\frac{t^k}{k!}, \qquad (50)$$

which gives:

$$S_k(p;\gamma) = \sum_{l=0}^{k} \gamma^l l^k.$$

For any $\gamma \in \mathbb{R}$ or $\mathbb{C}$, the multiple power sums $S_k^{(l)}(m;\gamma)$ are given by:

$$\left(\frac{1-\gamma^m e^{mt}}{1-\gamma e^t}\right)^l = \frac{1}{\gamma^l}\sum_{n=0}^{\infty}\left\{\sum_{p=0}^{n}\binom{n}{p}(-1)^{n-p}S_k^{(l)}(m;\gamma)\right\}\frac{t^n}{n!}. \qquad (51)$$

To prove the symmetry identities for the 3VHATFEP, we have the following theorems:

Theorem 12. *For all integers $c, d > 0$ and $n \geq 0$, $\alpha \geq 1$, $\lambda, u \in \mathbb{C}$, the following symmetry relation between the 3VHATFEP of order α and the generalized integer power sums holds true:*

$$\sum_{k=0}^{n} \binom{n}{k} c^{n-k} {}_H\mathfrak{F}_{n-k}^{(\alpha)}(dx, d^2y, d^3z; \lambda; u) \sum_{l=0}^{k} \binom{k}{l} d^k u^{c-1} \mathcal{S}_l(c-1; \tfrac{\lambda}{u}) {}_H\mathfrak{F}_{k-l}^{(\alpha-1)}(cX, c^2Y, c^3Z; \lambda; u)$$
$$= \sum_{k=0}^{n} \binom{n}{k} d^{n-k} u^{d-1} {}_H\mathfrak{F}_{n-k}^{(\alpha)}(cx, c^2y, c^3z; \lambda; u) \sum_{l=0}^{k} \binom{k}{l} c^k \mathcal{S}_l(d-1; \tfrac{\lambda}{u}) {}_H\mathfrak{F}_{k-l}^{(\alpha-1)}(dX, d^2Y, d^3Z; \lambda; u). \tag{52}$$

Proof. Let

$$G(t) := \frac{(1-u)^{2\alpha-1} e^{cdxt + y(cdt)^2 + z(cdt)^3} (\lambda^c e^{cdt} - u^c) e^{cdXt + Y(cdt)^2 + Z(cdt)^3}}{(\lambda e^{ct} - u)^\alpha (\lambda e^{dt} - u)^\alpha}, \tag{53}$$

which, upon rearranging the powers and then using Equations (4) and (50) in the resultant equation, yields:

$$G(t) = \left(\sum_{n=0}^{\infty} {}_H\mathfrak{F}_n^{(\alpha)}(dx, d^2y, d^3z; \lambda; u) \frac{(ct)^n}{n!} \right) \left(u^{c-1} \sum_{l=0}^{\infty} \mathcal{S}_l(c-1; \tfrac{\lambda}{u}) \frac{(dt)^l}{l!} \right)$$
$$\times \left(\sum_{k=0}^{\infty} {}_H\mathfrak{F}_k^{(\alpha-1)}(cX, c^2Y, c^3Z; \lambda; u) \frac{(dt)^k}{k!} \right). \tag{54}$$

Upon applying the Cauchy product rule in the above equation, we get:

$$G(t) = \sum_{n=0}^{\infty} \left(\sum_{k=0}^{n} \binom{n}{k} c^{n-k} d^k u^{c-1} {}_H\mathfrak{F}_{n-k}^{(\alpha)}(dx, d^2y, d^3z; \lambda; u) \sum_{l=0}^{k} \binom{k}{l} \mathcal{S}_l(c-1; \tfrac{\lambda}{u}) \right.$$
$$\left. \times {}_H\mathfrak{F}_{k-l}^{(\alpha-1)}(cX, c^2Y, c^3Z; \lambda; u) \right) \frac{t^n}{n!}. \tag{55}$$

In a similar manner, we obtain:

$$G(t) = \sum_{n=0}^{\infty} \left(\sum_{k=0}^{n} \binom{n}{k} d^{n-k} c^k u^{d-1} {}_H\mathfrak{F}_{n-k}^{(\alpha)}(cx, c^2y, c^3z; \lambda; u) \sum_{l=0}^{k} \binom{k}{l} \mathcal{S}_l(d-1; \tfrac{\lambda}{u}) \right.$$
$$\left. \times {}_H\mathfrak{F}_{k-l}^{(\alpha-1)}(dX, d^2Y, d^3Z; \lambda; u) \right) \frac{t^n}{n!}. \tag{56}$$

Equating the coefficients of the like powers of t in the r.h.s. of Expansions (55) and (56), we are led to Identity (52). □

Theorem 13. *For each pair of positive integers c, d and for all integers $n \geq 0$, $\alpha \geq 1$, $\lambda, u \in \mathbb{C}$, the following symmetry identity for the 3VHATFEP of order α holds true:*

$$\sum_{k=0}^{n} \binom{n}{k} \sum_{i=0}^{c-1} \sum_{j=0}^{d-1} u^{c+d-2} (\tfrac{\lambda}{u})^{i+j} c^{n-k} d^k {}_H\mathfrak{F}_k^{(\alpha)}\left(cX + \tfrac{c}{d}j, c^2Y, c^3Z; \lambda; u\right) {}_H\mathfrak{F}_{n-k}^{(\alpha)}\left(dx + \tfrac{d}{c}i, d^2y, d^3z; \lambda; u\right)$$
$$= \sum_{k=0}^{n} \binom{n}{k} \sum_{i=0}^{d-1} \sum_{j=0}^{c-1} u^{c+d-2} (\tfrac{\lambda}{u})^{i+j} d^{n-k} c^k {}_H\mathfrak{F}_k^{(\alpha)}\left(dX + \tfrac{d}{c}j, d^2Y, d^3Z; \lambda; u\right) {}_H\mathfrak{F}_{n-k}^{(\alpha)}\left(cx + \tfrac{c}{d}i, c^2y, c^3z; \lambda; u\right). \tag{57}$$

Proof. Let

$$H(t) := \frac{(1-u)^{2\alpha} e^{cdxt + y(cdt)^2 + z(cdt)^3} (\lambda^c e^{cdt} - u^c)(\lambda^d e^{cdt} - u^d) e^{cdXt + Y(cdt)^2 + Z(cdt)^3}}{(\lambda e^{ct} - u)^{\alpha+1} (\lambda e^{dt} - u)^{\alpha+1}}, \tag{58}$$

from which, upon rearranging the powers and using the series expansions for $\left(\frac{\lambda^c e^{cdt}-u^c}{\lambda e^{dt}-u}\right)$ and $\left(\frac{\lambda^d e^{cdt}-u^d}{\lambda e^{ct}u}\right)$ in the resultant equation, it follows that:

$$H(t) = \left(\frac{1-u}{\lambda e^{ct}-u}\right)^\alpha e^{dx(ct)+d^2y(ct)^2+d^3z(ct)^3} u^{c-1} \sum_{i=0}^{c-1} \left(\frac{\lambda}{u}\right)^i e^{dti}$$

$$\times \left(\frac{1-u}{\lambda e^{dt}-u}\right)^\alpha e^{cX(dt)+c^2Y(dt)^2+c^3Z(dt)^3} u^{d-1} \sum_{j=0}^{d-1} \left(\frac{\lambda}{u}\right)^j e^{ctj}. \tag{59}$$

Now, by making use of Equation (4) and the application of the Cauchy product rule in the resultant equation, we have:

$$H(t) = \sum_{k=0}^{n} \binom{n}{k} \sum_{i=0}^{c-1}\sum_{j=0}^{d-1} u^{c+d-2}\left(\frac{\lambda}{u}\right)^{i+j} c^{n-k} d^k {}_H\mathfrak{F}_k^{(\alpha)}\left(cX+\frac{c}{d}j, c^2Y, c^3Z; \lambda; u\right)$$
$${}_H\mathfrak{F}_{n-k}^{(\alpha)}\left(dx+\frac{d}{c}i, d^2y, d^3z; \lambda; u\right). \tag{60}$$

Following the same lines of proof as above gives another identity:

$$H(t) = \sum_{k=0}^{n} \binom{n}{k} \sum_{i=0}^{d-1}\sum_{j=0}^{c-1} u^{d+c-2}\left(\frac{\lambda}{u}\right)^{i+j} d^{n-k} c^k {}_H\mathfrak{F}_k^{(\alpha)}\left(dX+\frac{d}{c}j, d^2Y, d^3Z; \lambda; u\right)$$
$${}_H\mathfrak{F}_{n-k}^{(\alpha)}\left(cx+\frac{c}{d}i, c^2y, c^3z; \lambda; u\right). \tag{61}$$

Comparing the coefficients of the same powers of t in the r.h.s. of Expressions (60) and (61) gives Identity (57). □

Theorem 14. *For each pair of positive integers c, d and for all integers $n \geq 0$, $\alpha \geq 1$, $\lambda, u \in \mathbb{C}$, the following symmetry identity for the 3VHATFEP holds true:*

$$\sum_{m=0}^{d-1} u^{d-1}\left(\frac{\lambda}{u}\right)^m \sum_{l=0}^{n} \binom{n}{l} {}_H\mathfrak{F}_{n-l}\left(cx, c^2y, c^3z; \lambda; u\right) d^{n-l} (cm)^l$$
$$= \sum_{m=0}^{c-1} u^{c-1}\left(\frac{\lambda}{u}\right)^m \sum_{l=0}^{n} \binom{n}{l} {}_H\mathfrak{F}_{n-l}\left(dx, d^2y, d^3z; \lambda; u\right) c^{n-l} (dm)^l. \tag{62}$$

Proof. Let

$$N(t) := \frac{(1-u)e^{cdxt+y(cdt)^2+z(cdt)^3}(\lambda^d e^{cdt}-u^d)}{(\lambda e^{ct}-u)(\lambda e^{dt}-u)}. \tag{63}$$

Proceeding on the same lines of proof as in Theorem 13, we get Identity (62). Thus, we omit the proof. □

Theorem 15. *For each pair of positive integers c, d and for all integers $n \geq 0$, $\alpha \geq 1$, $\lambda, u \in \mathbb{C}$, the following symmetry relation between the 3VHATFEP and multiple power sums holds true:*

$$\sum_{l=0}^{n} \binom{n}{l} {}_H\mathfrak{F}_{n-l}(dx, d^2y, d^3z; \lambda; u) u^{d\alpha}\lambda^{-\alpha} \sum_{m=0}^{l} \binom{l}{m} \sum_{r=0}^{m} \binom{m}{r}(-\alpha)^{m-r} S_k^{(\alpha)}\left(d; \frac{\lambda}{u}\right)$$
$$\times {}_H\mathfrak{F}_{l-m}^{(\alpha+1)}(cX, c^2Y, c^3Z; \lambda; u) c^{n-l+m} d^{l-m}$$
$$= \sum_{l=0}^{n} \binom{n}{l} {}_H\mathfrak{F}_{n-l}(cx, c^2y, c^3z; \lambda; u) u^{c\alpha}\lambda^{-\alpha} \sum_{m=0}^{l} \binom{l}{m} \sum_{r=0}^{m} \binom{m}{r}(-\alpha)^{m-r} S_k^{(\alpha)}\left(c; \frac{\lambda}{u}\right)$$
$$\times {}_H\mathfrak{F}_{l-m}^{(\alpha+1)}(dX, d^2Y, d^3Z; \lambda; u) d^{n-l+m} c^{l-m}. \tag{64}$$

Proof. Let:
$$F(t) := \frac{(1-u)^{\alpha+2} e^{dx(ct)+d^2y(ct)^2+d^3z(ct)^3} \left(\lambda^d e^{dct} - u^d\right)^\alpha e^{cX(dt)+c^2Y(dt)^2+c^3Z(dt)^3}}{(\lambda e^{dt} - u)^{\alpha+1} (\lambda e^{ct} - u)^{\alpha+1}}, \tag{65}$$

which, upon rearranging the powers and use of Equations (4) and (51) in the resultant equation, yields:

$$F(t) := \sum_{n=0}^{\infty} {}_H\mathfrak{F}_n(dx, d^2y, d^3z; \lambda; u) c^n \frac{t^n}{n!} u^{d\alpha} \lambda^{-\alpha} \sum_{m=0}^{\infty} \sum_{r=0}^{m} \binom{m}{r} (-\alpha)^{m-r} S_k^{(\alpha)}\left(d; \frac{\lambda}{u}\right) c^m \frac{t^m}{m!}$$
$$\sum_{l=0}^{\infty} {}_H\mathfrak{F}_l^{(\alpha+1)}(cX, c^2Y, c^3Z; \lambda; u) d^l \frac{t^l}{l!}. \tag{66}$$

Now, appropriately applying the using Cauchy product rule in the above equation leads to:

$$F(t) := \sum_{n=0}^{\infty} \sum_{l=0}^{n} \binom{n}{l} {}_H\mathfrak{F}_{n-l}(dx, d^2y, d^3z; \lambda; u) c^{n-l} u^{d\alpha} \lambda^{-\alpha} \sum_{m=0}^{l} \binom{l}{m} \sum_{r=0}^{m} \binom{m}{r} (-\alpha)^{m-r} S_k^{(\alpha)}\left(d; \frac{\lambda}{u}\right)$$
$$ {}_H\mathfrak{F}_{l-m}^{(\alpha+1)}(cX, c^2y, c^3z; \lambda; u) c^m d^{l-m} \frac{t^n}{n!}. \tag{67}$$

Similarly, we can find:

$$F(t) := \sum_{n=0}^{\infty} \sum_{l=0}^{n} \binom{n}{l} {}_H\mathfrak{F}_{n-l}(cx, c^2y, c^3z; \lambda; u) d^{n-l} u^{c\alpha} \lambda^{-\alpha} \sum_{m=0}^{l} \binom{l}{m} \sum_{r=0}^{m} \binom{m}{r} (-\alpha)^{m-r} S_k^{(\alpha)}\left(c; \frac{\lambda}{u}\right)$$
$$ {}_H\mathfrak{F}_{l-m}^{(\alpha+1)}(dx, d^2y, d^3z; \lambda; u) d^m c^{l-m} \frac{t^n}{n!}. \tag{68}$$

Equating the coefficients of the like powers of $t^n/n!$ in the r.h.s. of Expansions (67) and (68) gives Identity (64). □

Theorem 16. *For each pair of positive integers c, d and for all integers $n \geq 0$, $\alpha \geq 1$, $\lambda, u \in \mathbb{C}$, the following symmetry relation between the 3VHATFEP of order α and multiple power sums holds true:*

$$\sum_{m=0}^{n} \binom{n}{m} {}_H\mathfrak{F}_{n-m}^{(\alpha)}(dx, d^2y, d^3z; \lambda; u) c^{n-m} u^{c\alpha} \lambda^{-\alpha} \sum_{r=0}^{m} \binom{m}{r} (-\alpha)^{m-r} S_k^{(\alpha)}\left(c; \frac{\lambda}{u}\right) d^m$$
$$= \sum_{m=0}^{n} \binom{n}{m} {}_H\mathfrak{F}_{n-m}^{(\alpha)}(cx, c^2y, c^3z; \lambda; u) d^{n-m} u^{d\alpha} \lambda^{-\alpha} \sum_{r=0}^{m} \binom{m}{r} (-\alpha)^{m-r} S_k^{(\alpha)}\left(d; \frac{\lambda}{u}\right) c^m. \tag{69}$$

Proof. Let:
$$M(t) := \frac{(1-u)^\alpha e^{dx(ct)+d^2y(ct)^2+d^3z(ct)^3} \left(\lambda^c e^{cdt} - u^c\right)^\alpha}{(\lambda e^{dt} - u)^\alpha (\lambda e^{ct} - u)^\alpha}. \tag{70}$$

Proceeding on the same lines of proof as in Theorem 15, we get Identity (69). Thus, we omit the proof. □

Theorem 17. *For each pair of positive integers c, d and for all integers $n \geq 0$, $\alpha \geq 1$, $\lambda, u \in \mathbb{C}$, the following symmetry relation between the 3VHATFEP of order α and the Hurwitz–Lerch Zeta function holds true:*

$$\left(\frac{1-u}{u}\right)^\alpha (-1)^\alpha \left(\sum_{p=0}^{n} \binom{n}{p} \sum_{s=0}^{n-p} \binom{n-p}{s} \Phi_\alpha\left(\frac{\lambda}{u}, s-n+p, cx\right) H_s(0, c^2y, c^3z) d^n u^c \lambda^{-1}\right)$$
$$\sum_{r=0}^{p} \binom{r}{p} \sum_{q=0}^{p-r} \binom{p-r}{q} (-1)^{p-r-q} S_q\left(c, \frac{\lambda}{u}\right) {}_H\mathfrak{F}_r^{(\alpha)}(dX, d^2Y, d^3Z; \lambda; u) c^r d^{p-r}\Big)$$
$$= \left(\frac{1-u}{u}\right)^\alpha (-1)^\alpha \left(\sum_{p=0}^{n} \binom{n}{p} \sum_{s=0}^{n-p} \binom{n-p}{s} \Phi_\alpha\left(\frac{\lambda}{u}, s-n+p, cx\right) H_s(0, d^2y, d^3z) c^n u^d \lambda^{-1}\right)$$
$$\sum_{r=0}^{p} \binom{r}{p} \sum_{q=0}^{p-r} \binom{p-r}{q} (-1)^{p-r-q} S_q\left(d, \frac{\lambda}{u}\right) {}_H\mathfrak{F}_r^{(\alpha)}(cX, c^2Y, c^3Z; \lambda; u) d^r c^{p-r}\Big). \tag{71}$$

Proof. Let:

$$P(t) := \frac{(1-u)^{2\alpha} e^{cx(dt)+c^2y(dt)^2+c^3z(dt)^3} \left(\lambda^c e^{cdt} - u^c\right) e^{dX(ct)+d^2Y(ct)^2+d^3Z(ct)^3}}{(\lambda e^{dt} - u)^{\alpha+1} (\lambda e^{ct} - u)^{\alpha}}, \tag{72}$$

which, upon rearranging the powers and after using Equations (4) and (51) (for $\alpha = 1$) and the following formula for the generalized binomial theorem:

$$(1+w)^{-\alpha} = \sum_{m=0}^{\infty} \binom{m+\alpha-1}{m}(-w)^m; \quad |w| < 1, \tag{73}$$

in the resultant equation becomes:

$$P(t) := \left(\frac{1-u}{u}\right)^{\alpha}(-1)^{\alpha} \sum_{m=0}^{\infty} \binom{m+\alpha-1}{m}\left(\frac{\lambda}{u}\right)^m e^{mdt} e^{cx(dt)+c^2y(dt)^2+c^3z(dt)^3} u^c \lambda^{-1} \sum_{p=0}^{\infty}\sum_{q=0}^{p} \binom{p}{q}(-1)^{p-q}$$
$$S_q\left(c, \frac{\lambda}{u}\right) d^p \frac{t^p}{p!} \sum_{r=0}^{\infty} {}_H\mathfrak{F}_r^{(\alpha)}(dX, d^2Y, d^3Z; \lambda; u) \frac{(ct)^r}{r!}. \tag{74}$$

Simplifying the above equation with the use of Equations (1) and (20) and then using the Cauchy product rule in the resultant equation, we get:

$$P(t) : = \left(\frac{1-u}{u}\right)^{\alpha}(-1)^{\alpha} \sum_{n=0}^{\infty}\left(\sum_{p=0}^{n}\binom{n}{p}\sum_{s=0}^{n-p}\binom{n-p}{s}\Phi_\alpha\left(\frac{\lambda}{u}, s-n+p, cx\right) H_s(0, c^2y, c^3z) d^n u^c \lambda^{-1}\right.$$
$$\left.\sum_{r=0}^{p}\binom{r}{p}\sum_{q=0}^{p-r}\binom{p-r}{q}(-1)^{p-r-q} S_q\left(c, \frac{\lambda}{u}\right) {}_H\mathfrak{F}_r^{(\alpha)}(dX, d^2Y, d^3Z; \lambda; u) c^r d^{p-r}\right) \frac{t^n}{n!}. \tag{75}$$

In a similar manner, we have:

$$P(t) : = \left(\frac{1-u}{u}\right)^{\alpha}(-1)^{\alpha} \sum_{n=0}^{\infty}\left(\sum_{p=0}^{n}\binom{n}{p}\sum_{s=0}^{n-p}\binom{n-p}{s}\Phi_\alpha\left(\frac{\lambda}{u}, s-n+p, dx\right) H_s(0, d^2y, d^3z) c^n u^d \lambda^{-1}\right.$$
$$\left.\sum_{r=0}^{p}\binom{r}{p}\sum_{q=0}^{p-r}\binom{p-r}{q}(-1)^{p-r-q} S_q\left(d, \frac{\lambda}{u}\right) {}_H\mathfrak{F}_r^{(\alpha)}(cX, c^2Y, c^3Z; \lambda; u) d^r c^{p-r}\right) \frac{t^n}{n!}. \tag{76}$$

Finally, canceling the coefficients of the same powers of t in the r.h.s. of Expansions (75) and (76), Identity (71) is proven. □

Note: The results established above for the 3VHATFEP can be reduced to the illustrative special cases mentioned in Table 1 simply by substituting special values of the variables or parameters. Therefore, we omit them.

5. Operational Representation

The classical and Apostol-type Frobenius–Euler numbers and polynomials are the generalization of Euler numbers and polynomials, and these are associated with the Brouwer fixed-point theorem and vector fields [28].

From generating Equation (4), we find that the 3VHATFEP are the solutions of the following equations:

$$\frac{\partial}{\partial y}{}_H\mathfrak{F}_n^{(\alpha)}(x, y, z; u; \lambda) = \frac{\partial^2}{\partial x^2}{}_H\mathfrak{F}_n^{(\alpha)}(x, y, z; u; \lambda), \tag{77}$$

$$\frac{\partial}{\partial z}{}_H\mathfrak{F}_n^{(\alpha)}(x, y, z; u; \lambda) = \frac{\partial^3}{\partial x^3}{}_H\mathfrak{F}_n^{(\alpha)}(x, y, z; u; \lambda), \tag{78}$$

under the following initial condition:

$${}_H\mathfrak{F}_n^{(\alpha)}(x, 0, 0; u; \lambda) = \mathfrak{F}_n^{(\alpha)}(x; u; \lambda). \tag{79}$$

Thus, in view of the above equation, we find that, for the 3VHATFEP, the following operational representation holds true:

$$_H\mathfrak{F}_n^{(\alpha)}(x,y,z;u;\lambda) = \exp\left(y\frac{\partial^2}{\partial x^2} + z\frac{\partial^3}{\partial x^3}\right)\{\mathfrak{F}_n^{(\alpha)}(x;u;\lambda)\}. \tag{80}$$

The operational formalism developed above can be used to obtain the corresponding identities for the 3VHATFEP and for their special cases. To give the applications of the operational representation (80), we apply the operation $\mathcal{O}$ given below:

$\mathcal{O}$: Operating $\exp\left(y\frac{\partial^2}{\partial x^2} + z\frac{\partial^3}{\partial x^3}\right)$ on both sides of a given result.

Consider the following identities for the FEP $\mathfrak{F}_n^{(\alpha)}(x;u)$ from [17]:

$$u\mathfrak{F}_n(x;u^{-1}) + \mathfrak{F}_n(x;u) = (1+u)\sum_{k=0}^{n}\binom{n}{k}\mathfrak{F}_{n-k}(u^{-1})\mathfrak{F}_k(x;u), \tag{81}$$

$$\frac{1}{n+1}\mathfrak{F}_k(x;u) + \mathfrak{F}_{n-k}(x;u) = \sum_{k=0}^{n-1}\frac{\binom{n}{k}}{n-k+1}\sum_{l=k}^{n}((-u)\mathfrak{F}_{l-k}(u)\mathfrak{F}_{n-l}(u) + 2u\mathfrak{F}_{n-k}(u)) \tag{82}$$
$$\mathfrak{F}_k(x;u)\,\mathfrak{F}_n(x;u),$$

$$\mathfrak{F}_n^{(\alpha)}(x;u) = \sum_{k=0}^{n}\binom{n}{k}\mathfrak{F}_{n-k}^{(\alpha-1)}(u)\mathfrak{F}_k(x;u) \quad (n \in \mathbb{Z}_+), \tag{83}$$

$$\mathfrak{F}_n(x;u) = \frac{1}{(1-u)^\alpha}\sum_{k=0}^{n}\binom{n}{k}\left(\sum_{j=0}^{\alpha}\binom{\alpha}{j}(-u)^{\alpha-j}\mathfrak{F}_{n-k}(j;u)\right)\mathfrak{F}_k^{(\alpha)}(x;u) \quad (n \in \mathbb{Z}_+), \tag{84}$$

which, upon using operation $(\mathcal{O})$ in both sides, yields the following identities for the polynomials $_H\mathfrak{F}_n^{(\alpha)}(x,y,z;u)$:

$$u_H\mathfrak{F}_n(x,y,z;u^{-1}) + _H\mathfrak{F}_n(x,y,z;u) = (1+u)\sum_{k=0}^{n}\binom{n}{k}\mathfrak{F}_{n-k}(u^{-1})_H\mathfrak{F}_k(x,y,z;u), \tag{85}$$

$$\frac{1}{n+1}{}_H\mathfrak{F}_k(x,y,z;u) + _H\mathfrak{F}_{n-k}(x,y,z;u) = \sum_{k=0}^{n-1}\frac{\binom{n}{k}}{n-k+1}\sum_{l=k}^{n}((-u)\mathfrak{F}_{l-k}(u)\mathfrak{F}_{n-l}(u) + 2u\mathfrak{F}_{n-k}(u)) \tag{86}$$
$$_H\mathfrak{F}_k(x,y,z;u)\,\mathfrak{F}_n(x;u),$$

$$_H\mathfrak{F}_n^{(\alpha)}(x,y,z;u) = \sum_{k=0}^{n}\binom{n}{k}\mathfrak{F}_{n-k}^{(\alpha-1)}(u)_H\mathfrak{F}_k(x,y,z;u) \quad (n \in \mathbb{Z}_+), \tag{87}$$

$$_H\mathfrak{F}_n(x,y,z;u) = \frac{1}{(1-u)^\alpha}\sum_{k=0}^{n}\binom{n}{k}\left(\sum_{j=0}^{\alpha}\binom{\alpha}{j}(-u)^{\alpha-j}\mathfrak{F}_{n-k}(j;u)\right)_H\mathfrak{F}_k^{(\alpha)}(x,y,z;u) \quad (n \in \mathbb{Z}_+). \tag{88}$$

Thus, we find that the aforementioned polynomials, which include the polynomials as their special cases given in Table 1 along with the underlying operational formalism, offer a powerful tool for the investigation of the properties of a wide class of polynomials. Thus, the combination of Hermite and Frobenius–Euler polynomials yields such interesting results.

Further, motivated by the ATFEP $\mathfrak{F}_n^{(\alpha)}(x;u;\lambda)$, we introduce the Apostol type Frobenius–Genocchi polynomials $\mathfrak{H}_n^{(\alpha)}(x;u;\lambda)$ (ATFGP). For $u \in \mathbb{C}$, $u \neq 1$, the ATFGP of order $\alpha \in \mathbb{C}$ are defined by:

$$\left(\frac{(1-u)t}{\lambda e^t - u}\right)^\alpha e^{xt} = \sum_{n=0}^{\infty}\mathfrak{H}_n^{(\alpha)}(x;u;\lambda)\frac{t^n}{n!}, \tag{89}$$

which, for $\lambda = \alpha = 1$, reduce to the Frobenius–Genocchi polynomials $G_n^F(x;u)$ [29].

Using the previous approach, we introduce the three-variable Hermite–Apostol-type Frobenius–Genocchi polynomials (3VHATFGP) ${}_H\mathfrak{H}_n^{(\alpha)}(x,y,z;u;\lambda)$ of order $\alpha \in \mathbb{C}$ defined by:

$$\left(\frac{(1-u)t}{\lambda e^t - u}\right)^\alpha e^{xt+yt^2+zt^3} = \sum_{n=0}^{\infty} {}_H\mathfrak{H}_n^{(\alpha)}(x,y,z;u;\lambda)\frac{t^n}{n!}. \tag{90}$$

The special members related to the 3VHATFGP ${}_H\mathfrak{H}_n^{(\alpha)}(x,y,z;u;\lambda)$ can be obtained, and corresponding results for these polynomials and for their special cases can be obtained easily. Thus, we omit them.

6. Conclusions

In this paper, a multi-variable hybrid class of the Hermite–Apostol-type Frobenius–Euler polynomials is introduced and their properties are explored using various generating function methods. Several explicit and recurrence relations, summation formulae and symmetry identities are established for these hybrid polynomials. A brief view of the operational approach is also given for these polynomials. The operational representations combined with integral transforms may lead to other interesting results, which may be helpful to the theory of fractional calculus. Several techniques and methods are used in [30,31], which are applicable to the other fields of mathematics. The applicability of these techniques to the hybrid polynomial families can also be explored. These aspects will be undertaken in further investigation.

Author Contributions: All authors contributed equally

Funding: Dr. S. Araci was supported by the Research Fund of Hasan Kalyoncu University in 2018. This work has been done under Post-Doctoral Fellowship (Office Memo No.2/40(38)/2016/R&D-II/1063) awarded to Dr. M. Riyasat by the National Board of Higher Mathematics, Department of Atomic Energy, Government of India, Mumbai.

Conflicts of Interest: The authors declare no conflict of interest.

References

1. Dattoli, G.; Lorenzutta, S.; Maino, G.; Torre, A.; Cesarano, C. Generalized Hermite polynomials and super-Gaussian forms. *J. Math. Anal. Appl.* **1996**, *203*, 597–609. [CrossRef]
2. Cesarano, C. Operational methods and new identities for Hermite polynomials. *Math. Model. Nat. Phenom.* **2017**, *12*, 44–50. [CrossRef]
3. Cesarano, C.; Fornaro, C.; Vázquez, L. A note on a special class of Hermite polynomials. *Int. J. Pure Appl. Math.* **2015**, *98*, 261–273. [CrossRef]
4. Dattoli, G. Generalized polynomials operational identities and their applications. *J. Comput. Appl. Math.* **2000**, *118*, 111–123. [CrossRef]
5. Appell, P.; de Fériet, J.K. *Fonctions Hypergéométriques et Hypersphériques: Polynômes d' Hermite*; Gauthier-Villars: Paris, France, 1926.
6. Andrews, L.C. *Special Functions for Engineers and Applied Mathematicians*; Macmillan Publishing Company: New York, NY, USA, 1985.
7. Özarslan, M.A. Unified Apostol-Bernoulli, Euler and Genocchi polynomials. *Comput. Math. Appl.* **2011**, *62*, 2452–2462. [CrossRef]
8. Luo, Q.M. Apostol–Euler polynomials of higher order and the Gaussian hypergeometric function. *Taiwan. J. Math.* **2006**, *10*, 917–925. [CrossRef]
9. Erdélyi, A.; Magnus, W.; Oberhettinger, F.; Tricomi, F.G. *Higher Transcendental Functions*; McGraw-Hill Book Company: New York, NY, USA; Toronto, ON, Canada; London, UK, 1955; Volume III.
10. Carlitz, L. Eulerian numbers and polynomials. *Math. Mag.* **1959**, *32*, 247–260. [CrossRef]
11. Khan, S.; Yasmin, G.; Khan, R.; Hassan, N.A.M. Hermite-based Appell polynomials: Properties and applications. *J. Math. Anal. Appl.* **2009**, *351*, 756–764. [CrossRef]

12. Khan, S.; Riyasat, M. A determinantal approach to Sheffer-Appell polynomials via monomiality principle. *J. Math. Anal. Appl.* **2015**, *421*, 806–829. [CrossRef]
13. Goyal, S.P.; Laddha, R.K. On the generalized Riemann zeta functions and the generalized Lambert transform. *Ganita Sandesh* **1997**, *11*, 99–108.
14. Jang, G.-W.; Kwon, H.-I.; Kim, T. A note on degenerate Apostol-Bernoulli numbers and polynomials. *Adv. Stud. Contemp. Math. (Kyungshang)* **2017**, *27*, 279–288.
15. Kim, T. An identity of the symmetry for the Frobenius–Euler polynomials associated with the fermionic p-adic invariant q-integrals on Z_p. *Rocky Mt. J. Math.* **2011**, *41*, 239–247. [CrossRef]
16. Kim, T. Identities involving Frobenius–Euler polynomials arising from non-linear differential equations. *J. Number Theory* **2012**, *132*, 2854–2865. [CrossRef]
17. Kim, D.S.; Kim, T. Some new identities of Frobenius–Euler numbers and polynomials. *J. Inequal. Appl.* **2012**, *307*, 1–10. [CrossRef]
18. Kim, D.S.; Kim, T. Higher-order Frobenius–Euler and poly-Bernoulli mixed-type polynomials. *Adv. Differ. Equ.* **2013**, *251*, 13. [CrossRef]
19. Kim, T.; Kim, D.S. Identities for degenerate Bernoulli polynomials and Korobov polynomials of the first kind. *Sci. China Math.* **2018**. [CrossRef]
20. Kim, T.; Kim, D.S. An identity of symmetry for the degenerate Frobenius–Euler polynomials. *Math. Slov.* **2018**, *68*, 239–243. [CrossRef]
21. Kim, T.; Kwon, H.-I.; Seo, J.J. Some identities of degenerate Frobenius–Euler polynomials and numbers. *Proc. Jangjeon Math. Soc.* **2016**, *19*, 157–163.
22. Kim, T.; Mansour, T. Umbral calculus associated with Frobenius-type Eulerian polynomials. *Russ. J. Math. Phys.* **2014**, *21*, 484–493. [CrossRef]
23. Kurt, V. Some symmetry identities for the Apostol-type polynomials related to multiple alternating sums. *Adv. Differ. Equ.* **2013**, *32*, 1–32. [CrossRef]
24. Bayad, A.; Kim, T. Identities for Apostol-type Frobenius–Euler polynomials resulting from the study of a nonlinear operator. *Russ. J. Math. Phys.* **2016**, *23*, 164–171. [CrossRef]
25. Duran, U.; Acikgoz, M.; Araci, S. Hermite based poly-Bernoulli polynomials with a q-parameter. *Adv. Stud. Contemp. Math. (Kyungshang)* **2018**, *28*, 285–296.
26. Yang, S.L. An identity of symmetry for the Bernoulli polynomials. *Discrete Math.* **2008**, *308*, 550–554. [CrossRef]
27. Zhang, Z.; Yang, H. Several identities for the generalized Apostol-Bernoulli polynomials. *Comput. Math. Appl.* **2008**, *56*, 2993–2999. [CrossRef]
28. Milnor, J.W. *Topology from the Differentiable View Point*; University of Virginia Press: Charlottesville, VA, USA, 1965.
29. Yilmaz, B.; Özarslan, M.A. Frobenius–Euler and Frobenius-Genocchi polynomials and their differential equations. *New Trends Math. Sci.* **2015**, *3*, 172–180.
30. Marin, M.; Florea, O. On temporal behaviour of solutions in thermoelasticity of porous micropolar bodies. *An. St. Univ. Ovidius Constanta-Ser. Math.* **2014**, *22*, 169–188. [CrossRef]
31. Marin, M. Weak solutions in elasticity of dipolar porous materials. *Math. Probl. Eng.* **2008**, *2008*, 158908. [CrossRef]

© 2018 by the authors. Licensee MDPI, Basel, Switzerland. This article is an open access article distributed under the terms and conditions of the Creative Commons Attribution (CC BY) license (http://creativecommons.org/licenses/by/4.0/).

Article

Symmetric Properties of Carlitz's Type q-Changhee Polynomials

Yunjae Kim [1], Byung Moon Kim [2] and Jin-Woo Park [3],*

[1] Department of Mathematics, Dong-A University, Busan 49315, Korea; kimholzi@gmail.com
[2] Department of Mechanical System Engineering, Dongguk University, Gyeongju 38066, Korea; kbm713@dongguk.ac.kr
[3] Department of Mathematics Education, Daegu University, Gyeongsan 38066, Korea
* Correspondence: a0417001@knu.ac.kr; Tel.: +82-53-850-4212

Received: 23 October 2018; Accepted: 10 November 2018; Published: 13 November 2018

Abstract: Changhee polynomials were introduced by Kim, and the generalizations of these polynomials have been characterized. In our paper, we investigate various interesting symmetric identities for Carlitz's type q-Changhee polynomials under the symmetry group of order n arising from the fermionic p-adic q-integral on $\mathbb{Z}_p$.

Keywords: fermionic p-adic q-integral on $\mathbb{Z}_p$; q-Euler polynomials; q-Changhee polynomials; symmetry group

MSC: 33E20; 05A30; 11B65; 11S05

---

1. Introduction

For an odd prime number p, $\mathbb{Z}_p$, $\mathbb{Q}_p$, and $\mathbb{C}_p$ denote the ring of p-adic integers, the field of p-adic rational numbers, and the completions of algebraic closure of $\mathbb{Q}_p$, respectively, throughout this paper.

The p-adic norm is normalized as $|p|_p = \frac{1}{p}$, and let q be an indeterminate in $\mathbb{C}_p$ with $|q-1|_p < p^{-\frac{1}{p-1}}$. The q-analogue of number x is defined as

$$[x]_q = \frac{1-q^x}{1-q}. \tag{1}$$

Note that $\lim_{q \to 1} [x]_q = x$ for each $x \in \mathbb{Z}_p$.

Let $C(\mathbb{Z}_p) = \{f | f : \mathbb{Z}_p \longrightarrow \mathbb{R} \text{ is continuous}\}$. Then, a fermionic p-adic q-integral of f ($\in C(\mathbb{Z}_p)$) is defined by Kim as [1–6]:

$$I_{-q}(f) = \int_{\mathbb{Z}_p} f(x) d\mu_{-q}(x) = \lim_{N \to \infty} \sum_{x=0}^{p^N - 1} f(x) \mu_{-q}\left(x + p^N \mathbb{Z}_p\right)$$

$$= \lim_{N \to \infty} \frac{1}{[p^N]_{-q}} \sum_{x=0}^{p^N - 1} f(x)(-q)^x \tag{2}$$

$$= \lim_{N \to \infty} \frac{[2]_q}{2} \sum_{x=0}^{p^N - 1} f(x)(-q)^x.$$

On the other hand, it is well known that the Euler polynomial $E_n(x)$ is given by the Appell sequence with $g(t) = \frac{1}{2}(e^t + 1)$, giving the the generating function

$$\frac{2}{e^t + 1} e^{xt} = \sum_{n=0}^{\infty} E_n(x) \frac{t^n}{n!},$$

(see [7–17]). In particular, if $x = 0$, $E_n = E_n(0)$ ($n \in \mathbb{N}$) is called the Euler number.

As a q-analogue of Euler polynomials, the Carlitz's type q-Euler polynomial $\mathcal{E}_{n,q}(x)$ is defined by

$$\sum_{n=0}^{\infty} \mathcal{E}_{n,q}(x) \frac{t^n}{n!} = \int_{\mathbb{Z}_p} e^{[x+y]_q t} d\mu_{-q}(y), \qquad (3)$$

(see [2,13–17]). In particular, if $x = 0$, $\mathcal{E}_{n,q} = \mathcal{E}_{n,q}(0)$ is called the q-Euler number.

By (3), the Carlitz's type q-Euler polynomial $\mathcal{E}_{n,q}(x)$ is obtained as

$$\mathcal{E}_{n,q}(x) = \int_{\mathbb{Z}_p} [x+y]_q^n d\mu_{-q}(y), \ (n \geq 0). \qquad (4)$$

From the fermionic p-adic q-integral on $\mathbb{Z}_p$, the degenerate q-Euler polynomial $\mathcal{E}_{n,\lambda,q}(x)$ is defined as [16]:

$$\sum_{n=0}^{\infty} \mathcal{E}_{n,\lambda,q}(x) \frac{t^n}{n!} = \int_{\mathbb{Z}_p} (1+\lambda t)^{\frac{[x+y]_q}{\lambda}} d\mu_{-q}(y). \qquad (5)$$

By the binomial expansion of $(1+\lambda t)^{\frac{[x+y]_q}{\lambda}}$, we get

$$\int_{\mathbb{Z}_p} (1+\lambda t)^{\frac{[x+y]_q}{\lambda}} d\mu_{-q}(y) = \sum_{n=0}^{\infty} \lambda^n \int_{\mathbb{Z}_p} \left(\frac{[x+y]_q}{\lambda}\right)_n d\mu_{-q}(y) \frac{t^n}{n!}, \qquad (6)$$

where $(\alpha)_n = \alpha(\alpha-1)\cdots(\alpha-n+1)$ for $n \in \mathbb{N}$, and by (5) and (6), we have

$$\mathcal{E}_{n,\lambda,q}(x) = \lambda^n \int_{\mathbb{Z}_p} \left(\frac{[x+y]_q}{\lambda}\right)_n d\mu_{-q}(y), \ (n \in \mathbb{N}). \qquad (7)$$

Since

$$(\alpha)_n = \alpha(\alpha-1)\cdots(\alpha-n+1) = \sum_{l=0}^{n} S_1(n,l) \alpha^l, \qquad (8)$$

$$\mathcal{E}_{n,\lambda,q}(x) = \lambda^n \sum_{l=0}^{n} S_1(n,l) \int_{\mathbb{Z}_p} \left(\frac{[x+y]_q}{\lambda}\right)^l d\mu_{-q}(y)$$

$$= \sum_{l=0}^{n} \lambda^{n-l} S_1(n,l) \mathcal{E}_{l,q}(x),$$

where $S_1(n,m)$ is the Stirling number of the first kind (see [2,7,8,12,17,18]).

Now, we apply these polynomials to Changhee polynomials, introduced by Kim et al. [19]. The Changhee polynomial of the first kind $Ch_n(x)$ is defined by the generating function to be

$$\sum_{n=0}^{\infty} Ch_n(x) \frac{t^n}{n!} = \int_{\mathbb{Z}_p} (1+t)^{x+y} d\mu_{-1}(y)$$

$$= \frac{2}{2+t} (1+t)^x. \qquad (9)$$

(see [20,21]).

In view point of (3) and (9), Carlitz's type q-Changhee polynomial $Ch_{n,q}(x)$ is defined by

$$\sum_{n=0}^{\infty} Ch_{n,q}(x)\frac{t^n}{n!} = \int_{\mathbb{Z}_p} (1+t)^{[x+y]_q} d\mu_{-q}(y), \qquad (10)$$

(see [18,22]).

By the binomial expansion of $(1+t)^{[x+y]_q}$,

$$\sum_{n=0}^{\infty} Ch_{n,q}(x)\frac{t^n}{n!} = \int_{\mathbb{Z}_p} (1+t)^{[x+y]_q} d\mu_{-q}(y)$$
$$= \sum_{n=0}^{\infty} \int_{\mathbb{Z}_p} ([x+y]_q)_n \, d\mu_{-q}(y)\frac{t^n}{n!}, \qquad (11)$$

and so the equation (10) and (11) yield the following:

$$Ch_{n,q}(x) = \int_{\mathbb{Z}_p} ([x+y]_q)_n \, d\mu_{-q}(y), \qquad (12)$$

(see [20,21]).

In the past decade, many different generalizations of Changhee polynomials have been studied (see [19,20,22–32]), and the relationship between important combinatorial polynomials and those polynomials was found.

Symmetric identities of special polynomials are important and interesting in number theory, pure and applied mathematics. Symmetric identities of many different polynomials were investigated in [5,10,14,16,32–39]. In particular, C. Cesarano [40] presented some techniques regarding the generating functions used, and these identities can be applicable to the theory of porous materials [41].

In current paper, we construct symmetric identities for the Carlitz's type q-Changhee polynomials under the symmetry group of order n arising from the fermionic p-adic q-integral on $\mathbb{Z}_p$, and the proof methods which was used in the Kim's previous researches are also used as good tools in this paper (see [5,10,14,16,32–39]).

2. Symmetric Identities for the Carlitz's Type q-Changhee Polynomials

Let $t \in \mathbb{C}_p$ with $|t|_p < p^{-\frac{1}{p-1}}$, and let S_n be the symmetry group of degree n. For positive integers $w_1, w_2, \ldots, w_n$ with $w_i \equiv 1 \pmod{2}$ for each $i = 1, 2, \ldots n$, we consider the following integral equation for the fermionic p-adic q-integral on $\mathbb{Z}_p$;

$$\int_{\mathbb{Z}_p} (1+t)^{\left[(\prod_{i=1}^{n-1} w_i)y + (\prod_{i=1}^{n} w_i)x + w_n \sum_{i=1}^{n-1} \left(\prod_{\substack{j=1 \\ j \neq i}}^{n-1} w_j\right) k_i\right]_q} d\mu_{-q^{w_1 w_2 \cdots w_{n-1}}}(y)$$

$$= \frac{[2]_{q^{w_1 \cdots w_{n-1}}}}{2} \lim_{N \to \infty} \sum_{m=0}^{w_n-1} \sum_{y=0}^{p^N-1} (1+t)^{\left[(\prod_{i=1}^{n-1} w_i)(m+w_n y) + (\prod_{i=1}^{n} w_i)x + w_n \sum_{i=1}^{n-1} \left(\sum_{\substack{j=1 \\ j \neq i}}^{n-1} w_j\right) k_i\right]_q} \qquad (13)$$

$$\times (-1)^{m+w_n y} q^{w_1 w_2 \cdots w_{n-1}(m+w_n y)}.$$

From (13), we get

$$\frac{2}{[2]_{q^{w_1 w_2 \cdots w_{n-1}}}} \prod_{m=1}^{n-1} \sum_{k_m=0}^{w_m-1} (-1)^{\sum_{i=1}^{n-1} k_i} q^{w_n \sum_{j=1}^{n-1} \left(\prod_{\substack{i=1\\i\neq j}}^{n-1} w_i\right) k_j}$$

$$\times \int_{\mathbb{Z}_p} (1+t)^{\left[\left(\prod_{i=1}^{n-1} w_i\right)y + \left(\prod_{i=1}^{n} w_i\right)x + w_n \sum_{i=1}^{n-1} \left(\prod_{\substack{j=1\\j\neq i}}^{n-1} w_j\right) k_i\right]} q \, d\mu_{-q^{w_1 w_2 \cdots w_{n-1}}}(y) \quad (14)$$

$$= \lim_{N\to\infty} \prod_{m=1}^{n-1} \sum_{k_m=0}^{w_m-1} \sum_{l=0}^{w_n-1} \sum_{y=0}^{p^N-1} (1+t)^{\left[\left(\prod_{i=1}^{n-1} w_i\right)(m+w_n y) + \left(\prod_{i=1}^{n} w_i\right)x + w_n \sum_{i=1}^{n} \left(\sum_{j=1}^{n-1} w_j\right) k_i\right]} q$$

$$\times (-1)^{\sum_{i=1}^{n-1} k_i + l + y} q^{\left(\sum_{j=1}^{n-1}\right)l + \left(\prod_{j=1}^{n} w_j\right) y + w_n \sum_{j=1}^{n-1} \left(\prod_{\substack{i=1\\i\neq j}}^{n-1} w_i\right) k_i}.$$

If we put

$$F(w_1, w_2, \ldots, w_n) = \frac{2}{[2]_{q^{w_1 w_2 \cdots w_{n-1}}}} \prod_{m=1}^{n-1} \sum_{k_m=0}^{w_m-1} (-1)^{\sum_{i=1}^{n-1} k_i} q^{w_n \sum_{j=1}^{n-1} \left(\prod_{\substack{i=1\\i\neq j}}^{n-1} w_i\right) k_j}$$

$$\times \int_{\mathbb{Z}_p} (1+t)^{\left[\left(\prod_{i=1}^{n-1} w_i\right)y + \left(\prod_{i=1}^{n} w_i\right)x + w_n \sum_{i=1}^{n-1} \left(\prod_{\substack{j=1\\j\neq i}}^{n-1} w_j\right) k_i\right]} q \, d\mu_{-q^{w_1 w_2 \cdots w_{n-1}}}(y), \quad (15)$$

then, by (14), we know that $F(w_1, w_2, \ldots, w_n)$ is invariant for any permutation $\sigma \in S_n$.

Hence, by (14) and (15), we obtain the following theorem.

Theorem 1. *Let $w_1, w_2, \ldots, w_n$ be positive odd integers. For any $\sigma \in S_n$, $F(w_{\sigma(1)}, w_{\sigma(2)}, \ldots, w_{\sigma(n)})$ have the same value.*

By (1), we know that

$$\left[\prod_{i=1}^{n-1} w_i\right]_q \left[y + w_n x + w_n \sum_{i=1}^{n-1} \frac{k_i}{w_i}\right]_{q^{w_1 w_2 \cdots w_{n-1}}} = \left[\left(\prod_{i=1}^{n-1} w_i\right) y + \left(\prod_{i=1}^{n} w_i\right) x + w_n \sum_{i=1}^{n-1} \left(\prod_{\substack{j=1\\j\neq i}}^{n-1} w_j\right) k_i\right]_q. \quad (16)$$

From (5) and (16), we derive the following identities.

$$\int_{\mathbb{Z}_p} (1+t)^{\left[(\prod_{i=1}^{n-1} w_i)y + (\prod_{i=1}^{n} w_i)x + w_n \sum_{i=1}^{n-1} \left(\prod_{\substack{j=1 \\ j \neq i}}^{n-1} w_j\right) k_i\right]_{q^{w_1 w_2 \cdots w_{n-1}}}} d\mu_{-q^{w_1 w_2 \cdots w_{n-1}}}(y)$$

$$= (1+t)^{[\prod_{i=1}^{n-1} w_i]_q} \int_{\mathbb{Z}_p} (1+t)^{\left[y + w_n x + w_n \sum_{i=1}^{n-1} \frac{k_i}{w_i}\right]_{q^{w_1 w_2 \cdots w_{n-1}}}} d\mu_{-q^{w_1 w_2 \cdots w_{n-1}}}(y)$$

$$= \left(\sum_{l=0}^{\infty} \binom{[\prod_{i=1}^{n-1} w_i]_q}{l} t^l\right) \left(\sum_{m=0}^{\infty} Ch_{m,q^{w_1 w_2 \cdots w_{n-1}}}\left(w_n x + w_n \sum_{i=1}^{n-1} \frac{k_i}{w_i}\right) \frac{t^m}{m!}\right) \quad (17)$$

$$= \sum_{m=0}^{\infty} \left(\sum_{r=0}^{m} \left(\left[\prod_{i=1}^{n-1} w_i\right]_q\right)_{m-r} \binom{m}{r} Ch_{r,q^{w_1 w_2 \cdots w_{n-1}}}\left(w_n x + w_n \sum_{i=1}^{n-1} \frac{k_i}{w_i}\right)\right) \frac{t^m}{m!},$$

for each positive integer n. Thus, by Theorem 1 and (17), we obtain the following corollary.

Corollary 1. Let $w_1, w_2, \ldots, w_n$ be positive integers with $w_i \equiv 1 \pmod{2}$ for each $i = 1, 2, \ldots, n$, and let m be a nonnegative integer. Then, for any permutation $\tau \in S_n$,

$$\frac{2}{[2]_{q^{w_{\tau(1)} w_{\tau(2)} \cdots w_{\tau(n)}}}} \sum_{r=0}^{m} \prod_{l=1}^{n-1} \sum_{k_l=0}^{w_{\tau(l)}-1} (-1)^{\sum_{i=1}^{n-1} k_i} q^{w_{\tau(n)} \sum_{i=1}^{n-1} \left(\sum_{\substack{j=1 \\ j\neq i}}^{n-1} w_{\tau(j)}\right) k_i}$$

$$\times \left(\left[\prod_{i=1}^{n-1} w_{\tau(i)}\right]_q\right)_{m-r} \binom{m}{r} Ch_{r,q^{w_{\tau(1)} w_{\tau(2)} \cdots w_{\tau(n-1)}}}\left(w_{\tau(n)} x + w_{\tau(n)} \sum_{i=1}^{n-1} \frac{k_i}{w_{\tau(i)}}\right)$$

have the same expressions.

Note that, by the definition of $[x]_q$,

$$\left[y + w_n x + w_n \sum_{i=1}^{n-1} \frac{k_i}{w_i}\right]_{q^{w_1 w_2 \cdots w_{n-1}}}$$

$$= \frac{[w_n]_q}{[\prod_{i=1}^{n-1} w_i]_q} \left[\sum_{i=1}^{n-1} \left(\prod_{\substack{j=1 \\ j\neq i}}^{n-1} w_j\right) k_i\right]_{q^{w_n}} + q^{w_n \sum_{i=1}^{n-1} \left(\prod_{\substack{j=1 \\ j\neq i}}^{n-1} w_j\right) k_i} [y + w_n x]_{q^{w_1 w_2 \cdots w_{n-1}}}. \quad (18)$$

By (12), we get

$$Ch_{m,q^{w_1 w_2 \cdots w_{n-1}}}\left(w_n x + w_n \sum_{i=1}^{n-1} \frac{k_i}{w_i}\right)$$

$$= \int_{\mathbb{Z}_p} \left(\left[y + w_n x + w_n \sum_{i=1}^{n-1} \frac{k_i}{w_i}\right]_{q^{w_1 w_2 \cdots w_{n-1}}}\right)_m d\mu_{-q^{w_1 w_2 \cdots w_{n-1}}}, \quad (19)$$

and by (8) and (18),

$$\left(\left[y + w_n x + w_n \sum_{i=1}^{n-1} \frac{k_i}{w_i}\right]_{q^{w_1 w_2 \cdots w_{n-1}}}\right)_m$$

$$= \left(\frac{[w_n]_q}{\left[\prod_{i=1}^{n-1} w_i\right]_q} \left[\sum_{i=1}^{n-1} \left(\prod_{\substack{j=1 \\ j \neq i}}^{n-1} w_j\right) k_i\right]_{q^{w_n}} + q^{w_n \sum_{i=1}^{n-1} \left(\prod_{\substack{j=1 \\ j \neq i}}^{n-1} w_j\right) k_i} [y + w_n x]_{q^{w_1 w_2 \cdots w_{n-1}}}\right)_m$$

$$= \sum_{l=0}^{m} S_1(m, l) \left(\frac{[w_n]_q}{\left[\prod_{i=1}^{n-1} w_i\right]_q} \left[\sum_{i=1}^{n-1} \left(\prod_{\substack{j=1 \\ j \neq i}}^{n-1} w_j\right) k_i\right]_{q^{w_n}} + q^{w_n \sum_{i=1}^{n-1} \left(\prod_{\substack{j=1 \\ j \neq i}}^{n-1} w_j\right) k_i} [y + w_n x]_{q^{w_1 w_2 \cdots w_{n-1}}}\right)^l \quad (20)$$

$$= \sum_{l=0}^{m} S_1(m, l) \sum_{i=1}^{l} \binom{l}{i} \left(\frac{[w_n]_q}{\left[\prod_{i=1}^{n-1} w_i\right]_q}\right)^{l-i} \left[\sum_{i=1}^{n-1} \left(\prod_{\substack{j=1 \\ j \neq i}}^{n-1} w_j\right) k_i\right]_{q^{w_n}}^{l-i}$$

$$\times q^{i w_n \sum_{i=1}^{n-1} \left(\prod_{\substack{j=1 \\ j \neq i}}^{n-1} w_j\right) k_i} [y + w_n x]_{q_1^w w_2 \cdots w_{n-1}}^i.$$

From (4), (19) and (20), we have

$$Ch_{m, q^{w_1 w_2 \cdots w_{n-1}}}\left(w_n x + w_n \sum_{i=1}^{n-1} \frac{k_i}{w_i}\right)$$

$$= \sum_{l=0}^{m} S_1(m, l) \sum_{i=1}^{l} \binom{l}{i} \left(\frac{[w_n]_q}{\left[\prod_{i=1}^{n-1} w_i\right]_q}\right)^{l-i} \left[\sum_{i=1}^{n-1} \left(\prod_{\substack{j=1 \\ j \neq i}}^{n-1} w_j\right) k_i\right]_{q^{w_n}}^{l-i}$$

$$\times q^{i w_n \sum_{i=1}^{n-1} \left(\prod_{\substack{j=1 \\ j \neq i}}^{n-1} w_j\right) k_i} \int_{\mathbb{Z}_p} [y + w_n x]_{q^{w_1 w_2 \cdots w_{n-1}}}^i d\mu_{-q^{w_1 w_2 \cdots w_{n-1}}}(y) \quad (21)$$

$$= \sum_{l=0}^{m} S_1(m, l) \sum_{i=1}^{l} \binom{l}{i} \left(\frac{[w_n]_q}{\left[\prod_{i=1}^{n-1} w_i\right]_q}\right)^{l-i} \left[\sum_{i=1}^{n-1} \left(\prod_{\substack{j=1 \\ j \neq i}}^{n-1} w_j\right) k_i\right]_{q^{w_n}}^{l-i}$$

$$\times q^{i w_n \sum_{i=1}^{n-1} \left(\prod_{\substack{j=1 \\ j \neq i}}^{n-1} w_j\right) k_i} Ch_{i, q^{w_1 w_2 \cdots w_{n-1}}}(w_n x).$$

From (21), we have

$$\frac{2}{[2]_{q^{w_1 w_2 \cdots w_n}}} \sum_{r=0}^{m} \prod_{l=1}^{n-1} \sum_{k_l=0}^{w_l-1} (-1)^{\sum_{i=1}^{n-1} k_i} q^{w_n \sum_{i=1}^{n-1} \left(\sum_{\substack{j=1 \\ j \neq i}}^{n-1} w_j \right) k_i}$$

$$\times \left(\left[\prod_{i=1}^{n-1} w_i \right]_q \right)_{m-r} \binom{m}{r} Ch_{r, q^{w_1 w_2 \cdots w_{n-1}}} \left(w_n x + w_n \sum_{i=1}^{n-1} \frac{k_i}{w_i} \right)$$

$$= \frac{2}{[2]_{q^{w_1 w_2 \cdots w_n}}} \sum_{r=0}^{m} \prod_{l=1}^{n-1} \sum_{k_l=0}^{w_l-1} (-1)^{\sum_{i=1}^{n-1} k_i} q^{w_n \sum_{i=1}^{n-1} \left(\sum_{\substack{j=1 \\ j \neq i}}^{n-1} w_j \right) k_i} \left(\left[\prod_{i=1}^{n-1} w_i \right]_q \right)_{m-r} \binom{m}{r}$$

$$\times \sum_{p=0}^{r} S_1(r, p) \sum_{i=1}^{l} \binom{l}{i} \left(\frac{[w_n]_q}{\left[\prod_{i=1}^{n-1} w_i \right]_q} \right)^{p-i} \left[\sum_{i=1}^{n-1} \left(\prod_{\substack{j=1 \\ j \neq i}}^{n-1} w_j \right) k_i \right]_{q^{w_n}}^{p-i}$$

$$\times q^{i w_n \sum_{i=1}^{n-1} \left(\prod_{\substack{j=1 \\ j \neq i}}^{n-1} w_j \right) k_i} Ch_{i, q^{w_1 w_2 \cdots w_{n-1}}}(w_n x) \quad (22)$$

$$= \sum_{r=0}^{m} \sum_{l=0}^{r} \sum_{i=1}^{l} S_1(r, l) \binom{m}{r} \binom{l}{i} \left(\frac{[w_n]_q}{\left[\prod_{i=1}^{n-1} w_i \right]_q} \right)^{p-i} \left(\left[\prod_{i=1}^{n-1} w_i \right]_q \right)_{m-r} Ch_{i, q^{w_1 w_2 \cdots w_{n-1}}}(w_n x)$$

$$\times \frac{2}{[2]_{q^{w_1 w_2 \cdots w_n}}} \prod_{l=1}^{n-1} \sum_{k_l=0}^{w_l-1} (-1)^{\sum_{i=1}^{n-1} k_i} q^{(1+i) w_n \sum_{i=1}^{n-1} \left(\sum_{\substack{j=1 \\ j \neq i}}^{n-1} w_j \right) k_i} \left[\sum_{i=1}^{n-1} \left(\prod_{\substack{j=1 \\ j \neq i}}^{n-1} w_j \right) k_i \right]_{q^{w_n}}^{p-i}$$

$$= \sum_{r=0}^{m} \sum_{l=0}^{r} \sum_{i=1}^{l} S_1(r, l) \binom{m}{r} \binom{l}{i}$$

$$\times \left(\frac{[w_n]_q}{\left[\prod_{i=1}^{n-1} w_i \right]_q} \right)^{p-i} \left(\left[\prod_{i=1}^{n-1} w_i \right]_q \right)_{m-r} Ch_{i, q^{w_1 w_2 \cdots w_{n-1}}}(w_n x) F_{n, q^{w_n}}(w_1, \ldots, w_{n-1} | i+1),$$

where

$$F_{n, q}(w_1, \ldots, w_{n-1} | i) = \frac{2}{[2]_{q^{w_1 w_2 \cdots w_n}}} \prod_{l=1}^{n-1} \sum_{k_l=0}^{w_l-1} (-1)^{\sum_{i=1}^{n-1} k_i} q^{i \sum_{i=1}^{n-1} \left(\sum_{\substack{j=1 \\ j \neq i}}^{n-1} w_j \right) k_i} \left[\sum_{t=1}^{n-1} \left(\prod_{\substack{j=1 \\ j \neq t}}^{n-1} w_j \right) k_t \right]_q^{p-i-1}$$

Theorem 2. *For each nonnegative odd integers $w_1, w_2, \ldots, w_n$ and for any permutation σ in the symmetry group of degree n, the expressions*

$$\sum_{r=0}^{m} \sum_{l=0}^{r} \sum_{i=1}^{l} S_1(r, l) \binom{m}{r} \binom{l}{i} \left(\frac{[w_{\sigma(n)}]_q}{\left[\prod_{i=1}^{n-1} w_{\sigma(i)} \right]_q} \right)^{p-i} \left(\left[\prod_{i=1}^{n-1} w_{\sigma(i)} \right]_q \right)_{m-r}$$

$$\times Ch_{i, q^{w_{\sigma(1)} w_{\sigma(2)} \cdots w_{\sigma(n-1)}}} \left(w_{\sigma(n)} x \right) F_{n, q^{w_{\sigma(n)}}} \left(w_{\sigma(1)}, \ldots, w_{\sigma(n-1)} | i+1 \right)$$

have the same.

3. Conclusion

The Changhee numbers are closely related with the Euler numbers, the Stirling numbers of the first kind and second kind and the harmonic numbers, and so on. Throughout this paper, we investigate that the function $F(w_{\sigma(1)}, w_{\sigma(2)}, \ldots, w_{\sigma(n)})$ for the Carlitz's type q-Changhee polynomials is invariant under the symmetry group $\sigma \in S_n$. From the invariance of $F(w_{\sigma(1)}, w_{\sigma(2)}, \ldots, w_{\sigma(n)})$, $\sigma \in S_n$, we construct symmetric identities of the Carlitz's type q-Changhee polynomials from the fermionic p-adic q-integral on $\mathbb{Z}_p$. As Bernoulli and Euler polynomials, our properties on the Carlitz's type q-Changhee polynomials play an crucial role in finding identities for numbers in algebraic number theory.

Author Contributions: All authors contributed equally to this work; All authors read and approved the final manuscript.

Funding: This research was supported by the Daegu University Research Grant, 2018.

Acknowledgments: The authors would like to thank the referees for their valuable and detailed comments which have significantly improved the presentation of this paper.

Conflicts of Interest: The authors declare no conflict of interest.

References

1. Kim, T. q-Euler numbers and polynomials associated with p-adic q-integrals. *J. Nonlinear Math. Phys.* **2007**, *14*, 15–27. [CrossRef]
2. Kim, T. Some identities on the q-Euler polynomials of higher order and q-Stirling numbers by the fermionic p-adic integral on $\mathbb{Z}_p$. *Russ. J. Math. Phys.* **2009**, *16*, 484–491. [CrossRef]
3. Kim, T. On q-analogue of the p-adic log gamma functions and related integral. *J. Number Theory* **1999**, *76*, 320–329. [CrossRef]
4. Kim, T. q-Volkenborn integration. *Russ. J. Math. Phys.* **2002**, *9*, 288–299.
5. Kim, T. Symmetry of power sum polynomials and multivariate fermionic p-adic invariant integral on $\mathbb{Z}_p$. *Russ. J. Math. Phys.* **2009**, *16*, 93–96. [CrossRef]
6. Kim, D.S.; Kim, T. Some p-adic integrals on $\mathbb{Z}_p$ associated with trigonometric Functions. *Russ. J. Math. Phys.* **2018**, *25*, 300–308. [CrossRef]
7. Kim, T. A study on the q-Euler numbers and the fermionic q-integrals of the product of several type q-Bernstein polynomials on $\mathbb{Z}_p$. *Adv. Stud. Contemp. Math.* **2013**, *23*, 5–11.
8. Bayad, A.; Kim, T. Identities involving values of Bernstein, q-Bernoulli, and q-Euler polynomials. *Russ. J. Math. Phys.* **2011**, *18*, 133–143. [CrossRef]
9. Gaboury, S.; Tremblay, R.; Fugere, B.-J. Some explicit formulas for certain new classes of Bernoulli, Euler and Genocchi polynomials. *Proc. Jangjeon Math. Soc.* **2014**, *17*, 115–123.
10. Kim, D.S.; Kim, T. Some symmetric identites for the higher-order q-Euler polynomials related to symmetry group S_3 arising from p-adic q-fermionic integral on $\mathbb{Z}_p$. *Filomat* **2016**, *30*, 1717–1721. [CrossRef]
11. Sharma, A. q-Bernoulli and Euler numbers of higher order. *Duke Math. J.* **1958**, *25*, 343–353. [CrossRef]
12. Srivastava, H. Some generalizations and basic (or q-)extensions of the Bernoulli, Euler and Genocchi polynomials. *Appl. Math. Inf. Sci.* **2011**, *5*, 390–444.
13. Zhang, Z.; Yang, H. Some closed formulas for generalized Bernoulli-Euler numbers and polynomials. *Proc. Jangjeon Math. Soc.* **2008**, *11*, 191–198.
14. Kim, T.; Kim, D.S. Identities of symmetry for degenerate Euler polynomials and alternating generalized falling factorial sums. *Iran. J. Sci. Technol. Trans. A Sci.* **2017**, *41*, 939–949. [CrossRef]
15. Carlitz, L. q-Bernoulli and Eulerian numbers. *Trans. Amer. Math. Soc.* **1954**, *76*, 332–350.
16. Kim, D.S.; Kim, T. Symmetric identities of higher-order degenerate q-Euler polynomials. *J. Nonlinear Sci. Appl.* **2016**, *9*, 443–451. [CrossRef]
17. Comtet, L. *Advanced Combinatorics*; Reidel: Dordrecht, The Netherlands, 1974.
18. Dolgy, D.V.; Jang, G.W.; Kwon, H.I.; Kim, T. A note on Carlitzs type q-Changhee numbers and polynomials. *Adv. Stud. Contemp. Math.* **2017**, *27*, 451–459.

19. Kim, D.S.; Kim, T.; Seo, J.J. A note on Changhee polynomials and numbers. *Adv. Stud. Theor. Phys.* **2013**, *7*, 993–1003. [CrossRef]
20. Kim, T.; Kim, D.S. A note on nonlinear Changhee differential equations. *Russ. J. Math. Phys.* **2016**, *23*, 88–92. [CrossRef]
21. Kim, T.; Kim, D.S. Identities for degenerate Bernoulli polynomials and Korobov polynomials of the first kind. *Sci. China Math.* **2018**. [CrossRef]
22. Kim, B.M.; Jang, L.C.; Kim, W.; Kwon, H.I. On Carlitz's Type Modified Degenerate Changhee Polynomials and Numbers. *Discrete Dyn. Nat. Soc.* **2018**, *2018*, 9520269. [CrossRef]
23. Kim, T.; Kim, D.S. Differential equations associated with degenerate Changhee numbers of the second kind. *Rev. R. Acad. Cienc. Exactas Fis. Nat. Ser. A Math. RACSAM* **2018**. [CrossRef]
24. Moon, E.J.; Park, J.W. A note on the generalized q-Changhee numbers of higher order. *J. Comput. Anal. Appl.* **2016**, *20*, 470–479.
25. Kwon, J.; Park, J.W. On modified degenerate Changhee polynomials and numbers. *J. Nonlinear Sci. Appl.* **2015**, *18*, 295–305. [CrossRef]
26. Kim, T.; Kwon, H.I.; Seo, J.J. Degenerate q-Changhee polynomials. *J. Nonlinear Sci. Appl.* **2016**, *9*, 2389–2393. [CrossRef]
27. Kwon, H.I.; Kim, T.; Seo, J.J. A note on degenerate Changhee numbers and polynomials. *Proc. Jangjeon Math. Soc.* **2015**, *18*, 295–305.
28. Arici, S.; Ağyüz, E.; Acikgoz, M. On a q-analogue of some numbers and polynomials. *J. Ineqal. Appl.* **2015**, *2015*, 19. [CrossRef]
29. Pak, H.K.; Jeong, J.; Kang, D.J.; Rim, S.H. Changhee-Genocchi numbers and their applications. *ARS Combin.* **2018**, *136*, 153–159.
30. Kim, B.M.; Jeong, J.; Rim, S.H. Some explicit identities on Changhee-Genocchi polynomials and numbers. *Adv. Differ. Equ.* **2016**, *2016*, 202. [CrossRef]
31. Rim, S.H.; Park, J.W.; Pyo, S.S.; Kwon, J. The n-th twisted Changhee polynomials and numbers. *Bull. Korean Math. Soc.* **2015**, *52*, 741–749. [CrossRef]
32. Simsek, Y. Identities on the Changhee numbers and Apostol-type Daehee polynomials. *Adv. Stud. Contemp. Math.* **2017**, *27*, 199–212.
33. Kim, T.; Kim, D.S. Degenerate Bernstein polynomials. *Rev. R. Acad. Cienc. Exactas Fis. Nat. Ser. A Math. RACSAM* **2018**. [CrossRef]
34. Liu, C.; Bao, W. Application of Probabilistic Method on Daehee Sequences. *Eur. J. Pure Appl. Math.* **2018**, *11*, 69–78. [CrossRef]
35. Kim, D.S.; Lee, N.; Na, J.; Park, K.H. Abundant symmetry for higher-order Bernoulli polynomials (I). *Adv. Stud. Contemp. Math.* **2013**, *23*, 461–482.
36. Kim, D.S.; Lee, N.; Na, J.; Pak, K.H. Identities of symmetry for higher-order Euler polynomials in three variables (I). *Adv. Stud. Contemp. Math.* **2012**, *22*, 51–74. [CrossRef]
37. Kim, D.S.; Lee, N.; Na, J.; Pak, K.H. Identities of symmetry for higher-order Bernoulli polynomials in three variables (II). *Proc. Jangjeon Math. Soc.* **2013**, *16*, 359–378.
38. Kim, D.S. Symmetry identities for generalized twisted Euler polynomials twisted by unramified roots of unity. *Proc. Jangjeon Math. Soc.* **2012**, *15*, 303–316.
39. Kim, T.; Kim, D.S. An identity of symmetry for the degenerate Frobenius-Euler polynomials. *Math. Slovaca* **2008**, *68*, 239–243. [CrossRef]
40. Cesarano, C. Operational methods and new identities for Hermit polynomials. *Math. Model. Nat. Phenom.* **2017**, *12*, 44–50. [CrossRef]
41. Marin, M. Weak solutions in elasticity of dipolar porous materials. *Math. Probl. Eng.* **2008**, *2008*, 158908. [CrossRef]

© 2018 by the authors. Licensee MDPI, Basel, Switzerland. This article is an open access article distributed under the terms and conditions of the Creative Commons Attribution (CC BY) license (http://creativecommons.org/licenses/by/4.0/).

Article

On Classical Gauss Sums and Some of Their Properties

Li Chen

School of Mathematics, Northwest University, Xi'an 710127, China; cl1228@stumail.nwu.edu.cn

Received: 18 October 2018; Accepted: 7 November 2018; Published: 11 November 2018

Abstract: The goal of this paper is to solve the computational problem of one kind rational polynomials of classical Gauss sums, applying the analytic means and the properties of the character sums. Finally, we will calculate a meaningful recursive formula for it.

Keywords: third-order character; classical Gauss sums; rational polynomials; analytic method; recursive formula

2010 Mathematics Subject Classification: 11L05, 11L07

1. Introduction

Let $q \geq 3$ be an integer. For any Dirichlet character χ mod q, according to the definition of classical Gauss sums $\tau(\chi)$, we can write

$$\tau(\chi) = \sum_{a=1}^{q} \chi(a) e\left(\frac{a}{q}\right),$$

where $e(y) = e^{2\pi i y}$.

Since this sum appears in numerous classical number theory problems, and it has a close connection with the trigonometric sums, we believe that classical Gauss sums play a crucial part in analytic number theory. Because of this phenomenon, plenty of experts have researched Gauss sums. Meanwhile, more conclusions have been obtained as regards their arithmetic properties. Such as the following results provided by Chen and Zhang [1]:

Let p be an odd prime with $p \equiv 1 \mod 4$, λ be any fourth-order character mod p. Then one has the identity

$$\tau^2(\lambda) + \tau^2\left(\overline{\lambda}\right) = \sqrt{p} \cdot \sum_{a=1}^{p-1} \left(\frac{a + \overline{a}}{p}\right) = 2\sqrt{p} \cdot \alpha,$$

where $\left(\frac{*}{p}\right) = \chi_2$ denotes the the Legendre's symbol mod p (please see Reference [1,2] for its definition and related properties), and $\alpha = \sum_{a=1}^{\frac{p-1}{2}} \left(\frac{a + \overline{a}}{p}\right)$.

If p is a prime with $p \equiv 1 \mod 3$, ψ is any third-order character mod p, then Zhang and Hu [3] had already obtained an analogous result (see Lemma 1). However, perhaps the most beautiful and important property of Gauss sums $\tau(\chi)$ is that $|\tau(\chi)| = \sqrt{q}$, for any primitive character χ mod q.

Reference [2] and References [4–13] have a good deal of various elementary properties of Gauss sums. In this paper, the following rational polynomials of Gauss sums attract our attention.

$$U_k(p,\chi) = \frac{\tau^{3k}(\chi)}{\tau^{3k}(\overline{\chi})} + \frac{\tau^{3k}(\overline{\chi})}{\tau^{3k}(\chi)}, \qquad (1)$$

where p is an odd prime, k is a non-negative integer, χ is any non-principal character mod p.

Observing the basic properties of Equation (1), we noticed that hardly anyone had published research in any academic papers to date. We consider that the question is significant. In addition, the regularity of the value distribution of classical Gauss sums could be better revealed. Presently, we will explain certain properties discovered in our investigation. See that $U_k(p,\chi)$ has some good properties. In fact, for some special character χ mod p, the second-order linear recurrence formula for $U_k(p,\chi)$ for all integers $k \geq 0$ may be found similarly.

The goal of this paper is to use the analytic method and the properties of the character sums to solve the computational problem of $U_k(p,\chi)$, and to calculate two recursive formulae, which are listed hereafter:

Theorem 1. *Let p be a prime with $p \equiv 1 \bmod 12$, ψ be any third-order character mod p. Then, for any positive integer k, we can deduce the following second-order linear recursive formulae*

$$U_{k+1}(p,\psi) = \frac{d^2 - 2p}{p} \cdot U_k(p,\psi) - U_{k-1}(p,\psi),$$

where the initial values $U_0(p,\psi) = 2$ and $U_1(p,\psi) = \frac{d^2-2p}{p}$, d is uniquely determined by $4p = d^2 + 27b^2$ and $d \equiv 1 \bmod 3$.

So we can deduce the general term

$$U_k(p,\psi) = \left(\frac{d^2 - 2p + 3dbi\sqrt{3}}{2p}\right)^k + \left(\frac{d^2 - 2p - 3dbi\sqrt{3}}{2p}\right)^k, \quad i^2 = -1.$$

Theorem 2. *Let p be a prime with $p \equiv 7 \bmod 12$, ψ be any third-order character mod p. Then, for any positive integer k, we will obtain the second-order linear recursive formulae*

$$U_{k+1}(p,\psi) = \frac{i(2p-d^2)}{p} \cdot U_k(p,\psi) - U_{k-1}(p,\psi),$$

where the initial values $U_0(p,\psi) = 2$, $U_1(p,\psi) = \frac{i(2p-d^2)}{p}$ and $i^2 = -1$.

Similarly, we can also deduce the general term

$$U_k(p,\psi) = i^k \left(\frac{2p - d^2 + \sqrt{8p^2 - 4pd^2 + d^4}}{2p}\right)^k + i^k \left(\frac{2p - d^2 - \sqrt{8p^2 - 4pd^2 + d^4}}{2p}\right)^k.$$

2. Several Lemmas

We have used five simple and necessary lemmas to prove our theorems. Hereafter, we will apply relevant properties of classical Gauss sums and the third-order character mod p, all of which can be found in books concerning elementary and analytic number theory, such as in References [2,10], so we will not duplicate the related contents.

Lemma 1. *If p is any prime with $p \equiv 1 \bmod 3$, ψ is any third-order character mod p, then, we have the equation*

$$\tau^3(\psi) + \tau^3(\overline{\psi}) = dp,$$

where $\tau(\psi)$ denotes the classical Gauss sums, d is uniquely determined by $4p = d^2 + 27b^2$ and $d \equiv 1 \bmod 3$.

Proof. See References [3] or [8]. □

Lemma 2. *Let p be a prime with $p \equiv 1 \mod 3$, ψ be any third-order character mod p, $\chi_2 = \left(\frac{*}{p}\right)$ denotes the Legendre's symbol mod p. The following identity holds*

$$\tau^2\left(\overline{\psi}\right) = \left(\frac{-1}{p}\right)\psi(4)\tau(\chi_2)\tau(\psi\chi_2).$$

Proof. Firstly, using the properties of Gauss sums, we get

$$\sum_{a=1}^{p-1} \overline{\psi}\left(a(a+1)\right) = \frac{1}{\tau(\psi)} \sum_{b=1}^{p-1} \psi(b) \sum_{a=1}^{p-1} \overline{\psi}(a) e\left(\frac{b(a+1)}{p}\right) \quad (2)$$

$$= \frac{\tau^2(\overline{\psi})}{\tau(\psi)} = \frac{\tau^3(\overline{\psi})}{p}.$$

On the other side, we get the sums

$$\sum_{a=1}^{p-1} \overline{\psi}\left(a(a+1)\right) = \psi(4) \sum_{a=0}^{p-1} \overline{\psi}\left(4a^2 + 4a\right)$$

$$= \psi(4) \sum_{a=0}^{p-1} \overline{\psi}\left((2a+1)^2 - 1\right) = \psi(4) \sum_{a=0}^{p-1} \overline{\psi}\left(a^2 - 1\right)$$

$$= \frac{\psi(4)}{\tau(\psi)} \sum_{b=1}^{p-1} \psi(b) \sum_{a=0}^{p-1} e\left(\frac{b(a^2-1)}{p}\right) = \frac{\psi(4)}{\tau(\psi)} \sum_{b=1}^{p-1} \psi(b) e\left(\frac{-b}{p}\right) \sum_{a=0}^{p-1} e\left(\frac{ba^2}{p}\right) \quad (3)$$

$$= \frac{\psi(4)\tau(\chi_2)}{\tau(\psi)} \sum_{b=1}^{p-1} \psi(b)\chi_2(b) e\left(\frac{-b}{p}\right) = \frac{\psi(4)\chi_2(-1)\tau(\chi_2)\tau(\psi\chi_2)}{\tau(\psi)}.$$

Combining Equations (2) and (3), we obtain

$$\tau^2\left(\overline{\psi}\right) = \left(\frac{-1}{p}\right)\psi(4)\tau(\chi_2)\tau(\psi\chi_2).$$

Now, Lemma 2 has been proved. □

Lemma 3. *Let p be a prime with $p \equiv 1 \mod 6$, χ be any sixth-order character mod p. Then, about classical Gauss sums $\tau(\chi)$, the following holds:*

$$\tau^3(\chi) + \tau^3\left(\overline{\chi}\right) = \begin{cases} p^{\frac{1}{2}}\left(d^2 - 2p\right) & \text{if } p = 12h+1, \\ -i \cdot p^{\frac{1}{2}}\left(d^2 - 2p\right) & \text{if } p = 12h+7, \end{cases}$$

where $i^2 = -1$, d is uniquely determined by $4p = d^2 + 27b^2$ and $d \equiv 1 \mod 3$.

Proof. Since $p \equiv 1 \mod 6$, ψ is a third-order character mod p. Any sixth-order character χ mod p can be denoted as $\chi = \psi\chi_2$ or $\chi = \overline{\psi}\chi_2$. Note that $\psi^3(4) = 1$, $\overline{\psi}^3(4) = 1$ and $\chi_2^3 = \chi_2$, from Lemma 2 we deduce

$$\tau^6\left(\overline{\psi}\right) = \left(\frac{-1}{p}\right)\tau^3(\chi_2)\tau^3(\psi\chi_2) \quad (4)$$

and

$$\tau^6(\psi) = \left(\frac{-1}{p}\right)\tau^3(\chi_2)\tau^3\left(\overline{\psi}\chi_2\right). \quad (5)$$

Adding Equations (4) and (5), and then applying Lemma 1 we have

$$\left(\frac{-1}{p}\right)\tau^3(\chi_2)\left(\tau^3(\psi\chi_2) + \tau^3\left(\overline{\psi}\chi_2\right)\right) = \tau^6\left(\overline{\psi}\right) + \tau^6(\psi)$$

$$= \left(\tau^3\left(\overline{\psi}\right) + \tau^3(\psi)\right)^2 - 2p^3 = d^2p^2 - 2p^3. \quad (6)$$

Note that χ_2 is a real character mod p, $\overline{\psi}\chi_2 = \overline{\psi\chi_2}$, and $\tau(\chi_2) = \sqrt{p}$. If $p \equiv 1 \bmod 4$; $\tau(\chi_2) = i \cdot \sqrt{p}$, $i^2 = -1$, if $p \equiv 3 \bmod 4$. From Equation (6) we may immediately prove the sum

$$\tau^3(\psi\chi_2) + \tau^3\left(\overline{\psi}\chi_2\right) = \begin{cases} p^{\frac{1}{2}}\left(d^2 - 2p\right) & \text{if } p = 12h + 1, \\ -i \cdot p^{\frac{1}{2}}\left(d^2 - 2p\right) & \text{if } p = 12h + 7. \end{cases} \tag{7}$$

Let $\chi = \psi\chi_2$, then χ is a sixth-order character mod p and $\overline{\psi}\chi_2 = \overline{\chi}$. From Equation (7) we can deduce the sum term

$$\tau^3(\chi) + \tau^3\left(\overline{\chi}\right) = \begin{cases} p^{\frac{1}{2}}\left(d^2 - 2p\right) & \text{if } p = 12h + 1, \\ -i \cdot p^{\frac{1}{2}}\left(d^2 - 2p\right) & \text{if } p = 12h + 7. \end{cases}$$

The proof of Lemma 3 has been completed. □

Lemma 4. *Let p be a prime with $p \equiv 7 \bmod 12$, ψ be any three-order character mod p. Then, we compute the sum term*

$$\frac{\tau^3\left(\overline{\psi}\right)}{\tau^3\left(\psi\right)} + \frac{\tau^3\left(\psi\right)}{\tau^3\left(\overline{\psi}\right)} = \frac{i \cdot (2p - d^2)}{p}.$$

Proof. Let ψ be a three-order character mod p. Then, for any six-order character χ mod p, we must have $\chi = \psi\chi_2$ or $\chi = \overline{\chi}\chi_2$. Without loss of generality we suppose that $\chi = \psi\chi_2$, then note that $\psi(-1) = 1$, $\chi_2(-1) = -1$ and Theorem 7.5.4 in Reference [10], we acquire

$$\sum_{a=0}^{p-1} e\left(\frac{ba^2}{p}\right) = \chi_2(b) \cdot \sqrt{p}, \quad (p, b) = 1.$$

Using the properties of Gauss sums we can write

$$\sum_{a=0}^{p-1} \chi\left(a^2 - 1\right) = \frac{1}{\tau(\overline{\chi})} \sum_{b=1}^{p-1} \overline{\chi}(b) \sum_{a=0}^{p-1} e\left(\frac{b(a^2-1)}{p}\right)$$

$$= \frac{1}{\tau(\overline{\chi})} \sum_{b=1}^{p-1} \overline{\chi}(b) e\left(\frac{-b}{p}\right) \sum_{a=0}^{p-1} e\left(\frac{ba^2}{p}\right) = \frac{\sqrt{p}}{\tau(\overline{\chi})} \sum_{b=1}^{p-1} \overline{\chi}(b)\chi_2(b) e\left(\frac{-b}{p}\right) \tag{8}$$

$$= \frac{\overline{\chi}(-1)\chi_2(-1)\sqrt{p}\,\tau(\overline{\chi}\chi_2)}{\tau(\overline{\chi})} = \frac{\sqrt{p}\,\tau(\overline{\chi}\chi_2)}{\tau(\overline{\chi})}.$$

Noting that $\overline{\chi}^2 = \overline{\psi}^2 = \psi$, we can deduce

$$\sum_{a=0}^{p-1} \chi\left(a^2 - 1\right) = \sum_{a=0}^{p-1} \chi\left((a+1)^2 - 1\right) = \sum_{a=1}^{p-1} \chi(a)\chi(a+2)$$

$$= \frac{1}{\tau(\overline{\chi})} \sum_{b=1}^{p-1} \overline{\chi}(b) \sum_{a=1}^{p-1} \chi(a) e\left(\frac{b(a+2)}{p}\right) = \frac{\tau(\chi)}{\tau(\overline{\chi})} \sum_{b=1}^{p-1} \overline{\chi}^2(b) e\left(\frac{2b}{p}\right) \tag{9}$$

$$= \frac{\overline{\psi}(2)\tau(\chi)\tau(\psi)}{\tau(\overline{\chi})}.$$

Obviously, $\overline{\chi}\chi_2 = \overline{\psi}$ and $\psi^3(2) = 1$, applying Equations (8) and (9) we have

$$\tau^3(\chi) = p^{\frac{3}{2}} \cdot \frac{\tau^3\left(\overline{\psi}\right)}{\tau^3\left(\psi\right)}. \tag{10}$$

Similarly, we can see

$$\tau^3\left(\overline{\chi}\right) = p^{\frac{3}{2}} \cdot \frac{\tau^3\left(\psi\right)}{\tau^3\left(\overline{\psi}\right)}. \tag{11}$$

Combining Equation (10), Equation (11) and Lemma 3 we compute

$$\frac{\tau^3(\psi)}{\tau^3(\bar{\psi})} + \frac{\tau^3(\bar{\psi})}{\tau^3(\psi)} = \frac{1}{p^{\frac{3}{2}}}\left(\tau^3(\chi) + \tau^3(\bar{\chi})\right) = \frac{i \cdot (2p - d^2)}{p}.$$

This completes the proof of Lemma 4. □

Lemma 5. *Let p be a prime with $p \equiv 1 \mod 12$, ψ be any three-order character mod p. Then, we obtain the sum term*

$$\frac{\tau^3(\bar{\psi})}{\tau^3(\psi)} + \frac{\tau^3(\psi)}{\tau^3(\bar{\psi})} = \frac{d^2 - 2p}{p}.$$

Proof. From Lemma 3 and the method of proving Lemma 4 we can easily deduce Lemma 5. □

3. Proofs of the Theorems

In this section, we prove our two theorems. For Theorem 1, since $p \equiv 1 \mod 12$, ψ is a third-order character mod p, then for any positive integer k, let

$$U_k(p) = \frac{\tau^{3k}(\psi)}{\tau^{3k}(\bar{\psi})} + \frac{\tau^{3k}(\bar{\psi})}{\tau^{3k}(\psi)}.$$

From Lemma 5 we have

$$U_1(p) = \frac{\tau^3(\bar{\psi})}{\tau^3(\psi)} + \frac{\tau^3(\psi)}{\tau^3(\bar{\psi})} = \frac{d^2 - 2p}{p} \tag{12}$$

and

$$\frac{d^2 - 2p}{p} \cdot U_k(p) = U_k(p)U_1(p) = \left(\frac{\tau^{3k}(\psi)}{\tau^{3k}(\bar{\psi})} + \frac{\tau^{3k}(\bar{\psi})}{\tau^{3k}(\psi)}\right) \cdot \left(\frac{\tau^3(\bar{\psi})}{\tau^3(\psi)} + \frac{\tau^3(\psi)}{\tau^3(\bar{\psi})}\right) \tag{13}$$

$$= \frac{\tau^{3k+3}(\bar{\psi})}{\tau^{3k+3}(\psi)} + \frac{\tau^{3k+3}(\psi)}{\tau^{3k+3}(\bar{\psi})} + \frac{\tau^{3k-3}(\bar{\psi})}{\tau^{3k-3}(\psi)} + \frac{\tau^{3k-3}(\psi)}{\tau^{3k-3}(\bar{\psi})} = U_{k+1}(p) + U_{k-1}(p).$$

Combining Equations (12) and (13) we may immediately compute the second-order linear recursive formula

$$U_{k+1}(p) = \frac{d^2 - 2p}{p} \cdot U_k(p) - U_{k-1}(p) \tag{14}$$

with initial values $U_0(p) = 2$ and $U_1(p) = \frac{d^2 - 2p}{p}$.

Note that the two roots of the equation $\lambda^2 - \frac{d^2 - 2p}{p}\lambda + 1 = 0$ are

$$\lambda_1 = \frac{d^2 - 2p + 3dbi\sqrt{3}}{2p} \quad \text{and} \quad \lambda_2 = \frac{d^2 - 2p - 3dbi\sqrt{3}}{2p}.$$

So from Equation (14) and its initial values we may immediately deduce the general term

$$U_k(p, \psi) = \left(\frac{d^2 - 2p + 3dbi\sqrt{3}}{2p}\right)^k + \left(\frac{d^2 - 2p - 3dbi\sqrt{3}}{2p}\right)^k,$$

where $i^2 = -1$. Now Theorem 1 has been finished.

Similarly, from Lemma 4 and the method of proving Theorem 1 we can easily obtain Theorem 2. Now, we have completed all the proofs of our Theorems.

4. Conclusions

The main results of this paper are Theorem 1 and 2. They give a new second-order linear recurrence formula for Equation (1) with the third-order character ψ mod p. Therefore, we can calculate the exact value of Equation (1). Note that $|\tau(\overline{\psi})/\tau(\psi)| = 1$, so $\tau(\overline{\psi})/\tau(\psi)$ is a unit root, thus, the results in this paper profoundly reveal the distributional properties of two different Gauss sums quotients on the unit circle.

For the other mod p characters, for example, the fifth-order character χ mod p with $p \equiv 1$ mod 5, we naturally ask whether there exists a similar formula as presented in our theorems. This is still an open problem. It will be the content of our future investigations.

Funding: This research was funded by [National Natural Science Foundation of China] Grant number [11771351].

Acknowledgments: The author wish to express her gratitude to the editors and the reviewers for their helpful comments.

Conflicts of Interest: The authors declare no conflict of interest.

References

1. Chen, Z.Y.; Zhang, W.P. On the fourth-order linear recurrence formula related to classical Gauss sums. *Open Math.* **2017**, *15*, 1251–1255.
2. Apostol, T.M. *Introduction to Analytic Number Theory*; Springer: New York, NY, USA, 1976.
3. Zhang, W.P.; Hu, J.Y. The number of solutions of the diagonal cubic congruence equation mod p. *Math. Rep.* **2018**, *20*, 73–80.
4. Chen, L.; Hu, J.Y. A linear recurrence formula involving cubic Gauss sums and Kloosterman sums. *Acta Math. Sin.* **2018**, *61*, 67–72.
5. Li, X.X.; Hu, J.Y. The hybrid power mean quartic Gauss sums and Kloosterman sums. *Open Math.* **2017**, *15*, 151–156.
6. Zhang, H.; Zhang, W.P. The fourth power mean of two-term exponential sums and its application. *Math. Rep.* **2017**, *19*, 75–83.
7. Zhang, W.P.; Liu, H.N. On the general Gauss sums and their fourth power mean. *Osaka J. Math.* **2005**, *42*, 189–199.
8. Berndt, B.C.; Evans, R.J. The determination of Gauss sums. *Bull. Am. Math. Soc.* **1981**, *5*, 107–128. [CrossRef]
9. Berndt, B.C.; Evans, R.J. Sums of Gauss, Jacobi, and Jacobsthal. *J. Number Theory* **1979**, *11*, 349–389. [CrossRef]
10. Hua, L.K. *Introduction to Number Theory*; Science Press: Beijing, China, 1979.
11. Kim, T. Power series and asymptotic series associated with the q analog of the two-variable p-adic L-function. *Russ. J. Math. Phys.* **2005**, *12*, 186–196.
12. Kim, H.S.; Kim, T. On certain values of p-adic q-L-function. *Rep. Fac. Sci. Eng. Saga Univ. Math.* **1995**, *23*, 1–2.
13. Chae, H.; Kim, D.S. L function of some exponential sums of finite classical groups. *Math. Ann.* **2003**, *326*, 479–487. [CrossRef]

© 2018 by the authors. Licensee MDPI, Basel, Switzerland. This article is an open access article distributed under the terms and conditions of the Creative Commons Attribution (CC BY) license (http://creativecommons.org/licenses/by/4.0/).

Article

Connection Problem for Sums of Finite Products of Chebyshev Polynomials of the Third and Fourth Kinds

Dmitry Victorovich Dolgy [1], Dae San Kim [2], Taekyun Kim [3,4] and Jongkyum Kwon [5,*]

1. Institute of National Sciences, Far Eastern Federal University, 690950 Vladivostok, Russia; d_dol@mail.ru
2. Department of Mathematics, Sogang University, Seoul 04107, Korea; dskim@sogang.ac.kr
3. Department of Mathematics, College of Science, Tianjin Polytechnic University, Tianjin 300160, China; tkkim@kw.ac.kr
4. Department of Mathematics, Kwangwoon University, Seoul 01897, Korea
5. Department of Mathematics Education and ERI, Gyeongsang National University, Jinju 52828, Gyeongsangnamdo, Korea
* Correspondence: mathkjk26@gnu.ac.kr; Tel.: +82-(0)10-8978-73576

Received: 8 October 2018; Accepted: 5 November 2018; Published: 9 November 2018

Abstract: This paper treats the connection problem of expressing sums of finite products of Chebyshev polynomials of the third and fourth kinds in terms of five classical orthogonal polynomials. In fact, by carrying out explicit computations each of them are expressed as linear combinations of Hermite, generalized Laguerre, Legendre, Gegenbauer, and Jacobi polynomials which involve some terminating hypergeometric functions $_2F_0, _2F_1$, and $_3F_2$.

Keywords: sums of finite products of Chebyshev polynomials of the third and fourth kinds; Hermite; generalized Laguerre; Legendre; Gegenbauer; Jacobi

MSC: 11B83; 33C05; 33C20; 33C45

1. Introduction and Preliminaries

In this section, we will recall some basic facts about relevant orthogonal polynomials that will be needed throughout this paper. For this, we will first fix some notations. For any nonnegative integer n, the falling factorial polynomials $(x)_n$ and the rising factorial polynomials $<x>_n$ are respectively given by

$$(x)_n = x(x-1)\cdots(x-n+1), \ (n \geq 1), \ (x)_0 = 1, \tag{1}$$

$$<x>_n = x(x+1)\cdots(x+n-1), \ (n \geq 1), \ <x>_0 = 1. \tag{2}$$

The two factorial polynomials are evidently related by

$$(-1)^n (x)_n = <-x>_n, \ (-1)^n <x>_n = (-x)_n. \tag{3}$$

$$\frac{(2n-2s)!}{(n-s)!} = \frac{2^{2n-2s}(-1)^s <\frac{1}{2}>_n}{<\frac{1}{2}-n>_s}, \ (n \geq s \geq 0). \tag{4}$$

$$B(x,y) = \int_0^1 t^{x-1}(1-t)^{y-1}dt = \frac{\Gamma(x)\Gamma(y)}{\Gamma(x+y)}, \ (\text{Re } x, \text{Re } y > 0). \tag{5}$$

$$\Gamma(n+\frac{1}{2}) = \frac{(2n)!\sqrt{\pi}}{2^{2n}n!}, \ (n \geq 0). \tag{6}$$

$$\frac{\Gamma(x+1)}{\Gamma(x+1-n)} = (x)_n, \frac{\Gamma(x+n)}{\Gamma(x)} = <x>_n, \ (n \geq 0), \tag{7}$$

where $\Gamma(x)$ and $B(x,y)$ are the gamma and beta functions respectively.

The hypergeometric function is defined by

$$\begin{aligned}
&_pF_q(a_1, \cdots, a_p; b_1, \cdots, b_q; x) \\
&= \sum_{n=0}^{\infty} \frac{<a_1>_n \cdots <a_p>_n}{<b_1>_n \cdots <b_q>_n} \frac{x^n}{n!}.
\end{aligned} \tag{8}$$

We are now ready to state some basic facts about Chebyshev polynomials of the third kind $V_n(x)$, those of the fourth kind $W_n(x)$, Hermite polynomials $H_n(x)$, generalized (extended) Laguerre polynomials $L_n^\alpha(x)$, Legendre polynomials $P_n(x)$, Gegenbauer polynomials $C_n^{(\lambda)}(x)$, and Jacobi polynomials $P_n^{(\alpha,\beta)}$. Chebyshev polynomials are diversely used in approximation theory and numerical analysis, Hermite polynomials appear as the eigenfunctions of the harmonic oscillator in quantum mechanics, Laguerre polynomials have important applications to the solution of Schrödinger's equation for the hydrogen atom, Legendre polynomials can be used to write the Coulomb potential as a series, Gegenbauer polynomials play an important role in the constructive theory of spherical functions and Jacobi polynomials occur in the solution to the equations of motion of the symmetric top. All the necessary facts on those special polynomials can be found in [1–9]. For the full accounts of this fascinating area of orthogonal polynomials, the reader may refer to [10–13].

The above special polynomials are given in terms of generating functions by

$$F(t,x) = \frac{1-t}{1-2xt+t^2} = \sum_{n=0}^{\infty} V_n(x) t^n, \tag{9}$$

$$G(t,x) = \frac{1+t}{1-2xt+t^2} = \sum_{n=0}^{\infty} W_n(x) t^n, \tag{10}$$

$$e^{2xt-t^2} = \sum_{n=0}^{\infty} H_n(x) \frac{t^n}{n!}, \tag{11}$$

$$(1-t)^{-\alpha-1} \exp\left(-\frac{xt}{1-t}\right) = \sum_{n=0}^{\infty} L_n^\alpha(x) t^n, \ (\alpha > -1), \tag{12}$$

$$(1-2xt+t^2)^{-\frac{1}{2}} = \sum_{n=0}^{\infty} P_n(x) t^n, \tag{13}$$

$$\frac{1}{(1-2xt+t^2)^\lambda} = \sum_{n=0}^{\infty} C_n^{(\lambda)}(x) t^n, \ (\lambda > -\frac{1}{2}, \lambda \neq 0, |t| < 1, |x| \leq 1), \tag{14}$$

$$\frac{2^{\alpha+\beta}}{R(1-t+R)^\alpha(1+t+R)^\beta} = \sum_{n=0}^{\infty} P_n^{(\alpha,\beta)}(x) t^n, \tag{15}$$

$(R = \sqrt{1-2xt+t^2}, \ \alpha, \beta > -1)$.

Explicit expressions for the above special polynomials are as in the following:

$$\begin{aligned}
V_n(x) &= {}_2F_1(-n, n+1; \tfrac{1}{2}; \tfrac{1-x}{2}) \\
&= \sum_{l=0}^{n} \binom{2n-l}{l} 2^{n-l} (x-1)^{n-l},
\end{aligned} \tag{16}$$

$$W_n(x) = (2n+1){}_2F_1(-n, n+1; \tfrac{3}{2}; \tfrac{1-x}{2})$$

$$= (2n+1) \sum_{l=0}^{n} \frac{2^{n-l}}{2n-2l+1} \binom{2n-l}{l} (x-1)^{n-l}, \tag{17}$$

$$H_n(x) = n! \sum_{l=0}^{[\frac{n}{2}]} \frac{(-1)^l}{l!(n-2l)!} (2x)^{n-2l}, \tag{18}$$

$$L_n^\alpha(x) = \frac{<\alpha+1>_n}{n!} {}_1F_1(-n, \alpha+1; x)$$

$$= \sum_{l=0}^{n} \frac{(-1)^l \binom{n+\alpha}{n-l}}{l!} x^l, \tag{19}$$

$$P_n(x) = {}_2F_1(-n, n+1; 1; \tfrac{1-x}{2})$$

$$= \frac{1}{2^n} \sum_{l=0}^{[\frac{n}{2}]} (-1)^l \binom{n}{l} \binom{2n-2l}{n} x^{n-2l}, \tag{20}$$

$$C_n^{(\lambda)}(x) = \binom{n+2\lambda-1}{n} {}_2F_1(-n, n+2\lambda; \lambda+\tfrac{1}{2}; \tfrac{1-x}{2})$$

$$= \sum_{k=0}^{[\frac{n}{2}]} (-1)^k \frac{\Gamma(n-k+\lambda)}{\Gamma(\lambda)k!(n-2k)!} (2x)^{n-2k}, \tag{21}$$

$$P_n^{(\alpha,\beta)}(x) = \frac{<\alpha+1>_n}{n!} {}_2F_1(-n, 1+\alpha+\beta+n; \alpha+1; \tfrac{1-x}{2})$$

$$= \sum_{k=0}^{n} \binom{n+\alpha}{n-k} \binom{n+\beta}{k} (\tfrac{x-1}{2})^k (\tfrac{x+1}{2})^{n-k}. \tag{22}$$

Next, we state Rodrigues-type formulas for Hermite and generalized Laguerre polynomials and Rodrigues' formulas for Legendre, Gegenbauer and Jacobi polynomials.

$$H_n(x) = (-1)^n e^{x^2} \frac{d^n}{dx^n} e^{-x^2}, \tag{23}$$

$$L_n^\alpha(x) = \frac{1}{n!} x^{-\alpha} e^x \frac{d^n}{dx^n} (e^{-x} x^{n+\alpha}), \tag{24}$$

$$P_n(x) = \frac{1}{2^n n!} \frac{d^n}{dx^n} (x^2-1)^n, \tag{25}$$

$$(1-x^2)^{\lambda-\frac{1}{2}} C_n^{(\lambda)}(x) = \frac{(-2)^n}{n!} \frac{<\lambda>_n}{<n+2\lambda>_n} \frac{d^n}{dx^n} (1-x^2)^{n+\lambda-\frac{1}{2}}, \tag{26}$$

$$(1-x)^\alpha (1+x)^\beta P_n^{(\alpha,\beta)}(x) = \frac{(-1)^n}{2^n n!} \frac{d^n}{dx^n} (1-x)^{n+\alpha} (1+x)^{n+\beta}. \tag{27}$$

The last thing we want to mention is the orthogonalities with respect to various weight functions enjoyed by Hermite, generalized Laguerre, Legendre, Gegenbauer and Jacobi polynomials.

$$\int_{-\infty}^{\infty} e^{-x^2} H_n(x) H_m(x) \, dx = 2^n n! \sqrt{\pi} \delta_{n,m}, \tag{28}$$

$$\int_0^{\infty} x^\alpha e^{-x} L_n^\alpha(x) L_m^\alpha(x) \, dx = \frac{1}{n!} \Gamma(\alpha+n+1) \delta_{n,m}, \tag{29}$$

$$\int_{-1}^{1} P_n(x) P_m(x) \, dx = \frac{2}{2n+1} \delta_{n,m}, \tag{30}$$

$$\int_{-1}^{1} (1-x^2)^{\lambda-\frac{1}{2}} C_n^{(\lambda)}(x) C_m^{(\lambda)}(x) \, dx = \frac{\pi 2^{1-2\lambda} \Gamma(n+2\lambda)}{n!(n+\lambda)\Gamma(\lambda)^2} \delta_{n,m}, \tag{31}$$

$$\int_{-1}^{1} (1-x)^\alpha (1+x)^\beta P_n^{(\alpha,\beta)}(x) P_m^{(\alpha,\beta)}(x)\, dx$$
$$= \frac{2^{\alpha+\beta+1}\Gamma(n+\alpha+1)\Gamma(n+\beta+1)}{(2n+\alpha+\beta+1)\Gamma(n+\alpha+\beta+1)\Gamma(n+1)} \delta_{n,m}. \tag{32}$$

For convenience, let us put

$$\gamma_{n,r}(x) = \sum_{l=0}^{n} \sum_{i_1+i_2+\cdots+i_{r+1}=l} \binom{r-1+n-l}{r-1} V_{i_1}(x) V_{i_2}(x) \cdots V_{i_{r+1}}(x), \ (n \geq 0, r \geq 1), \tag{33}$$

$$\mathcal{E}_{n,r}(x) = \sum_{l=0}^{n} \sum_{i_1+i_2+\cdots+i_{r+1}=l} (-1)^{n-l} \binom{r-1+n-l}{r-1} W_{i_1}(x) W_{i_2}(x) \cdots W_{i_{r+1}}(x), (n \geq 0, r \geq 1). \tag{34}$$

We observe here that both $\gamma_{n,r}(x)$ and $\mathcal{E}_{n,r}(x)$ have degree n.

In this paper, we will consider the connection problem of expressing the sums of finite products in (33) and (34) as linear combinations of $H_n(x)$, $L_n^\alpha(x)$, $P_n(x)$, $C_n^{(\lambda)}(x)$, and $P_n^{(\alpha,\beta)}(x)$. These will be done by performing explicit computations based on Proposition 1. We observe here that the formulas in Proposition 1 follow from their orthogonalities, Rodrigues' and Rodrigues-type formulas and integration by parts.

Our main results are the following Theorems 1 and 2.

Theorem 1. *Let n, r be any integers with $n \geq 0, r \geq 1$. Then we have the following.*

$$\sum_{l=0}^{n} \sum_{i_1+i_2+\cdots+i_{r+1}=l} \binom{r-1+n-l}{r-1} V_{i_1}(x) V_{i_2}(x) \cdots V_{i_{r+1}}(x)$$

$$= \frac{(2n+2r)!}{r! 4^{n+r}(n+r-\frac{1}{2})_{n+r}} \sum_{k=0}^{n} \frac{(-2)^k}{(n-k)!}$$
$$\times \sum_{j=0}^{[\frac{k}{2}]} \frac{{}_2F_1(2j-k, \frac{1}{2}-n-r; -2n-2r; 2)}{j! 4^j (k-2j)!} H_{n-k}(x) \tag{35}$$

$$= \frac{1}{r!} \sum_{k=0}^{n} \sum_{l=0}^{k} \frac{(-2)^{n-l}(2n+2r-1)!(n+r-1)!}{l!(2n+2r-2l)!(k-l)!}$$
$$\times {}_2F_0(l-k, n-k+\alpha+1; -; 1) L_{n-k}^\alpha(x) \tag{36}$$

$$= \frac{(-1)^n n!(2n+2r)!}{r! 4^r (n+r-\frac{1}{2})_{n+r}} \sum_{k=0}^{n} \frac{(-1)^k(2k+1)}{(n-k)!(n+k+1)!}$$
$$\times {}_3F_2(k-n, \frac{1}{2}-n-r, -n-k-1; -2n-2r, -n; 1) P_k(x) \tag{37}$$

$$= \frac{(-1)^n(2n+2r)! 4^{\lambda-r} \Gamma(\lambda) \Gamma(n+\lambda+\frac{1}{2})}{\sqrt{\pi} r!(n+r-\frac{1}{2})_{n+r}} \sum_{k=0}^{n} \frac{(-1)^k(k+\lambda)}{\Gamma(n+k+2\lambda+1)(n-k)!}$$
$$\times {}_3F_2(k-n, \frac{1}{2}-n-r, -n-k-2\lambda; -2n-2r, -n-\lambda+\frac{1}{2}; 1) C_k^{(\lambda)}(x) \tag{38}$$

$$= \frac{(-1)^n(2n+2r)!\Gamma(n+\alpha+1)}{r! 4^r(n+r-\frac{1}{2})_{n+r}} \sum_{k=0}^{n} \frac{(-1)^k(2k+\alpha+\beta+1)\Gamma(k+\alpha+\beta+1)}{(n-k)!\Gamma(\alpha+k+1)\Gamma(n+k+\alpha+\beta+2)}$$
$$\times {}_3F_2(k-n, \frac{1}{2}-n-r, -n-k-\alpha-\beta-1; -2n-2r, -n-\alpha; 1) P_k^{(\alpha,\beta)}(x). \tag{39}$$

Theorem 2. Let n, r be any integers with $n \geq 0, r \geq 1$. Then we have the following.

$$\sum_{l=0}^{n} \sum_{i_1+i_2+\cdots+i_{r+1}=l} (-1)^{n-l} \binom{r-1+n-l}{r-1} W_{i_1}(x) W_{i_2}(x) \cdots W_{i_{r+1}}(x)$$

$$= \frac{(2n+1)(2n+2r)!}{r! 2^{2n+2r+1}(n+r+\frac{1}{2})_{n+r+1}} \sum_{k=0}^{n} \frac{(-2)^k}{(n-k)!}$$

$$\times \sum_{j=0}^{[\frac{k}{2}]} \frac{{}_2F_1(2j-k, -n-r-\frac{1}{2}; -2n-2r; 2)}{j! 4^j (k-2j)!} H_{n-k}(x) \quad (40)$$

$$= \frac{(2n+1)}{r!} \sum_{k=0}^{n} \sum_{l=0}^{k} \frac{(-2)^{n-l}(2n+2r-l)!(n+r-l)!}{l!(2n+2r-2l+1)!(k-l)!}$$

$$\times {}_2F_0(l-k, n-k+\alpha+1; -; 1) L_{n-k}^{\alpha}(x) \quad (41)$$

$$= \frac{(-1)^n n!(2n+1)(2n+2r)!}{r! 2^{2r+1}(n+r+\frac{1}{2})_{n+r+1}} \sum_{k=0}^{n} \frac{(-1)^k(2k+1)}{(n-k)!(n+k+1)!}$$

$$\times {}_3F_2(k-n, -n-r-\frac{1}{2}, -n-k-1; -2n-2r, -n; 1) P_k(x) \quad (42)$$

$$= \frac{(-1)^n (2n+2r)! 2^{2\lambda-2r-1}(2n+1)\Gamma(\lambda)\Gamma(n+\lambda+\frac{1}{2})}{\sqrt{\pi} r!(n+r+\frac{1}{2})_{n+r+1}}$$

$$\times \sum_{k=0}^{n} \frac{(-1)^k(k+\lambda)}{\Gamma(n+k+2\lambda+1)(n-k)!}$$

$$\times {}_3F_2(k-n, -n-r-\frac{1}{2}, -n-k-2\lambda; -2n-2r, -n-\lambda+\frac{1}{2}; 1) C_k^{(\lambda)}(x) \quad (43)$$

$$= \frac{(-1)^n (2n+2r)!(2n+1)\Gamma(n+\alpha+1)}{r! 2^{2r+1}(n+r+\frac{1}{2})_{n+r+1}}$$

$$\times \sum_{k=0}^{n} \frac{(-1)^k(2k+\alpha+\beta+1)\Gamma(k+\alpha+\beta+1)}{(n-k)!\Gamma(\alpha+k+1)\Gamma(n+k+\alpha+\beta+2)}$$

$$\times {}_3F_2(k-n, -n-r-\frac{1}{2}, -n-k-\alpha-\beta-1; -2n-2r, -n-\alpha; 1) P_k^{(\alpha,\beta)}(x). \quad (44)$$

Before closing the section, we are going to mention some of previous results on the related connection problems. The papers [14–16] treat the connection problem of expressing sums of finite products of Bernoulli, Euler and Genocchi polynomials in terms of Bernoulli polynomials. In fact, they were carried out by deriving Fourier series expansions for the functions closely related to those sums of finite products. Moreover, the same was done for the sums of finite products of Chebyshev polynomials of the second kind and of Fibonacci polynomials in [17].

Along the same line as the present paper, sums of finite products of Chebyshev polynomials of the second kind and Fibonacci polynomials were expressed in [18] as linear combinations of the orthogonal polynomials $H_n(x), L_n^{\alpha}(x), P_n(x), C_n^{(\lambda)}(x)$, and $P_n^{(\alpha,\beta)}(x)$. Also, the connection problem of expressing in terms of all kinds of Chebyshev polynomials were done for sums of finite products of Chebyshev polynomials of the second, third and fourth kinds and of Fibonacci, Legendre and Laguerre polynomials in [19–21].

Finally, we let the reader refer to [22,23] for some applications of Chebyshev polynomials.

2. Proof of Theorem 1

First, we will state Propositions 1 and 2 that will be needed in showing Theorems 1 and 2.

The results in (a), (b), (c), (d) and (e) in Proposition 1 follow respectively from (3.7) of [4], (2.3) of [4] (see also (2.4) of [2]), (2.3) of [5], (2.3) of [3] and (2.7) of [7]. They can be derived from their orthogonalities in (28)–(32), Rodrigues-type and Rodrigues' formulas in (23)–(27) and integration by parts.

Proposition 1. *Let $q(x) \in \mathbb{R}[x]$ be a polynomial of degree n. Then we have the following.*

(a) $q(x) = \sum_{k=0}^{n} C_{k,1} H_k(x)$, where

$$C_{k,1} = \frac{(-1)^k}{2^k k! \sqrt{\pi}} \int_{-\infty}^{\infty} q(x) \frac{d^k}{dx^k} e^{-x^2} dx,$$

(b) $q(x) = \sum_{k=0}^{n} C_{k,2} L_k^\alpha(x)$, where

$$C_{k,2} = \frac{1}{\Gamma(\alpha + k + 1)} \int_0^\infty q(x) \frac{d^k}{dx^k} (e^{-x} x^{k+\alpha}) dx,$$

(c) $q(x) = \sum_{k=0}^{n} C_{k,3} P_k(x)$, where

$$C_{k,3} = \frac{2k+1}{2^{k+1} k!} \int_{-1}^{1} q(x) \frac{d^k}{dx^k} (x^2 - 1)^k dx,$$

(d) $q(x) = \sum_{k=0}^{n} C_{k,4} C_k^{(\lambda)}(x)$, where

$$C_{k,4} = \frac{(k+\lambda)\Gamma(\lambda)}{(-2)^k \sqrt{\pi} \Gamma(k + \lambda + \frac{1}{2})} \int_{-1}^{1} q(x) \frac{d^k}{dx^k} (1-x^2)^{k+\lambda - \frac{1}{2}} dx,$$

(e) $q(x) = \sum_{k=0}^{n} C_{k,5} P_n^{(\alpha,\beta)}(x)$, where

$$C_{k,5} = \frac{(-1)^k (2k + \alpha + \beta + 1) \Gamma(k + \alpha + \beta + 1)}{2^{\alpha + \beta + k + 1} \Gamma(\alpha + k + 1) \Gamma(\beta + k + 1)}$$

$$\times \int_{-1}^{1} q(x) \frac{d^k}{dx^k} (1-x)^{k+\alpha} (1+x)^{k+\beta} dx.$$

The following proposition will be used in showing Theorems 1 and 2. In fact, (a) is needed for (35) and (40), (b) for (39) and (44), and (b) or (c) for (37), (38), (42) and (43).

Proposition 2. *The following holds true.*

(a) For any nonnegative integer m,

$$\int_{-\infty}^{\infty} x^m e^{-x^2} dx = \begin{cases} 0, & \text{if } m \equiv 1 \pmod{2}, \\ \frac{m! \sqrt{\pi}}{(\frac{m}{2})! 2^m}, & \text{if } m \equiv 0 \pmod{2}, \end{cases}$$

(b) For any real numbers $r, s > -1$, we have

$$\int_{-1}^{1} (1-x)^r (1+x)^s dx = 2^{r+s+1} \frac{\Gamma(r+1)\Gamma(s+1)}{\Gamma(r+s+2)},$$

(c) For any real numbers r, s with $r + s > -1, s > -1$, we have

$$\int_{-1}^{1} (1-x)^r (1-x^2)^s dx = 2^{r+2s+1} \frac{\Gamma(r+s+1)\Gamma(s+1)}{\Gamma(r+2s+2)}.$$

Proof.
(a) This is an easy exercise.
(c) This follows from (b) with r replaced by $r + s$.
(b) This follows from the change of variable $1 + x = 2y$ and (5). □

The following lemma can be obtained by differentiating (9), as was shown in [24].

Lemma 1. *Let n, r be integers with $n \geq 0, r \geq 1$. Then we have the following identity.*

$$\sum_{l=0}^{n} \sum_{i_1+i_2+\cdots+i_{r+1}=l} \binom{r-1+n-l}{r-1} V_{i_1}(x) V_{i_2}(x) \cdots V_{i_{r+1}}(x) = \frac{1}{2^r r!} V_{n+r}^{(r)}(x), \tag{45}$$

where the inner sum runs over all nonnegative integers $i_1, i_2, \cdots i_{r+1}$, with $i_1 + i_2 + \cdots + i_{r+1} = l$.

From (16), we see that the rth derivative of $V_n(x)$ is given by

$$V_n^{(r)}(x) = \sum_{l=0}^{n-r} \binom{2n-l}{l} 2^{n-l}(n-l)_r (x-1)^{n-l-r}. \tag{46}$$

Especially, we have

$$V_{n+r}^{(r+k)}(x) = \sum_{l=0}^{n-k} \binom{2n+2r-l}{l} 2^{n+r-l}(n+r-l)_{r+k} (x-1)^{n-k-l}. \tag{47}$$

Now, we are ready to prove Theorem 1. As (38) and (39) can be shown similarly to (43) and (44) in the next section, we will show only (35), (36) and (37). With $\gamma_{n,r}(x)$ as in (33), we let

$$\gamma_{n,r}(x) = \sum_{k=0}^{n} C_{k,1} H_k(x). \tag{48}$$

Then, from (a) of Proposition 1, (45), (47), and integration by parts k times, we obtain

$$\begin{aligned}
C_{k,1} &= \frac{(-1)^k}{2^k k! \sqrt{\pi}} \int_{-\infty}^{\infty} \gamma_{n,r}(x) \frac{d^k}{dx^k} e^{-x^2} dx \\
&= \frac{(-1)^k}{2^{k+r} r! k! \sqrt{\pi}} \int_{-\infty}^{\infty} V_{n+r}^{(r)}(x) \frac{d^k}{dx^k} e^{-x^2} dx \\
&= \frac{1}{2^{k+r} r! k! \sqrt{\pi}} \int_{-\infty}^{\infty} V_{n+r}^{(r+k)}(x) e^{-x^2} dx \\
&= \frac{1}{2^{k+r} r! k! \sqrt{\pi}} \sum_{l=0}^{n-k} \binom{2n+2r-l}{l} 2^{n+r-l}(n+r-l)_{r+k} \\
&\quad \times \int_{-\infty}^{\infty} (x-1)^{n-k-l} e^{-x^2} dx.
\end{aligned} \tag{49}$$

Before proceeding further, by making use of (a) in Proposition 2, we note that

$$\int_{-\infty}^{\infty} (x-1)^m e^{-x^2} dx$$

$$= \sum_{s=0}^{m} \binom{m}{s}(-1)^{m-s} \int_{-\infty}^{\infty} x^s e^{-x^2} dx$$

$$= \sum_{\substack{0 \le s \le m \\ s \equiv 0 \pmod 2}} \binom{m}{s}(-1)^{m-s} \frac{s!\sqrt{\pi}}{(\frac{s}{2})! 2^s} \tag{50}$$

$$=(-1)^m \sqrt{\pi} \sum_{j=0}^{[\frac{m}{2}]} \binom{m}{2j} \frac{(2j)!}{j! 2^{2j}}, \quad (m \ge 0).$$

From (48)–(50), and after simplifications, we have

$$\gamma_{n,r}(x) = \frac{1}{r!} \sum_{k=0}^{n} \frac{(-2)^k}{(n-k)!} \sum_{l=0}^{k} \sum_{j=0}^{[\frac{k-l}{2}]} \frac{(-\frac{1}{2})^l (2n+2r-1)!(n+r-1)!}{l!(2n+2r-2l)!(k-l-2j)!j!4^j} H_{n-k}(x)$$

$$= \frac{1}{r!} \sum_{k=0}^{n} \frac{(-2)^k}{(n-k)!} \sum_{j=0}^{[\frac{k}{2}]} \frac{1}{j!4^j} \sum_{l=0}^{k-2j} \frac{(-\frac{1}{2})^l (2n+2r-1)!(n+r-1)!}{l!(2n+2r-2l)!(k-l-2j)!} H_{n-k}(x)$$

$$= \frac{(2n+2r)!}{r!4^{n+r} <\frac{1}{2}>_{n+r}} \sum_{k=0}^{n} \frac{(-2)^k}{(n-k)!} \sum_{j=0}^{[\frac{k}{2}]} \frac{1}{j!4^j(k-2j)!} \tag{51}$$

$$\times \sum_{l=0}^{k-2j} \frac{2^l <2j-k>_l <\frac{1}{2}-n-r>_l}{l! <-2n-2r>_l} H_{n-k}(x)$$

$$= \frac{(2n+2r)!}{r!4^{n+r}(n+r-\frac{1}{2})_{n+r}} \sum_{k=0}^{n} \frac{(-2)^k}{(n-k)!}$$

$$\times \sum_{j=0}^{[\frac{k}{2}]} \frac{{}_2F_1(2j-k, \frac{1}{2}-n-r; -2n-2r; 2)}{j!4^j(k-2j)!} H_{n-k}(x).$$

This shows (35) of Theorem 1.

Next, we let

$$\gamma_{n,r}(x) = \sum_{k=0}^{n} C_{k,2} L_k^{\alpha}(x). \tag{52}$$

Then, from (b) of Proposition 1, (45), (47) and integration by parts k times, we get

$$\begin{aligned}
C_{k,2} &= \frac{1}{2^r r! \Gamma(\alpha+k+1)} \int_0^\infty V_{n+r}^{(r)}(x) \frac{d^k}{dx^k}(e^{-x} x^{k+\alpha}) dx \\
&= \frac{(-1)^k}{2^r r! \Gamma(\alpha+k+1)} \int_0^\infty V_{n+r}^{(r+k)}(x) e^{-x} x^{k+\alpha} dx \\
&= \frac{(-1)^k}{2^r r! \Gamma(\alpha+k+1)} \sum_{l=0}^{n-k} \binom{2n+2r-l}{l} 2^{n+r-l}(n+r-l)_{r+k} \\
&\quad \times \int_0^\infty (x-1)^{n-k-l} e^{-x} x^{k+\alpha} dx \\
&= \frac{(-1)^k}{2^r r! \Gamma(\alpha+k+1)} \sum_{l=0}^{n-k} \binom{2n+2r-l}{l} 2^{n+r-l}(n+r-l)_{r+k} \\
&\quad \times \sum_{s=0}^{n-k-l} \binom{n-k-l}{s} (-1)^{n-k-l-s} \Gamma(s+k+\alpha+1) \\
&= \frac{(-1)^k}{2^r r!} \sum_{l=0}^{n-k} \binom{2n+2r-l}{l} 2^{n+r-l}(n+r-l)_{r+k} \\
&\quad \times \sum_{s=0}^{n-k-l} \binom{n-k-l}{s} (-1)^{n-k-l-s} <k+\alpha+1>_s \\
&= \frac{1}{r!} \sum_{l=0}^{n-k} \frac{(2n+2r-l)!(-2)^{n-l}(n+r-l)!}{l!(2n+2r-2l)!(n-k-l)!} \\
&\quad \times \sum_{s=0}^{n-k-l} \frac{1}{s!} <k+l-n>_s <k+\alpha+1>_s \\
&= \frac{1}{r!} \sum_{l=0}^{n-k} \frac{(2n+2r-l)!(-2)^{n-l}(n+r-l)!}{l!(2n+2r-2l)!(n-k-l)!} \\
&\quad \times {}_2F_0(k+l-n, k+\alpha+1; -; 1).
\end{aligned}$$
(53)

Combining (52)–(53), we finally have

$$\gamma_{n,r}(x) = \frac{1}{r!} \sum_{k=0}^n \sum_{l=0}^k \frac{(2n+2r-l)!(-2)^{n-l}(n+r-l)!}{l!(2n+2r-2l)!(k-l)!} \\
\times {}_2F_0(l-k, n-k+\alpha+1; -; 1) L_{n-k}^\alpha(x). \tag{54}$$

This completes the proof for (36) of Theorem 1.

Finally, let us put

$$\gamma_{n,r}(x) = \sum_{k=0}^n C_{k,3} P_k(x). \tag{55}$$

Then, from (c) of Proposition 1, (45), (47) and integration by parts k times, we have

$$\begin{aligned}
C_{k,3} &= \frac{2k+1(-1)^k}{2^{k+r+1} k! r!} \int_{-1}^1 V_{n+r}^{(r+k)}(x)(x^2-1)^k dx \\
&= \frac{(2k+1)(-1)^k}{2^{k+r+1} k! r!} \sum_{l=0}^{n-k} \binom{2n+2r-l}{l} 2^{n+r-l}(n+r-l)_{r+k} \\
&\quad \times \int_{-1}^1 (x-1)^{n-k-l}(x^2-1)^k dx.
\end{aligned} \tag{56}$$

By making use of (c) in Proposition 2 and after simplifications, from (56) we obtain

$$C_{k,3} = \frac{(-1)^{n+k}(2k+1)}{r!} \sum_{l=0}^{n-k}$$

$$\times \frac{(-1)^l 4^{n-l}(2n+2r-l)!(n+r-1)!(n-l)!}{l!(2n+2r-2l)!(n-k-l)!(n+k-l+1)!}$$

$$= \frac{(-1)^n(2n+2r)!n!}{r!4^r(n+r-\frac{1}{2})_{n+r}} \frac{(-1)^k(2k+1)}{(n-k)!(n+k+1)!} \tag{57}$$

$$\times \sum_{l=0}^{n-k} \frac{< k-n >_l < \frac{1}{2} - n - r >_l < -n - k - 1 >_l}{l! < -2n - 2r >_l < -n >_l}$$

$$= \frac{(-1)^n(2n+2r)!n!}{r!4^r(n+r-\frac{1}{2})_{n+r}} \frac{(-1)^k(2k+1)}{(n-k)!(n+k+1)!}$$

$$\times {}_3F_2(k-n, \frac{1}{2} - n - r, -n - k - 1; -2n - 2r, -n; 1).$$

From (55) and (57), we get

$$\gamma_{n,r}(x) = \frac{(-1)^n(2n+2r)!n!}{r!4^r(n+r-\frac{1}{2})_{n+r}} \sum_{k=0}^{n} \frac{(-1)^k(2k+1)}{(n-k)!(n+k+1)!} \tag{58}$$

$$\times {}_3F_2(k-n, \frac{1}{2} - n - r, -n - k - 1; -2n - 2r, -n; 1) P_k(x).$$

This proves (37) of Theorem 1.

3. Proof of Theorem 2

Here we will show only (43) and (44) in Theorem 2, as (40)–(42) can be shown analogously to the proofs for (35)–(37), respectively. The following can be derived by differentiating the Equation (10) and is stated in [24].

Lemma 2. *Let n, r be integers with $n \geq 0, r \geq 1$. Then we have the following identity.*

$$\sum_{l=0}^{n} \sum_{i_1+i_2+\cdots+i_{r+1}=l} (-1)^{n-l} \binom{r-1+n-l}{r-1} W_{i_1}(x) W_{i_2}(x) \cdots W_{i_{r+1}}(x) = \frac{1}{2^r r!} W_{n+r}^{(r)}(x), \tag{59}$$

where the inner sum runs over all nonnegative integers $i_1, i_2, \cdots, i_{r+1}$, with $i_1 + i_2 + \cdots + i_{r+1} = l$.

From (17), the rth derivative of $W_n(x)$ is given by

$$W_n^{(r)}(x) = (2n+1) \sum_{l=0}^{n-r} \frac{2^{n-l}}{2n+1-2l} \binom{2n-l}{l} (n-l)_r (x-1)^{n-l-r}. \tag{60}$$

In particular, we have

$$W_{n+r}^{(r+k)}(x) = (2n+1) \sum_{l=0}^{n-k} \frac{2^{n+r-l}}{2n+2r+1-2l} \binom{2n+2r-l}{l} (n+r-l)_{r+k} (x-1)^{n-k-l}. \tag{61}$$

With $\mathcal{E}_{n,r}(x)$ as in (34), we let

$$\mathcal{E}_{n,r}(x) = \sum_{k=0}^{n} C_{k,4} C_k^{(\alpha)}(x). \tag{62}$$

Then, from (d) of Proposition 1, (59), (61) and integration by parts k times, we get

$$C_{k,4} = \frac{(k+\lambda)\Gamma(\lambda)}{2^{k+r}r!\sqrt{\pi}\Gamma(k+\lambda+\frac{1}{2})}$$
$$\times \int_{-1}^{1} W_{n+r}^{(r+k)}(x)(1-x^2)^{k+\lambda-\frac{1}{2}}dx$$
$$= \frac{(k+\lambda)\Gamma(\lambda)(2n+1)}{2^{k+r}r!\sqrt{\pi}\Gamma(k+\lambda+\frac{1}{2})}$$
$$\times \sum_{l=0}^{n-k} \frac{2^{n+r-l}}{2n+2r+1-2l}\binom{2n+2r-l}{l}(n+r-l)_{r+k}$$
$$\times \int_{-1}^{1} (x-1)^{n-k-l}(1-x^2)^{k+\lambda-\frac{1}{2}}dx$$
$$= \frac{(k+\lambda)\Gamma(\lambda)(2n+1)(-2)^{n-k}}{r!\sqrt{\pi}\Gamma(k+\lambda+\frac{1}{2})}$$
$$\times \sum_{l=0}^{n-k} \frac{(-\frac{1}{2})^l(2n+2r-l)!(n+r-l)!}{(2n+2r-2l+1)l!(2n+2r-2l)!(n-k-l)!}$$
$$\times \int_{-1}^{1} (1-x)^{n-k-l}(1-x^2)^{k+\lambda-\frac{1}{2}}dx. \quad (63)$$

Invoking (c) of Proposition 2 and after simplifications, from (63) we obtain

$$C_{k,4} = \frac{(-1)^{n-k}(k+\lambda)\Gamma(\lambda)(2n+1)2^{2n+2\lambda+1}\Gamma(n+\lambda+\frac{1}{2})(2n+2r)!}{\Gamma(n+k+2\lambda+1)(n-k)!r!\sqrt{\pi}}$$
$$\times \sum_{l=0}^{n-k} \frac{(-\frac{1}{4})^l(2n+2r-l)!(n+r-l+1)!(n-k)!(n+k+2\lambda)_l}{l!(2n+2r)!(2n+2r-2l+2)!(n-k-l)!(n+\lambda-\frac{1}{2})_l}$$
$$= \frac{(-1)^k(k+\lambda)\Gamma(\lambda)(2n+1)2^{2\lambda-2r-1}(-1)^n\Gamma(n+\lambda+\frac{1}{2})(2n+2r)!}{\Gamma(n+k+2\lambda+1)(n-k)!r!\sqrt{\pi}(n+r+\frac{1}{2})_{n+r+1}}$$
$$\times \sum_{l=0}^{n-k} \frac{<k-n>_l<-n-r-\frac{1}{2}>_l<-n-k-2\lambda>_l}{l!<-2n-2r>_l<-n-\lambda+\frac{1}{2}>_l} \quad (64)$$
$$= \frac{(-1)^k(k+\lambda)\Gamma(\lambda)(2n+1)2^{2\lambda-2r-1}(-1)^n\Gamma(n+\lambda+\frac{1}{2})(2n+2r)!}{\Gamma(n+k+2\lambda+1)(n-k)!r!\sqrt{\pi}(n+r+\frac{1}{2})_{n+r+1}}$$
$$\times {}_3F_2(k-n,-n-r-\frac{1}{2},-n-k-2\lambda;-2n-2r,-n-\lambda+\frac{1}{2};1).$$

From (62) and (64), we have

$$\mathcal{E}_{n,r}(x) = \frac{\Gamma(\lambda)(2n+1)2^{2\lambda-2r-1}(-1)^n\Gamma(n+\lambda+\frac{1}{2})(2n+2r)!}{r!\sqrt{\pi}(n+r+\frac{1}{2})_{n+r+1}}$$
$$\times \sum_{k=0}^{n} \frac{(-1)^k(k+\lambda)}{\Gamma(n+k+2\lambda+1)(n-k)!} \quad (65)$$
$$\times {}_3F_2(k-n,-n-r-\frac{1}{2},-n-k-2\lambda;-2n-2r,-n-\lambda+\frac{1}{2};1)C_k^{(\alpha)}(x).$$

This shows (43) of Theorem 2.
Next, we let

$$\mathcal{E}_{n,r}(x) = \sum_{k=0}^{n} C_{k,5} P_n^{(\alpha,\beta)}(x). \quad (66)$$

Then, from (e) of Proposition 1, and (59), (61), and integrating by parts k times, we obtain

$$C_{k,5} = \frac{(2k+\alpha+\beta+1)\Gamma(k+\alpha+\beta+1)}{2^{\alpha+\beta+k+r+1}r!\Gamma(\alpha+k+1)\Gamma(\beta+k+1)}$$
$$\times \int_{-1}^{1} W_{n+r}^{(r+k)}(x)(1-x)^{k+\alpha}(1+x)^{k+\beta}dx$$
$$= \frac{(2k+\alpha+\beta+1)\Gamma(k+\alpha+\beta+1)(2n+1)}{2^{\alpha+\beta+k+r+1}r!\Gamma(\alpha+k+1)\Gamma(\beta+k+1)} \quad (67)$$
$$\times \sum_{l=0}^{n-k} \frac{2^{n+r-l}}{2n+2r-2l+1}\binom{2n+2r-l}{l}(n+r-l)_{r+k}(-1)^{n-k-l}$$
$$\times \int_{-1}^{1}(1-x)^{n+\alpha-l}(1+x)^{k+\beta}dx.$$

By exploiting (b) in Proposition 2 and after simplifications, from (67) we get

$$C_{k,5} = \frac{(-1)^{n-k}(2k+\alpha+\beta+1)\Gamma(k+\alpha+\beta+1)2^{2n+1}(2n+1)\Gamma(n+\alpha+1)}{\Gamma(\alpha+k+1)\Gamma(n+k+\alpha+\beta+2)r!}$$
$$\times \sum_{l=0}^{n-k} \frac{(-\frac{1}{4})^l(2n+2r-l)!(n+r-l+1)!(n+k+\alpha+\beta+1)_l}{l!(2n+2r-2l+2)!(n-k-l)!(n+\alpha)_l}$$
$$= \frac{(-1)^{n-k}(2k+\alpha+\beta+1)\Gamma(k+\alpha+\beta+1)(2n+1)\Gamma(n+\alpha+1)(2n+2r)!}{\Gamma(\alpha+k+1)\Gamma(n+k+\alpha+\beta+2)(n-k)!r!2^{2r+1}(n+r+\frac{1}{2})_{n+r+1}} \quad (68)$$
$$\times \sum_{l=0}^{n-k} \frac{<k-n>_l<-n-r-\frac{1}{2}>_l<-n-k-\alpha-\beta-1>_l}{l!<-2n-2r>_l<-n-\alpha>_l}$$
$$= \frac{(-1)^{n-k}(2k+\alpha+\beta+1)\Gamma(k+\alpha+\beta+1)(2n+1)\Gamma(n+\alpha+1)(2n+2r)!}{\Gamma(\alpha+k+1)\Gamma(n+k+\alpha+\beta+2)(n-k)!r!2^{2r+1}(n+r+\frac{1}{2})_{n+r+1}}$$
$$\times {}_3F_2(k-n,-n-r-\frac{1}{2},-n-k-\alpha-\beta-1;-2n-2r,-n-\alpha;1).$$

Thus, from (66) and (68), we have

$$\mathcal{E}_{n,r}(x) = \frac{(-1)^n(2n+1)\Gamma(n+\alpha+1)(2n+2r)!}{r!2^{2r+1}(n+r+\frac{1}{2})_{n+r+1}}$$
$$\times \sum_{k=0}^{n} \frac{(-1)^k(2k+\alpha+\beta+1)\Gamma(k+\alpha+\beta+1)}{\Gamma(\alpha+k+1)\Gamma(n+k+\alpha+\beta+2)(n-k)!}$$
$$\times {}_3F_2(k-n,-n-r-\frac{1}{2},-n-k-\alpha-\beta-1;-2n-2r,-n-\alpha;1)P_n^{(\alpha,\beta)}(x).$$

4. Conclusions

In this paper, we considered sums of finite products of Chebyshev polynomials of the third and fourth kinds and expressed each of them in terms of five orthogonal polynomials. Indeed, by explicit computations we expressed each of them as linear combinations of Hermite, generalized Laguerre, Legendre, Gegenbauer and Jacobi polynomials which involve some terminating hypergeometric functions. This can be viewed as a generalization of the classical linearization problem. In general, the linearization problem deals with determining the coefficients in the expansion of the products of two polynomials $a_m(x)$ and $b_n(x)$ in terms of an arbitrary polynomial sequence $\{p_k(x)\}_{k\geq 0}$:

$$a_m(x)b_n(x) = \sum_{k=0}^{m+n} c_{mn}(k)p_k(x).$$

Those sums of finite products were also represented by all kinds of Chebyshev polynomials in [20]. In addition, the same had been done for sums of finite products of Chebyshev polynomials of the first and second kinds, Fibonacci polynomials, Legendre polynomials, Laguerre polynomials and Lucas polynomials.

Author Contributions: T.K. and D.S.K. conceived the framework and structured the whole paper; T.K. wrote the paper; J.K. typed; D.V.D. checked for typos; D.S.K.completed the revision of the article.

Funding: This work was supported by the National Research Foundation of Korea (NRF) grant funded by the Korea government (MEST) (No. 2017R1E1A1A03070882).

Acknowledgments: We would like to thank the referees for their valuable comments and suggestions.

Conflicts of Interest: The authors declare no conflict of interest.

References

1. Kim, D.S.; Dolgy, D.V.; Kim, T.; Rim, S.H. Identities involving Bernoulli and Euler polynomials arising from Chebyshev polynomials. *Proc. Jangjeon Math. Soc.* **2012**, *15*, 361–370.
2. Kim, D.S.; Kim, T.; Dolgy, D.V. Some identities on Laguerre polynomials in connection with Bernoulli and Euler numbers. *Discret. Dyn. Nat. Soc.* **2012**, *2012*, 619197. [CrossRef]
3. Kim, D.S.; Kim, T.; Rim, S.-H. Some identities involving Gegenbauer polynomials. *Adv. Differ. Equ.* **2012**, *2012*, 219. [CrossRef]
4. Kim, D.S.; Kim, T.; Rim, S.-H.; Lee, S.H. Hermite polynomials and their applications associated with Bernoulli and Euler numbers. *Discret. Dyn. Nat. Soc.* **2012**, *2012*, 974632. [CrossRef]
5. Kim, D.S.; Rim, S.-H.; Kim, T. Some identities on Bernoulli and Euler polynomials arising from orthogonality of Legendre polynomials. *J. Inequal. Appl.* **2012**, *2012*, 227. [CrossRef]
6. Kim, T.; Kim, D.S. Extended Laguerre polynomials associated with Hermite, Bernoulli, and Euler numbers and polynomials. *Abstr. Appl. Anal.* **2012**, *2012*, 957350. [CrossRef]
7. Kim, T.; Kim, D.S.; Dolgy, D.V. Some identities on Bernoulli and Hermite polynomials associated with Jacobi polynomials. *Discret. Dyn. Nat. Soc.* **2012**, *2012*, 584643. [CrossRef]
8. Kim, T.; Kim, D.S. Identities for degenerate Bernoulli polynomials and Korobov polynomials of the first kind. *Sci. China Math.* **2018**. [CrossRef]
9. Kim, T.; Kim, D.S.; Kwon, J.; Gang, -W. J. Sums of finite products of Legendre and Laguerre polynomials by Chebyshev polynomials. *Adv. Stud. Contemp. Math. (Kyungshang)* **2018**, *28*, 551–565.
10. Andrews, G.E.; Askey, R.; Roy, R. *Special Functions*; Encyclopedia of Mathematics and its Applications 71; Cambridge University Press: Cambridge, UK, 1999.
11. Beals, R.; Wong, R. *Special Functions and Orthogonal Polynomials*; Cambridge Studies in Advanced Mathematics 153; Cambridge University Press: Cambridge, UK, 2016.
12. Wang, Z.X.; Guo, D.R. *Special Functions*; Translated by Guo, D.R., Xia, X.J.; World Scientific Publishing Co., Inc.: Teaneck, NJ, USA, 1989.
13. Mason, J.C.; Handscomb, D.C. *Chebyshev Polynomials*; Chapman & Hall/CRC: Boca Raton, FL, USA, 2003.
14. Agarwal, R.P.; San Kim, D.; Kim, T.; Kwon, J. Sums of finite products of Bernoulli functions. *Adv. Differ. Equ.* **2017**, *2017*, 237. [CrossRef]
15. Kim, T.; San Kim, D.; Jang, G.W.; Kwon, J. Sums of finite products of Euler functions. In *Advances in Real and Complex Analysis with Applications*; Trends in Mathematics; Birkhäuser: Singapore, 2017; pp. 243–260.
16. Kim, T.; Kim, D.S.; Jang, L.C.; Jang, G.-W. Sums of finite products of Genocchi functions. *Adv. Differ. Equ.* **2017**, *2017*, 268. [CrossRef]
17. Kim, T.; San Kim, D.; Dolgy, D.V.; Park, J.W. Sums of finite products of Chebyshev polynomials of the second kind and Fibonacci polynomials. *J. Inequal. Appl.* **2018**, *2018*, 148. [CrossRef] [PubMed]
18. Kim, T.; Kim, D.S.; Kwon, J.; Dolgy, D.V. Expressing sums of finite products of Chebyshev polynomials of the second kind and Fibonacci polynomials by several orthogonal polynomials. *Mathematics* **2018**, *6*, 210. [CrossRef]
19. Kim, T.; Dolgy, D.V.; Kim, D.S. Representing sums of finite products of Chebyshev polynomials of the second kind and Fibonacci polynomials in terms of Chebyshev polynomials. *Adv. Stud. Contemp. Math. (Kyungshang)* **2018**, *28*, 321–335.

20. Kim, T.; Kim, D.S.; Dolgy, D.V.; Ryoo, C.-S. Representing sums of finite products of Chebyshev polynomials of the third and fourth kinds by Chebyshev polynomials. *Symmetry* **2018**, *10*, 258. [CrossRef]
21. Kim, T.; Kim, D.; Victorovich, D.; Ryoo, C. Representing sums of finite products of Legendre and Laguerre polynomials by Chebyshev polynomials. *Adv. Stud. Contemp. Math. (Kyungshang)* **2018**, *28*, in press.
22. Doha, E.H.; Abd-Elhameed, W.M.; Alsuyuti, M.M. On using third and fourth kinds Chebyshev polynomials for solving the integrated forms of high odd-order linear boundary value problems. *J. Egyptian Math. Soc.* **2015**, *23*, 397–405. [CrossRef]
23. Mason, J.C. Chebyshev polynomials of the second, third and fourth kinds in approximation, indefinite integration, and integral transforms. *J. Comput. Appl. Math.* **1993**, *49*, 169–178. [CrossRef]
24. Kim, T.; San Kim, D.; Dolgy, D.V.; Kwon, J. Sums of finite products of Chebyshev polynomials of the third and fourth kinds. *Adv. Differ. Equ.* **2018**, *2018*, 283. [CrossRef]

© 2018 by the authors. Licensee MDPI, Basel, Switzerland. This article is an open access article distributed under the terms and conditions of the Creative Commons Attribution (CC BY) license (http://creativecommons.org/licenses/by/4.0/).

Article

Fibonacci and Lucas Numbers of the Form $2^a + 3^b + 5^c + 7^d$

Yunyun Qu [1,2], Jiwen Zeng [1,*] and Yongfeng Cao [3]

1. School of Mathematical Sciences, Xiamen University, Xiamen 361005, China; qucloud@163.com
2. School of Mathematical Sciences, Guizhou Normal University, Guiyang 550001, China
3. School of Big Data and Computer Science, Guizhou Normal University, Guiyang 550001, China; cyfeis@whu.edu.cn
* Correspondence: jwzeng@xmu.edu.cn

Received: 22 September 2018; Accepted: 14 October 2018; Published: 16 October 2018

Abstract: In this paper, we find all Fibonacci and Lucas numbers written in the form $2^a + 3^b + 5^c + 7^d$, in non-negative integers a, b, c, d, with $0 \leq \max\{a, b, c\} \leq d$.

Keywords: Fibonacci; Lucas; linear form in logarithms; continued fraction; reduction method

MSC: 11B39; 11J86; 11D61

1. Introduction

Let $\{F_n\}_{n \geq 0}$ be the Fibonacci sequence which is a second-order linear recursive sequence given by $F_{n+2} = F_{n+1} + F_n$, its initial values are $F_0 = 0$ and $F_1 = 1$, and its companion Lucas sequence $\{L_n\}_{n \geq 0}$ follows the same recursive pattern as the Fibonacci numbers, but with initial values $L_0 = 2$ and $L_1 = 1$. Fibonacci and Lucas numbers are very famous because they have amazing features (consult [1–3]). The problem of looking for a specific form of second-order recursive sequence has a very rich history. Bugeaud, Mignotte and Siksek [4] showed that $0, 1, 8, 144$ and $1, 4$ are the only Fibonacci and Lucas numbers, respectively, of the form y^t with $t > 1$ (perfect power). Other related papers searched for Fibonacci numbers of forms such as $px^2 + 1, px^3 + 1$ [5], $k^2 + k + 2$ [6], $p^a \pm p^b + 1$ [7]. In 1989, Luo [8] solved Vern Hoggatt's conjecture and proved that the only triangle numbers in the Fibonacci sequence $\{F_n\}$ are $1, 3, 21, 55$. In 1991, Luo [9] found all triangular numbers in the Lucas sequence $\{L_n\}$. In [10], Eric F. Bravo and Jhon J. Bravo found all positive integer solutions of the Diophantine equation $F_n + F_m + F_l = 2^a$ in non-negative integers n, m, l, and a with $n \geq m \geq l$. In [11], Normenyo, Luca and Togbé determined all base-10 repdigits that are expressible as sums of four Fibonacci or Lucas numbers. In [12], Marques and Togbé searched for Fibonacci numbers of the form $2^a + 3^b + 5^c$ which are sum of three perfect powers of some prescribed distinct bases.

In this paper, we are interested in Fibonacci numbers and Lucas numbers which are sum of four perfect powers of several prescribed distinct bases. The number of perfect powers involved in the Diophantine equation solved by the literature [12] is one less than the perfect powers involved in the equation solved by us and the amount of computation in the literature [12] is relatively small. More precisely, our results are the following.

Theorem 1. *The solutions of the Diophantine equation*

$$F_n = 2^a + 3^b + 5^c + 7^d \tag{1}$$

in non-negative integers n, a, b, c, d with $0 \leq \max\{a, b, c\} \leq d$ are $(n, a, b, c, d) \in \{(7, 1, 1, 0, 1), (10, 1, 1, 0, 2), (10, 2, 0, 0, 2), (14, 1, 3, 1, 3), (14, 3, 0, 2, 3)\}$.

Theorem 2. *The solutions of the Diophantine equation*

$$L_n = 2^a + 3^b + 5^c + 7^d \qquad (2)$$

in non-negative integers n, a, b, c, d with $0 \leq \max\{a, b, c\} \leq d$ are $(n, a, b, c, d) \in \{(3, 0, 0, 0, 0), (5, 1, 0, 0, 1), (9, 0, 0, 2, 2)\}$.

2. Preliminaries

Before proceeding further, we recall some facts and tools which will be used later.
First, we recall the Binet's formulae for Fibonacci and Lucas sequences:

$$F_n = \frac{\gamma^n - \mu^n}{\gamma - \mu}$$

and

$$L_n = \gamma^n + \mu^n,$$

where $\gamma = \frac{1+\sqrt{5}}{2}$ and $\mu = \frac{1-\sqrt{5}}{2}$ are the roots of $F_n's$ characteristic polynomial $x^2 - x - 1 = 0$. For all positive integers n, the inequalities

$$\gamma^{n-2} \leq F_n \leq \gamma^{n-1}, \quad \gamma^{n-1} \leq L_n \leq 2\gamma^n \qquad (3)$$

hold.

In order to prove our theorem, one tool used is a Baker type lower bound for a linear form in logarithms of algebraic numbers, and such a bound was given by the following result of Matveev (see [13]).

Lemma 1. *Let $\gamma_1, \gamma_2, \cdots, \gamma_t$ be real algebraic numbers and let $b_1, \cdots, b_t$ be non-zero rational integers. Let D be the degree of the number field $\mathbb{Q}(\gamma_1, \gamma_2, \cdots, \gamma_t)$ over $\mathbb{Q}$ and let A_j be a real number satisfying*

$$A_j \geq \max\{Dh(\gamma_j), |\log \gamma_j|, 0.16\}$$

for $j = 1, \cdots, t$. Assume that

$$B \geq \max\{|b_1|, \cdots, |b_t|\}.$$

If $\gamma_1^{b_1} \cdots \gamma_t^{b_t} \neq 1$, then

$$|\gamma_1^{b_1} \cdots \gamma_t^{b_t} - 1| \geq \exp(-1.4 \times 30^{t+3} \times t^{4.5} \times D^2(1 + \log D)(1 + \log B) A_1 \cdots A_t).$$

As usual, in the above statement, the logarithmic height of an s-degree algebraic number γ is defined as

$$h(\gamma) = \frac{1}{s}(\log|a| + \sum_{j=1}^{s} \log\max\{1, |\gamma^{(j)}|\}),$$

where a is the leading coefficient of the minimal polynomial of γ (over $\mathbb{Z}$) and $\gamma^{(j)}, 1 \leq j \leq s$ are the conjugates of γ (over $\mathbb{Q}$).

After finding an upper bound on n which is in general too large, the next step is to reduce it. For that, we need a variant of the famous Baker–Davenport lemma which was developed by Dujella and Pethő [14]. For a real number x, we use $\|x\| = \min\{|x - n| : n \in \mathbb{Z}\}$ for the distance from x to the nearest integer.

Lemma 2. *(see [10]) Let M be a positive integer, let $\frac{p}{q}$ be a convergent of the continued fraction of the irrational number α such that $q > 6M$, and let A, B, τ be some real numbers with $A > 0$ and $B > 1$.*

Let $\epsilon := \|\tau q\| - M\|\alpha q\|$, where $\|\cdot\|$ denotes the distance from the nearest integer. If $\epsilon > 0$, then no solution to the inequality

$$0 < |u\alpha - v + \tau| < AB^{-w}$$

exists in positive integers u, v, and w with $u \leq M$ and $w \geq \frac{\log(Aq/\epsilon)}{\log B}$.

Next, we are ready to handle the proofs of our results.

3. Proof of Theorem 1

3.1. Bounding n

By combining the Binet formula together with (1), we get

$$\frac{\gamma^n}{\sqrt{5}} - 7^d = 2^a + 3^b + 5^c + \frac{\mu^n}{\sqrt{5}} > 0, \qquad (4)$$

because $|\mu| < 1$ while $2^a \geq 1$. Thus,

$$\frac{\gamma^n 7^{-d}}{\sqrt{5}} - 1 = \frac{2^a}{7^d} + \frac{3^b}{7^d} + \frac{5^c}{7^d} + \frac{\mu^n}{7^d\sqrt{5}} > 0 \qquad (5)$$

yields

$$\left|\frac{\gamma^n 7^{-d}}{\sqrt{5}} - 1\right| < \frac{4}{7^{0.1d}}. \qquad (6)$$

From the first inequality of (3), we obtain the estimate $\gamma^{n-2} < 4 \times 7^d$ and $7^d < \gamma^{n-1}$, which implies that $0.24n - 1.9 < d < 0.25(n-1)$; also, this yields $d < n$.

We are in a situation where we can apply Matveev's result Lemma 1 to the left side of (6). The left expression of (6) is nonzero, since, if this expression is zero, it means that $\gamma^{2n} = 7^{2d} \times 5 \in \mathbb{Z}$, so $\gamma^{2n} \in \mathbb{Z}$ for some positive integer n, which is false. We take $t := 3$, $\gamma_1 := \gamma$, $\gamma_2 := 7$, $\gamma_3 := \sqrt{5}$ and $b_1 := n, b_2 := -d, b_3 := -1$. Then we have $D = [\mathbb{Q}(\sqrt{5}) : \mathbb{Q}] = 2$. Note that $h(\gamma_1) = \frac{1}{2}\log\gamma$, $h(\gamma_2) = \log 7$ and $h(\gamma_3) = \log\sqrt{5}$. Thus, we can take $A_1 := 0.5$, $A_2 := 3.9$ and $A_3 := 1.7$. Note that $\max\{|b_1|, |b_2|, |b_3|\} = \max\{n, d, 1\} = n$. We are in position to apply Matveev's result Lemma 1. This lemma together with a straightforward calculation gives

$$\left|\frac{\gamma^n 7^{-d}}{\sqrt{5}} - 1\right| > \exp(-C(1 + \log n)), \qquad (7)$$

where $C = 3.22 \times 10^{12}$. Thus, from (6), (7) and $d > 0.24n - 1.9$, taking logarithms in the inequalities (6), (7) and comparing the resulting inequalities, we get

$$0.046n - 1.8 < 3.22 \times 10^{12} \times (1 + \log n),$$

giving $n < 2.56 \times 10^{15}$. We summarize the conclusions of this section as follows.

Lemma 3. *If (n, a, b, c, d) is a solution in positive integers to Equation (1) with $0 \leq \max\{a, b, c\} \leq d$, then*

$$d < n < 2.56 \times 10^{15}.$$

3.2. Reducing the Bound on n

We use Lemma 2 several times to reduce the bound for n. We return to (6). Put

$$\Lambda_F := n\log\gamma - d\log 7 - \log\sqrt{5}.$$

Then (5), (6) implies that
$$0 < \Lambda_F < e^{\Lambda_F} - 1 < \frac{4}{7^{0.1d}}. \tag{8}$$

Dividing across by log 7, we get
$$0 < |n\frac{\log \gamma}{\log 7} - d - \frac{\log \sqrt{5}}{\log 7}| < \frac{2.1}{7^{0.1d}}. \tag{9}$$

We are now ready to apply Lemma 2 with the obvious parameters,
$$\alpha := \frac{\log \gamma}{\log 7}, v := d, \tau := -\frac{\sqrt{5}}{\log 7}, A := 2.1, B := 1.2.$$

It is easy to see that α is irrational. In fact, we assume that $\alpha = \frac{p}{q}$, where $p, q \in \mathbb{Z}^+$ and $\gcd(p, q) = 1$. Then $\gamma^q = 7^p$, hence $\overline{\gamma}^q = 7^p$, where $\overline{\gamma}$ is the conjugate of γ. Thus, we can get $\gamma^q \overline{\gamma}^q = 7^{2p}$; hence, $(-1)^q = 7^{2p}$ which is an absurdity. We can take $M := 2.56 \times 10^{15}$. Let $\frac{p_k}{q_k}$ be the kth convergent of the continued fraction of α. By applying Lemma 2 and performing the calculations with $q_{39} > 6M$ and $\epsilon = \|\tau q_{39}\| - M\|\alpha q_{39}\| = 0.42904\cdots$, we get that if (n, a, b, c, d) is a solution in positive integers of Equation (1), then $d < 225$, which implies that
$$n < \frac{226.9}{0.24} = 945.417 < 946.$$

Then we can take $M := 946$. By applying Lemma 2 again and performing the calculations with $q_8 > 6M$ and $\epsilon = \|\tau q_8\| - M\|\alpha q_8\| = 0.07417\cdots$, we get that if (n, a, b, c, d) is a solution in positive integers of Equation (1), then $d < 73$, which implies that
$$n < 313.$$

Finally, we apply a program written in Mathematica to determine the solutions to (1) in the range $0 \le \max\{a, b, c\} \le d < 73$ and $n < 313$. Quickly, the program returns the following solutions: $(n, a, b, c, d) \in \{(7, 1, 1, 0, 1), (10, 1, 1, 0, 2), (10, 2, 0, 0, 2), (14, 1, 3, 1, 3), (14, 3, 0, 2, 3)\}$. This proof has been completed.

4. Proof of Theorem 2

4.1. Bounding n

By combining Binet formula together with (2), we get
$$\gamma^n - 7^d = 2^a + 3^b + 5^c - \mu^n > 0, \tag{10}$$

because $|\mu| < 1$ while $2^a \ge 1$. Thus,
$$\gamma^n 7^{-d} - 1 = \frac{2^a}{7^d} + \frac{3^b}{7^d} + \frac{5^c}{7^d} - \mu^n 7^{-d} > 0 \tag{11}$$

yields
$$|\gamma^n 7^{-d} - 1| < \frac{4}{7^{0.1d}}. \tag{12}$$

From the second inequality of (3) and (2), we obtain the estimate $\gamma^{n-1} < 4 \times 7^d$ and $7^d < 2 \times \gamma^n$, which implies that $4.04d - 1.45 < n < 4.05d + 3.89$; also, this yields $d \le n$.

We are also in a situation where we can apply Matveev's result Lemma 1 to the left side of (12). The left expression of (12) is nonzero, since, if this expression is zero, it means that $\gamma^n = 7^d \in \mathbb{Z}$, so $\gamma^n \in \mathbb{Z}$ for some positive integer n, which is false. We take $t := 2$, $\gamma_1 := \gamma, \gamma_2 := 7$ and $b_1 := n, b_2 := -d$.

Then we have $D = [\mathbb{Q}(\sqrt{5}) : \mathbb{Q}] = 2$. Note that $h(\gamma_1) = \frac{1}{2}\log\gamma$, $h(\gamma_2) = \log 7$. Thus, we can take $A_1 := 0.5$, $A_2 := 3.9$. Note that $B = \max\{|b_1|, |b_2|\} = \max\{n, d\} = n$. We are in position to apply Matveev's result Lemma 1. This lemma together with a straightforward calculation gives

$$|\gamma^n 7^{-d} - 1| > \exp(-C(1 + \log n)), \tag{13}$$

where $C = 1.02 \times 10^{10}$. Thus, from (12), (13) and $d > \frac{n-3.89}{4.05}$, taking logarithms in the inequalities (12), (13) and comparing the resulting inequalities, we get

$$0.1(n-1)\log\gamma - 1.1 \times \log 4 < C \times (1 + \log n),$$

giving $n < 6.47 \times 10^{12}$. The conclusions of this section are as follows.

Lemma 4. *If (n, a, b, c, d) is a solution in positive integers to Equation (2) with $0 \le \max\{a, b, c\} \le d$, then*

$$d \le n < 6.47 \times 10^{12}.$$

4.2. Reducing the Bound on n

We use the extremality property of continued fraction to reduce the bound for n. We return to (12) and put

$$\Lambda_L := n\log\gamma - d\log 7.$$

Then (11), (12) implies that

$$0 < \Lambda_L < e^{\Lambda_L} - 1 < \frac{4}{7^{0.1d}}. \tag{14}$$

Dividing by $\log 7$, we get

$$0 < n\frac{\log\gamma}{\log 7} - d < \frac{2.1}{1.2^d}. \tag{15}$$

Let $[a_0, a_1, a_2, a_3, a_4, \cdots,] = [0, 4, 22, 1, 5, 1, 1, 17, \cdots]$ be the continued fraction of $\frac{\log\gamma}{\log 7}$, and let $\frac{p_k}{q_k}$ be its kth convergent. Recall that $n < 6.47 \times 10^{12}$ by Lemma 4. A quick inspection using Mathematica reveals that $q_{19} < 1.662 \times 10^{12} < q_{20}$. Furthermore, $a_M := \max\{a_i : i = 0, 1, \cdots, 27\} = a_{14} = 35$. So, in accordance with the extremality property of continued fraction, we obtain that

$$\left|n\frac{\log\gamma}{\log 7} - d\right| > \frac{1}{(a_M + 2)n} = \frac{1}{37n}. \tag{16}$$

By comparing estimates (15) and (16), we get right away that

$$\frac{1}{37n} < \frac{2.1}{1.2^d}.$$

This leads to

$$d < \frac{\log(2.1 \times 37n)}{\log 1.2} < 186,$$

which implies that

$$n < 757.$$

This can lead to

$$d < \frac{\log(2.1 \times 37n)}{\log 1.2} < 61,$$

which implies that

$$n < 251.$$

Finally, we use a program written in Mathematica to find the solutions to (2) in the range $0 \leq \max\{a,b,c\} \leq d < 61$ and $n < 251$. Quickly, the program returns the following solutions: $(n,a,b,c,d) \in \{(3,0,0,0,0),(5,1,0,0,1),(9,0,0,2,2)\}$. This completes the proof.

5. Conclusions

In this paper, we find all the solutions of the Diophantine equation (1) by using a Baker type lower bound for a nonzero linear form in logarithms of algebraic numbers and the Lemma 2 from Diophantine approximation to reduce the upper bounds on the variables of the equation. For the Diophantine equation (2), we solve the equation by using the lower bound for a nonzero linear form in logarithms of algebraic numbers and the extremality properties of continued fraction to reduce the upper bounds on the variables of the equation.

6. Future Developments

We remark that we can further take advantage of our method to prove that there are only finitely many solutions (and all of them are effectively computable) for the Diophantine equation $F_n = -2^a - 3^b - 5^c + 7^d$, $L_n = -2^a - 3^b - 5^c + 7^d$ in non-negative integers n,a,b,c,d with $0 \leq \max\{a,b,c\} \leq d$. We leave this as a problem for other researchers.

Author Contributions: All authors contributed equally to this work. All authors read and approved the final manuscript.

Funding: This research was supported by the Youth Science and Technology Talent Growth Program of Guizhou Provincial Education Department(No. QIANJIAOHEKYZI[2016]130), the Natural Science Foundation of Educational Commission of Guizhou Province (No. KY[2016]027), the Natural Science Foundation of Science and Technology Department of Guizhou Province (No. GZKJ[2017]1128) and the National Natural Science Foundation of China (No. 11261060).

Acknowledgments: The authors would like to express their sincere gratitude to the referees for their valuable comments which have significantly improved the presentation of this paper.

Conflicts of Interest: The authors declare no conflict of interest.

References

1. Koshy, T. *Fibonacci and Lucas Numbers with Applications*; Pure and Applied Mathematics (New York); Wiley-Interscience: New York, NY, USA, 2001.
2. Kim, T.; Kim, D.S.; Dolgy, D.V.; Park, J.-W. Sums of finite products of Chebyshev polynomials of the second kind and of Fibonacci polynomials. *J. Inequal. Appl.* **2018**, *2018*, 148. [CrossRef] [PubMed]
3. Kim, T.; Dolgy, D.V.; Kim, D.S.; Seo, J.J. Convolved Fibonacci numbers and their applications. *ARS Comb.* **2017**, *135*, 119–131.
4. Bugeaud, Y.; Mignotte, M.; Siksek, S. Classical and modular approaches to exponential Diophantine equations. I. Fibonacci and Lucas perfect powers. *Ann. Math.* **2006**, *163*, 969–1018. [CrossRef]
5. Robbins, N. Fibonacci numbers of the forms $pX^2 + 1$, $pX^3 + 1$, where p is prime. In Proceedings of the Applications of Fibonacci Numbers, San Jose, CA, USA, 13–16 August 1986; Kluwer Academic Publishers: Dordrecht, The Netherlands, 1988; pp. 77–88.
6. Luca, F. Fibonacci numbers of the form $k^2 + k + 2$. In Proceedings of the Applications of Fibonacci Numbers, Rochester, NY, USA, 22–26 June 1998; Kluwer Academic Publishers: Dordrecht, The Netherlands, 1999; Volume 8, pp. 241–249.
7. Luca, F.; Szalay, L. Fibonacci numbers of the form $p^a \pm p^b + 1$. *Fibonacci Quart.* **2007**, *45*, 98–103.
8. Luo, M. On triangular Fibonacci numbers. *Fibonacci Quart.* **1989**, *27*, 98–108.
9. Luo, M. On triangular Lucas numbers. In *Applications of Fibonacci Numbers*; Springer: Dordrecht, The Netherlands, 1991; pp. 231–240.
10. Bravo, E.F.; Bravo, J.J. Powers of two as sums of three Fibonacci numbers. *Lith. Math. J.* **2015**, *55*, 301–311. [CrossRef]
11. Normenyo, B.V.; Luca, F.; Togbé, A. Repdigits as Sums of Four Fibonacci or Lucas Numbers. *J. Integer Seq.* **2018**, *21*, 1–30.

12. Marques, D.; Togbé, A. Fibonacci and Lucas numbers of the form $2^a + 3^b + 5^c$. *Proc. Jpn. Acad. Ser. A Math. Sci.* **2013**, *89*, 47–50. [CrossRef]
13. Matveev, E.M. An explicit lower bound for a homogeneous rational linear form in logarithms of algebraic numbers, II. *Izv. Ross. Akad. Nauk Ser. Mat.* **2000**, *64*, 125–180. *English Translation in Izv. Math.* **2000**, *64*, 1217–1269. [CrossRef]
14. Dujella, A.; Pethő, A. A generalization of a theorem of Baker and Davenport. *Q. J. Math.* **1998**, *49*, 291–306. [CrossRef]

© 2018 by the authors. Licensee MDPI, Basel, Switzerland. This article is an open access article distributed under the terms and conditions of the Creative Commons Attribution (CC BY) license (http://creativecommons.org/licenses/by/4.0/).

Article

A Note on Modified Degenerate Gamma and Laplace Transformation

YunJae Kim [1], Byung Moon Kim [2], Lee-Chae Jang [3] and Jongkyum Kwon [4],*

[1] Department of Mathematics, Dong-A University, Busan 49315, Korea; kimholzi@gmail.com
[2] Department of Mechanical System Engineering, Dongguk University, Gyungju-si, Gyeongsangbukdo 38066, Korea; kbm713@dongguk.ac.kr
[3] Graduate School of Education, Konkuk University, Seoul 139-701, Korea; lcjang@konkuk.ac.kr
[4] Department of Mathematics Education and ERI, Gyeongsang National University, Jinju, Gyeongsangnamdo 52828, Korea
* Correspondence: mathkjk26@gnu.ac.kr

Received: 8 September 2018; Accepted: 3 October 2018; Published: 10 October 2018

Abstract: Kim-Kim studied some properties of the degenerate gamma and degenerate Laplace transformation and obtained their properties. In this paper, we define modified degenerate gamma and modified degenerate Laplace transformation and investigate some properties and formulas related to them.

Keywords: the degenerate gamma function; the modified degenerate gamma function; the degenerate Laplace transform; the modified degenerate Laplace transform

1. Introduction

It is well known that gamma function is defied by

$$\Gamma(s) = \int_0^\infty e^{-t} t^{s-1} dt, \text{ where } s \in \mathbf{C} \text{ with } Re(s) > 0, \qquad (1)$$

(see [1,2]). From (1), we note that

$$\Gamma(s+1) = s\Gamma(s), \text{ and } \Gamma(n+1) = n!, \text{ where } n \in \mathbf{N}. \qquad (2)$$

Let $f(t)$ be a function defined for $t \geq 0$. Then, the integral

$$L(f(t)) = \int_0^\infty e^{-st} f(t) dt, \qquad (3)$$

(see [1–4]), is said to be the Laplace transform of f, provided that the integral converges. For $\lambda \in (0, \infty)$. Kim-Kim [2] introduced the degenerate gamma function for the complex variable s with $0 < Re(s) < \frac{1}{\lambda}$ as follows:

$$\Gamma_\lambda(s) = \int_0^\infty (1+\lambda t)^{-\frac{1}{\lambda}} t^{s-1} dt, \qquad (4)$$

(see [2]) and degenerate Laplace transformation which was defined by

$$L_\lambda(f(t)) = \int_0^\infty (1+\lambda t)^{-\frac{s}{\lambda}} f(t) dt, \qquad (5)$$

(see [2,5]), if the integral converges. The authors obtained some properties and interesting formulas related to the degenerate gamma function. For examples, For $\lambda \in (0,1)$ and $0 < Re(s) < \frac{1-\lambda}{\lambda}$,

$$\Gamma_\lambda(s+1) = \frac{s}{(1-\lambda)^{s-1}} \Gamma_{\frac{\lambda}{1-\lambda}}(s), \qquad (6)$$

and $\lambda \in (0, \frac{1}{k+s})$ with $k \in \mathbf{N}$ and $0 < Re(s) < \frac{1-\lambda}{\lambda}$,

$$\Gamma_\lambda(s+1) = \frac{s(s-1)\cdots(s-(k+1)+1)}{(1-\lambda)(1-2\lambda)\cdots(1-k\lambda)(1-(k+1)\lambda)} \Gamma_{\frac{\lambda}{1-(k+1)\lambda}}(s-k), \qquad (7)$$

and for $k \in \mathbf{N}$ and $\lambda \in (0, \frac{1}{k})$,

$$\Gamma_\lambda(k) = \frac{(k-1)!}{(1-\lambda)(1-2\lambda)\cdots(1-k\lambda)}. \qquad (8)$$

The authors obtained some formulas related to the degenerate Laplace transformation. For examples,

$$L_\lambda(1) = \frac{1}{s-\lambda}, \text{ if } s > \lambda, \qquad (9)$$

and

$$L_\lambda((1+\lambda t)^{-\frac{a}{\lambda}}) = \frac{1}{s+a-\lambda}, \text{ if } s > -a+\lambda, \qquad (10)$$

and

$$L_\lambda(\cos_\lambda(at)) = \frac{s-\lambda}{(s-\lambda)^2+a^2}, \qquad (11)$$

and

$$L_\lambda(\sin_\lambda(at)) = \frac{a}{(s-\lambda)^2+a^2}, \qquad (12)$$

where $\cos_\lambda(t) = \frac{1}{2}\left((1+\lambda t)^{\frac{it}{\lambda}} + (1+\lambda t)^{-\frac{it}{\lambda}}\right)$ and $\sin_\lambda(t) = \frac{1}{2i}\left((1+\lambda t)^{\frac{it}{\lambda}} - (1+\lambda t)^{-\frac{it}{\lambda}}\right)$.

Furthermore, the authors obtained that

$$L_\lambda(t^n) = \frac{n!}{(s-\lambda)(s-2\lambda)\cdots(s-n\lambda)(s-(n+1)\lambda)}, \qquad (13)$$

for $n \in \mathbf{N}$ and $s > (n+1)\lambda$, and

$$L_\lambda(f^{(n)}(t)) = s(s+\lambda)(s+2\lambda)\cdots(s+(n-1)\lambda)L_\lambda((1+\lambda t)^{-n}f(t)) \\ - \sum_{i=0}^{n-1} f^{(i)}(0)\left(\prod_{t=1}^{n-i-1} s+(l-1)\lambda\right). \qquad (14)$$

where $f, f^{(1)}, \cdots, f^{(n-1)}$ are continuous on $(0, \infty)$ and are of degenerate exponential order and $f^{(n)}(t)$ is piecewise continuous on $(0, \infty)$, and

$$L_\lambda((\log(1+\lambda t))^n f(t)) = (-1)^n \lambda^n \left(\frac{d}{ds}\right)^n L_\lambda(s), \qquad (15)$$

for $n \in \mathbf{N}$.

At first, L. Carlitz introduced the degenerate special polynomials (see [6,7]). The recently works which can be cited in this and researchers have studied the degenerate special polynomials and numbers (see [2,8–19]). Recently, the concept of degenerate gamma function and degenerate Laplace transformation was introduced by Kim-Kim [2]. They studied some properties of the degenerate gamma and degenerate Laplace transformation and obtained their properties. We observe whether or not that holds. Thus, we consider the modified degenerate Laplace transform which are satisfied (16). The degenerate gamma and degenerate Laplace transformation applied to engineer's mathematical toolbox as they make solving linear ODEs and related initial value problems. This paper consists of two sections. The first section contains the modified degenerate gamma function and investigate the properties of the modified gamma function. The second part of the paper provide the modified degenerate Laplace transformation and investigate interesting results of the modified degenerate Laplace transformation.

$$L_\lambda(f * g) = L_\lambda(f) L_\lambda(g) \tag{16}$$

2. Modified Degenerate Gamma Function

In this section, we will define modified degenerate gamma functions which are different to degenerate gamma functions. For each $\lambda \in (0, \infty)$, we define modified degenerate gamma function for the complex variable s with $0 < Re(s)$ as follows:

$$\Gamma_\lambda^*(s) = \int_0^\infty (1+\lambda)^{-\frac{t}{\lambda}} t^{s-1} dt. \tag{17}$$

Let $\lambda \in (0,1)$. Then, for $0 < Re(s)$, we have

$$\Gamma_\lambda^*(s+1) = \int_0^\infty (1+\lambda)^{-\frac{t}{\lambda}} t^s dt$$

$$= \frac{1}{(\log(1+\lambda))^{-\frac{1}{\lambda}}} (1+\lambda)^{-\frac{t}{\lambda}} t^s \Big|_0^\infty + \frac{\lambda}{\log(1+\lambda)} \int_0^\infty s(1+\lambda)^{-\frac{t}{\lambda}} t^{s-1} dt \tag{18}$$

$$= \frac{\lambda}{\log(1+\lambda)} s \Gamma_\lambda^*(s).$$

Therefore, by (18), we obtain the following theorem.

Theorem 1. Let $\lambda \in (0,1)$. Then, for $0 < Re(s)$, we have

$$\Gamma_\lambda^*(s+1) = \frac{\lambda s}{\log(1+\lambda)} \Gamma_\lambda^*(s). \tag{19}$$

Then, for $0 < Re(s)$ and $\lambda \in (0,1)$, repeatly we calculate

$$\Gamma_\lambda^*(s+1) = \frac{\lambda s}{\log(1+\lambda)} \Gamma_\lambda^*(s) = \frac{\lambda^2 (s-1)}{(\log(1+\lambda))^2} \Gamma_\lambda^*(s-1). \tag{20}$$

Thus, continuing this process, for $0 < Re(s)$ and $\lambda \in (0,1)$, we have

$$\Gamma_\lambda^*(s+1) = \frac{\lambda^k (s-1) \cdots (s-k+1)}{(\log(1+\lambda))^k} \Gamma_\lambda^*(s-k). \tag{21}$$

Therefore, by (21), we obtain the following theorem.

Theorem 2. Let $\lambda \in (0,1)$. Then, for $0 < Re(s)$, we have

$$\Gamma_\lambda^*(s+1) = \frac{\lambda^k (s-1) \cdots (s-k+1)}{(\log(1+\lambda))^k} \Gamma_\lambda^*(s-k). \tag{22}$$

Let us take $s = k+1$. Then, by Theorem 2, we get

$$\Gamma_\lambda^*(k+2) = \frac{\lambda^{k+1} k \cdots 2}{(\log(1+\lambda))^{k+1}} \Gamma_\lambda^*(1)$$

$$= \frac{\lambda^{k+1} k!}{(\log(1+\lambda))^{k+1}} \Gamma_\lambda^*(1) \tag{23}$$

and

$$\Gamma_\lambda^*(1) = \int_0^\infty (1+\lambda)^{-\frac{t}{\lambda}} dt$$
$$= -\frac{\lambda}{(\log(1+\lambda))}(1+\lambda)^{-\frac{t}{\lambda}} \Big|_0^\infty \quad (24)$$
$$= \frac{\lambda}{(\log(1+\lambda))}.$$

Therefore, by (23) and (24), we obtain the following theorem.

Theorem 3. *For $k \in \mathbb{N}$ and $\lambda \in (0,1)$, we have*

$$\Gamma_\lambda^*(k+1) = \frac{\lambda^{k+1} k!}{(\log(1+\lambda))^{k+1}}. \quad (25)$$

3. Modified Degenerate Laplace Transformation

In this section, we will define modified Laplace transformation which are different to degenerate Laplace transformation. Let $\lambda \in (0, \infty)$ and let $f(t)$ be a function defined for $t \geq 0$. Then the integral

$$\mathcal{L}_\lambda^*(f(t)) = \int_0^\infty (1+\lambda s)^{-\frac{t}{\lambda}} f(t) dt. \quad (26)$$

is said to be the modified degenerate Laplace transformation of f if the integral converges which is also defined by $\mathcal{L}_\lambda^*(f(t)) = F_\lambda(s)$.

From (26), we get

$$\mathcal{L}_\lambda^*(\alpha f(t) + \beta g(t)) = \alpha \mathcal{L}_\lambda^*(f(t)) + \beta \mathcal{L}_\lambda^*(g(t)), \quad (27)$$

where α and β are constant real numbers.

First, we observe that for $n \in \mathbb{N}$,

$$\mathcal{L}_\lambda^*(t^n) = \int_0^\infty (1+\lambda s)^{-\frac{t}{\lambda}} t^n dt$$
$$= -\frac{\lambda}{\log(1+\lambda s)}(1+\lambda s)^{-\frac{t}{\lambda}} t^n \Big|_0^\infty + \frac{\lambda n}{\log(1+\lambda s)} \int_0^\infty (1+\lambda s)^{-\frac{t}{\lambda}} t^{n-1} dt$$
$$= \frac{\lambda n}{\log(1+\lambda s)} \mathcal{L}_\lambda^*(t^{n-1})$$
$$= \frac{\lambda n}{\log(1+\lambda s)} \left(-\frac{\lambda}{\log(1+\lambda s)}(1+\lambda s)^{-\frac{t}{\lambda}} t^{n-1} \Big|_0^\infty + \frac{\lambda(n-1)}{\log(1+\lambda s)} \int_0^\infty (1+\lambda s)^{-\frac{t}{\lambda}} t^{n-2} dt \right) \quad (28)$$
$$= \left(\frac{\lambda}{\log(1+\lambda s)} \right)^2 n(n-1) \mathcal{L}_\lambda^*(t^{n-2})$$
$$= \cdots$$
$$= \left(\frac{\lambda}{\log(1+\lambda s)} \right)^n n! \mathcal{L}_\lambda^*(1)$$
$$= \left(\frac{\lambda}{\log(1+\lambda s)} \right)^{n+1} n!.$$

Therefore, by (28), we obtain the following theorem.

Theorem 4. *For $k \in \mathbb{N}$ and $\lambda \in (0,1)$, we have*

$$\mathcal{L}_\lambda^*(t^n) = \left(\frac{\lambda}{\log(1+\lambda s)} \right)^{n+1} n!. \quad (29)$$

Secondly, we note that if f is a periodic function with a period T.

$$\begin{aligned}\mathcal{L}_\lambda^*(f(t)) &= \int_0^\infty (1+\lambda s)^{-\frac{t}{\lambda}} f(t) dt \\ &= \int_0^T (1+\lambda s)^{-\frac{t}{\lambda}} f(t) dt + \int_T^\infty (1+\lambda s)^{-\frac{t}{\lambda}} f(t) dt \\ &= \int_0^T (1+\lambda s)^{-\frac{t}{\lambda}} f(t) dt + \int_0^\infty (1+\lambda s)^{-\frac{t+T}{\lambda}} f(t+T) dt \\ &= \int_0^T (1+\lambda s)^{-\frac{t}{\lambda}} f(t) dt + (1+\lambda s)^{-\frac{T}{\lambda}} \int_0^\infty (1+\lambda s)^{-\frac{t}{\lambda}} f(t) dt\end{aligned} \quad (30)$$

By (30), we get

$$\left(1 - (1+\lambda s)^{-\frac{T}{\lambda}}\right) \mathcal{L}_\lambda^*(f(t)) = \int_0^T (1+\lambda s)^{-\frac{t}{\lambda}} f(t) dt. \quad (31)$$

Thus, by (31), we get

$$\mathcal{L}_\lambda^*(f(t)) = \frac{1}{\left(1-(1+\lambda s)^{-\frac{T}{\lambda}}\right)} \int_0^T (1+\lambda s)^{-\frac{t}{\lambda}} f(t) dt. \quad (32)$$

We recall that the degenerate Bernoulli numbers are introduced as

$$\frac{t}{(1+\lambda)^{-\frac{t}{\lambda}}} = \sum_{n=0}^\infty B_{n,\lambda} \frac{t^n}{n!}, \quad (33)$$

Thus, by (32) and (33), we have

$$\begin{aligned}\frac{1}{1-(1+\lambda S)^{-\frac{T}{\lambda}}} &= -\frac{1}{TS} \frac{ST}{(1+\lambda s)^{-\frac{TS}{\lambda S}} - 1} \\ &= -\frac{1}{TS} \sum_{n=0}^\infty B_{n,\lambda S}(-1)^n S^n \frac{T^n}{n!}.\end{aligned} \quad (34)$$

Therefore, by (33) and (34), we obtain the following theorem.

Theorem 5. *If f is a function defined $t \geq 0$ and $\mathcal{L}_\lambda^*(f(t))$ exists, then we have*

$$\begin{aligned}\mathcal{L}_\lambda^*(f(t)) &= -\frac{1}{TS} \sum_{n=0}^\infty B_{n,\lambda S}(-1)^n S^n \int_0^T (1+\lambda s)^{-\frac{t}{\lambda}} f(t) dt \frac{T^n}{n!} \\ &= -\frac{1}{TS} \sum_{n=0}^\infty B_{n,\lambda S}(-1)^n S^n \mathcal{L}_\lambda^*(U(t-T)f(t)),\end{aligned} \quad (35)$$

where $U(t-a) = \begin{cases} 0, \text{for } 0 \geq t \geq a, \\ 1, \text{for } t \leq a. \end{cases}$ *is the Heviside function.*

Thirdly, we observe the modified degenerate Laplace transformation of $f(t-a)U(t-a)$ as follows:

$$\begin{aligned}\mathcal{L}_\lambda^*(f(t-a)U(t-a)) &= \int_0^\infty (1+\lambda s)^{-\frac{t}{\lambda}} f(t-a) U(t-a) dt \\ &= \int_a^\infty (1+\lambda s)^{-\frac{t}{\lambda}} f(t-a) dt \\ &= \int_0^\infty (1+\lambda s)^{-\frac{t+a}{\lambda}} f(t) dt \\ &= (1+\lambda s)^{-\frac{a}{\lambda}} \int_0^\infty (1+\lambda s)^{-\frac{t}{\lambda}} f(t) dt \\ &= (1+\lambda s)^{-\frac{a}{\lambda}} \mathcal{L}_\lambda^*(f(t)).\end{aligned} \quad (36)$$

Therefore, by (36), we obtain the following theorem.

Theorem 6. *For $\lambda \in (0,1)$ and $a \in (0,\infty)$ we have*

$$\mathcal{L}_\lambda^*(f(t-a)U(t-a)) = (1+\lambda s)^{-\frac{a}{\lambda}} \mathcal{L}_\lambda^*(f(t)), \tag{37}$$

where $U(t-a)$ is the Heviside function.

Fourthly, we observe the modified degenerate Laplace transformation of the convolution $f * g$ of two function f, g as follows:

$$\begin{aligned}
\mathcal{L}_\lambda^*(f)\mathcal{L}_\lambda^*(g) &= \left(\int_0^\infty (1+\lambda s)^{-\frac{t}{\lambda}} f(t) dt\right)\left(\int_0^\infty (1+\lambda s)^{-\frac{\tau}{\lambda}} g(\tau) d\tau\right) \\
&= \int_0^\infty \int_0^\infty (1+\lambda s)^{-\frac{t+\tau}{\lambda}} f(t)g(\tau) dt d\tau \\
&= \int_0^\infty f(t) \int_\tau^\infty (1+\lambda s)^{-\frac{\mu}{\lambda}} g(\mu-\tau) d\mu d\tau \\
&= \int_0^\infty \int_\tau^\infty f(t)(1+\lambda s)^{-\frac{\mu}{\lambda}} g(\mu-\tau) d\mu d\tau \\
&= \int_0^\infty (f * g)(1+\lambda s)^{-\frac{\mu}{\lambda}} d\mu \\
&= \mathcal{L}_\lambda(f * g).
\end{aligned} \tag{38}$$

Therefore, by (38), we obtain the following theorem.

Theorem 7. *For $\lambda \in (0,1]$, we have*

$$\mathcal{L}_\lambda^*(f * g) = \mathcal{L}_\lambda^*(f)\mathcal{L}_\lambda^*(g). \tag{39}$$

We note that

$$\begin{aligned}
\mathcal{L}_\lambda^*(1) &= \int_0^\infty (1+\lambda s)^{-\frac{t}{\lambda}} 1 dt \\
&= -\frac{\lambda}{\log(1+\lambda s)}(1+\lambda s)^{-\frac{t}{\lambda}} \big|_0^\infty \\
&= \frac{\lambda}{\log(1+\lambda s)}.
\end{aligned} \tag{40}$$

By (40), we have

$$\mathcal{L}_\lambda^*(f * 1) = \mathcal{L}_\lambda^*(f)\mathcal{L}_\lambda^*(1) = \mathcal{L}_\lambda^*(f)\frac{\lambda}{\log(1+\lambda s)}. \tag{41}$$

Therefore, by (41), we obtain the following theorem.

Theorem 8. *For $\lambda \in (0,1]$, we have*

$$\mathcal{L}_\lambda^{*-1}(\mathcal{L}_\lambda^*(f)\frac{\lambda}{\log(1+\lambda s)}) = f * 1(t) = \int_0^t f(t) dt. \tag{42}$$

Fifthly, we observe that the modified degenerate Laplace transformation of derivative of f which is $f(t) = 0((1+\lambda s)^{-\frac{t}{\lambda}})$, where $f(t) = 0(u(t))$ means

$$\mathcal{L}_\lambda^*(f') = \int_0^\infty (1+\lambda s)^{-\frac{t}{\lambda}} f' dt$$

$$= (1+\lambda s)^{-\frac{t}{\lambda}} f(t) \Big|_0^\infty + \int_0^\infty \frac{\log(1+\lambda s)}{\lambda}(1+\lambda s)^{-\frac{t}{\lambda}} f(t) dt \quad (43)$$

$$= -f(0) + \frac{\log(1+\lambda s)}{\lambda} \mathcal{L}_\lambda^*(f).$$

and

$$\mathcal{L}_\lambda^*(f^{(2)}) = \int_0^\infty (1+\lambda s)^{-\frac{t}{\lambda}} f^{(2)} dt$$

$$= (1+\lambda s)^{-\frac{t}{\lambda}} f'(t) \Big|_0^\infty + \frac{\log(1+\lambda s)}{\lambda} \int_0^\infty (1+\lambda s)^{-\frac{t}{\lambda}} f'(t) dt$$

$$= -f'(0) + \frac{\log(1+\lambda s)}{\lambda} \left(-f(0) + \frac{\log(1+\lambda s)}{\lambda} \mathcal{L}_\lambda^*(f) \right) \quad (44)$$

$$= \left(\frac{\log(1+\lambda s)}{\lambda} \right)^2 \mathcal{L}_\lambda^*(f) - \frac{\log(1+\lambda s)}{\lambda} f(0) - f'(0).$$

By using mathematical induction, we obtain the following theorem.

Theorem 9. *For $\lambda \in (0,1]$, we have*

$$\mathcal{L}_\lambda^*(f^{(n)}) = \left(\frac{\log(1+\lambda s)}{\lambda} \right)^n \mathcal{L}_\lambda^*(f) - \sum_{i=0}^{n-1} \left(\frac{\log(1+\lambda s)}{\lambda} \right)^{n-1-i} f^{(i)}(0). \quad (45)$$

Finally, we observe

$$\frac{dF_\lambda^*}{ds} = \int_0^\infty \frac{\lambda}{1+\lambda s} \left(-\frac{t}{\lambda} \right)(1+\lambda s)^{-\frac{t}{\lambda}} f(t) dt$$

$$= -\frac{1}{1+\lambda s} \int_0^\infty (1+\lambda s)^{-\frac{t}{\lambda}} t f(t) dt \quad (46)$$

$$= -\frac{1}{1+\lambda s} \mathcal{L}_\lambda^*(t f(t)).$$

By (46), we obtain the following theorem.

Theorem 10. *For $\lambda \in (0,1]$ and $0 < Re(s)$, we have*

$$\frac{dF_\lambda^*}{ds} = -\frac{1}{1+\lambda s} \mathcal{L}_\lambda^*(t f(t)). \quad (47)$$

4. Conclusions

Kim-Kim ([9]) defined a degenerate gamma function and a degenerate Laplace transformation. The motivation of this paper is to define modified degenerate gamma functions and modified degenerate Laplace transformations which are different to degenerate gamma function and degenerate Laplace transformation and to obtain more useful results which are Theorems 7 and 8 for the modified degenerate Laplace transformation. We do not obtain these result from the degenerate Laplace transformation. Also, we investigated some results which are Theorems 1 and 3 for modified degenerate gamma functions. Furthermore, Theorems 6 and 9 are some interesting properties which are applied to differential equations in engineering mathematics.

Author Contributions: All authors contributed equally to this work. All authors read and approved the final manuscript.

Funding: This work was supported by the National Research Foundation of Korea (NRF) grant funded by the Korea government (MEST) (No. 2017R1E1A1A03070882).

Conflicts of Interest: The authors declare that they have no competing interests.

References

1. Kreyszig, E.; Kreyszig, H.; Norminton, E.J. *Advanced Engineering Mathematics*; John Wiley & Sons Inc.: New Jersey, NJ, USA, 2011.
2. Kim, T.; Kim, D.S. Degenerate Laplace transform and degenerate Gamma function. *Russ. J. Math. Phys.* **2017**, *24*, 241–248. [CrossRef]
3. Chung, W.S.; Kim, T.; Kwon, H.I. On the q-analog of the Laplace transform. *Russ. J. Math. Phys.* **2014**, *21*, 156–168. [CrossRef]
4. Spiegel, M.R. *Laplace Transforms (Schaum's Outlines)*; McGraw Hill: New York, NY, USA, 1965.
5. Upadhyaya, L.M. On the degenerate Laplace transform. *Int. J. Eng. Sci. Res.* **2018**, *6*, 198–209. [CrossRef]
6. Carlitz, L. A degenerate Staudt-Clausen Theorem. *Arch. Math. (Basel)* **1956**, *7*, 28–33. [CrossRef]
7. Carlitz, L. Degenerate Stirling, Bernoulli and Eulerian numbers. *Util. Math.* **1979**, *15*, 51–88.
8. Dolgy, D.V.; Kim, T.; Seo, J.-J. On the symmetric identities of modified degenerate Bernoulli polynomials. *Proc. Jangjeon Math. Soc.* **2016**, *19*, 301–308.
9. Kim, D.S.; Kim, T. Some identities of degenerate Euler polynomials arising from p-adic fermionic integral on $\mathbb{Z}_p$. *Integral Transf. Spec. Funct.* **2015**, *26*, 295–302. [CrossRef]
10. Kim, T. Degenerate Euler Zeta function. *Russ. J. Math. Phys.* **2015**, *22*, 469–472. [CrossRef]
11. Kim, T. On the degenerate q-Bernoulli polynomials. *Bull. Korean Math. Soc.* **2016**, *53*, 1149–1156. [CrossRef]
12. Kim, T.; Kim, D.S.; Dolgy, D.V. Degenerate q-Euler polynomials. *Adv. Difference Equ.* **2015**, *2015*, 13662. [CrossRef]
13. Kim, T.; Dolgy, D.V.; Kim, D.S. Symmetric identities for degenerate generalized Bernoulli polynomials. *J. Nonlinear Sci. Appl.* **2016**, *9*, 677–683. [CrossRef]
14. Kim, T.; Kim, D.S.; Seo, J.J. Differential equations associated with degenerate Bell polynomials. *Int. J. Pure Appl. Math.* **2016**, *108*, 551–559.
15. Kwon, H.I.; Kim, T.; Seo, J.-J. A note on degenerate Changhee numbers and polynomials. *Proc. Jangjeon Math. Soc.* **2015**, *18*, 295–305.
16. Kim, T.; Jang, G.-W. A note on degenerate gamma function and degenerate Stirling number of the second kind. *Adv. Stud. Contemp. Math. (Kyungshang)* **2018**, *28*, 207–214.
17. Kim, T.; Yao, Y.; Kim, D.S.; Jang, G.-W. Degenerate r-Stirling numbers and r-Bell polynomials. *Russ. J. Math. Phys.* **2018**, *25*, 44–58. [CrossRef]
18. Kim, T.; Kim, D.S. A new approach to Catalan numbers using differential equations. *Russ. J. Math. Phys.* **2017**, *24*, 465–475. [CrossRef]
19. Jang, G.-W.; Kim, T.; Kwon, H.I. On the extension of degenerate Stirling polynomials of the second kind and degenerate Bell polynomials. *Adv. Stud. Contemp. Math. (Kyungshang)* **2018**, *28*, 305–316.

© 2018 by the authors. Licensee MDPI, Basel, Switzerland. This article is an open access article distributed under the terms and conditions of the Creative Commons Attribution (CC BY) license (http://creativecommons.org/licenses/by/4.0/).

Article

On p-adic Integral Representation of q-Bernoulli Numbers Arising from Two Variable q-Bernstein Polynomials

Dae San Kim [1], Taekyun Kim [2,3], Cheon Seoung Ryoo [4,*] and Yonghong Yao [2,5]

1 Department of Mathematics, Sogang University, Seoul 121-742, Korea; dskim@sogang.ac.kr
2 Department of Mathematics, Tianjin Polytechnic University, Tianjin 300387, China; tkkim@kw.ac.kr (T.K.); yaoyonghong@aliyun.com (Y.Y.)
3 Department of Mathematics, Kwangwoon University, Seoul 139-701, Korea
4 Department of Mathematics, Hannam University, Daejeon 306-791, Korea
5 Institute of Fundamental and Frontier Sciences, University of Electronic Science and Technology of China, Chengdu 610054, China
* Correspondence: ryoocs@hnu.kr

Received: 28 July 2018; Accepted: 20 September 2018; Published: 1 October 2018

Abstract: The q-Bernoulli numbers and polynomials can be given by Witt's type formulas as p-adic invariant integrals on $\mathbb{Z}_p$. We investigate some properties for them. In addition, we consider two variable q-Bernstein polynomials and operators and derive several properties for these polynomials and operators. Next, we study the evaluation problem for the double integrals on $\mathbb{Z}_p$ of two variable q-Bernstein polynomials and show that they can be expressed in terms of the q-Bernoulli numbers and some special values of q-Bernoulli polynomials. This is generalized to the problem of evaluating any finite product of two variable q-Bernstein polynomials. Furthermore, some identities for q-Bernoulli numbers are found.

Keywords: q-Bernoulli numbers; q-Bernoulli polynomials; two variable q-Bernstein polynomials; two variable q-Bernstein operators; p-adic integral on $\mathbb{Z}_p$

1. Introduction

Let p be a fixed prime number. Throughout this paper, $\mathbb{N}$, $\mathbb{Z}_p$, $\mathbb{Q}_p$, and $\mathbb{C}_p$ will denote the set of natural numbers, the ring of p-adic integers, the field of p-adic rational numbers, and the completion of the algebraic closure of $\mathbb{Q}_p$, respectively. The p-adic norm $|\cdot|_p$ is normalized as $|p|_p = \frac{1}{p}$. Assume that q is an indeterminate in $\mathbb{C}_p$ such that $|1-q|_p < p^{-\frac{1}{p-1}}$.

It is known that the q-number is defined by

$$[x]_q = \frac{1-q^x}{1-q},$$

see [1–20].

Please note that $\lim_{q \to 1}[x]_q = x$. Let $UD(\mathbb{Z}_p)$ be the space of uniformly differentiable functions on $\mathbb{Z}_p$. For $f \in UD(\mathbb{Z}_p)$, the p-adic q-integral on $\mathbb{Z}_p$ is defined by Kim as

$$I_q(f) = \int_{\mathbb{Z}_p} f(x)d\mu_q(x) = \lim_{N \to \infty} \frac{1}{[p^N]_q} \sum_{x=0}^{p^N-1} f(x)q^x, \qquad (1)$$

see [9,10].

As $q \to 1$ in (1), we have the p-adic integral on $\mathbb{Z}_p$ which is given by

$$I_1(f) = \lim_{q \to 1} I_q(f) = \int_{\mathbb{Z}_p} f(x) d\mu_1(x) = \lim_{N \to \infty} \frac{1}{p^N} \sum_{x=0}^{p^N-1} f(x), \qquad (2)$$

see [7–11,17].

From (2), we note that

$$I_1(f_1) - I_1(f) = f'(0), \qquad (3)$$

see [9]. Where $f_1(x) = f(x+1)$ and $f'(0) = \frac{df(x)}{dx}\big|_{x=0}$.

Thus, by (3), we get

$$\int_{\mathbb{Z}_p} e^{(x+y)t} d\mu_1(y) = \frac{t}{e^t - 1} e^{xt} = \sum_{n=0}^{\infty} B_n(x) \frac{t^n}{n!}, \qquad (4)$$

see [6,9], where $B_n(x)$ are the ordinary Bernoulli polynomials.

From (4), we note that

$$\int_{\mathbb{Z}_p} (x+y)^n d\mu_1(y) = B_n(x), \ (n \geq 0), \qquad (5)$$

see [7–11,17,18].

When $x = 0$, $B_n = B_n(0)$, $(n \geq 0)$, are called the ordinary Bernoulli numbers.
The Equation (4) implies the following recurrence relation for Bernoulli numbers:

$$B_0 = 1, \quad (B+1)^n - B_n = \begin{cases} 1 & \text{if } n = 1, \\ 0 & \text{if } n > 1, \end{cases} \qquad (6)$$

with the usual convention about replacing B^n by B_n (see [21]).

In [3,4], L. Carlitz introduced the q-Bernoulli numbers given by the recurrence relation

$$\beta_{0,q} = 1, \quad q(q\beta_q + 1)^n - \beta_{n,q} = \begin{cases} 1 & \text{if } n = 1, \\ 0 & \text{if } n > 1, \end{cases} \qquad (7)$$

with the usual convention about replacing β_q^n by $\beta_{n,q}$.

He also defined q-Bernoulli polynomials as

$$\beta_{n,q}(x) = (q^x \beta_q + [x]_q)^n = \sum_{l=0}^{n} \binom{n}{l} [x]_q^{n-l} q^{lx} \beta_{l,q}, \qquad (8)$$

see [3,4].

In 1999, Kim proved the following formula.

$$\int_{\mathbb{Z}_p} [x+y]_q^n d\mu_q(y) = \beta_{n,q}(x), \ (n \geq 0), \qquad (9)$$

see [10].

In the view of (5) and (9), we define the q-Bernoulli polynomials, different from Carlitz's q-Bernoulli polynomials, as

$$B_{n,q}(x) = \int_{\mathbb{Z}_p} [x+y]_q^n d\mu_1(y), \ (n \geq 0), \qquad (10)$$

see [8,9].

When $x = 0$, $B_{n,q} = B_{n,q}(0)$ are called the q-Bernoulli numbers.

From (3) and (10), we have

$$B_{0,q} = 1, \ (qB_q + 1)^n - B_{n,q} = \begin{cases} \frac{\log q}{q-1} & \text{if } n = 1, \\ 0 & \text{if } n > 1, \end{cases} \quad (11)$$

with the usual convention about replacing B_q^n by $B_{n,q}$.

By (10), we easily get

$$B_{n,q}(x) = (q^x B_q + [x]_q)^n = \sum_{l=0}^{n} \binom{n}{l} [x]_q^{n-l} q^{lx} B_{l,q}, \quad (12)$$

see [9].

As is known, the p-adic q-Bernstein operator is given by

$$\mathbb{B}_{n,q}(f|x) = \sum_{k=0}^{n} \binom{n}{k} f\left(\frac{k}{n}\right) [x]_q^k [1-x]_{q^{-1}}^{n-k} = \sum_{k=0}^{n} f\left(\frac{k}{n}\right) B_{k,n}(x|q),$$

where $n, k \in \mathbb{N} \cup \{0\}$, $x \in \mathbb{Z}_p$, and f is a continuous function on $\mathbb{Z}_p$ (see [7]). Here

$$B_{k,n}(x|q) = \binom{n}{k} [x]_q^k [1-x]_{q^{-1}}^{n-k}, \ (n, k \geq 0),$$

are called the p-adic q-Bernstein polynomials of degree n (see [7]). Please note that $\lim_{q \to 1} B_{k,n}(x|q) = B_{k,n}(x)$, where $B_{k,n}$ are the Bernstein polynomials (see [1,2,18–20,22]).

Here we cannot go without mentioning that Phillips (see [16]) introduced earlier in 1997 a different version of q-Bernstein polynomials from Kim's. Let f be a function defined on $[0,1]$, q any positive real number, and let

$$[n]_q! = [1]_q [2]_q \cdots [n]_q, \ (n \geq 1), \quad [0]_q! = 1, \quad \begin{bmatrix} n \\ k \end{bmatrix}_q = \frac{[n]_q!}{[k]_q! [n-k]_q!}.$$

Then Phillips' q-Bernstein polynomial of order n for f is given by

$$\mathbb{B}_n(f, q; x) = \sum_{k=0}^{n} f\left(\frac{[k]_q}{[n]_q}\right) \begin{bmatrix} n \\ k \end{bmatrix}_q x^k \prod_{s=0}^{n-1-k} (1 - q^s x),$$

Many results of Phillips' q-Bernstein polynomials for $q > 1$ were obtained for instance in [14,15], while those for $q \in (0,1)$ were derived for example in [12,13]. However, all of these and other related papers deal only with analytic properties of those q-Bernstein polynomials and some applications of them.

The Volkenborn integral and the fermionic p-adic, the p-adic q-invariant and the fermionic p-adic q-invariant integrals introduced by Kim have been studied for more than twenty years. Numerous results of arithmetic or combinatorial nature have been found by Kim and his colleagues around the world.

The present and related paper (see [5,6]) concern about Kim's q-Bernstein polynomials which have some merits over Phillips'. Indeed, by considering p-adic integrals on $\mathbb{Z}_p$ of them we can easily derive integral representations of q-Bernoulli numbers in the present paper, those of a q-analogue of Euler numbers in [5] and those of q-Euler numbers in [6]. These approaches also yield some identities for q-Bernoulli numbers, q-analogue of Euler numbers and q-Euler numbers. In conclusion, the Phillips' q-Bernstein polynomials are more analytic nature, while the Kim's are more arithmetic and combinatorial nature.

In this paper, we will study q-Bernoulli numbers and polynomials, which is introduced as p-adic invariant integrals on $\mathbb{Z}_p$, and investigate some properties for these numbers and polynomials. Also, we will consider two variable q-Bernstein polynomials and operators and derive several properties for these polynomials and operators. Next, we will consider p-adic integrals on $\mathbb{Z}_p$ of any finite product of two variable q-Bernstein polynomials and show that they can be expressed in terms of the q-Bernoulli numbers and some special values of q-Bernoulli polynomials. Furthermore, some identities for q-Bernoulli numbers will be found.

2. Some Integral Representations of q-Bernoulli Numbers and Polynomials

First, we consider the two variable q-Bernstein operator of order n which is given by

$$\mathbb{B}_{n,q}(f|x_1,x_2) = \sum_{k=0}^{n} \binom{n}{k} f\left(\frac{k}{n}\right) [x_1]_q^k [1-x_2]_{q^{-1}}^{n-k} = \sum_{k=0}^{n} f\left(\frac{k}{n}\right) B_{k,n}(x_1,x_2|q),$$

where $n \in \mathbb{N}$, and $x_1, x_2 \in \mathbb{Z}_p$.

Here, for $n, k \geq 0$,

$$B_{k,n}(x_1,x_2|q) = \binom{n}{k} [x_1]_q^k [1-x_2]_{q^{-1}}^{n-k} \tag{13}$$

are called two variable q-Bernstein polynomials of degree n (see [6,7]). In particular, this implies that $B_{k,n}(x_1,x_2|q) = 0$, for $0 \leq n < k$. In (13), if $x_1 = x_2 = x$, then $B_{k,n}(x,x|q) = B_{k,n}(x|q)$ are the q-Bernstein polynomials. It is not difficult to show that the generating function of $B_{k,n}(x_1,x_2|q)$ is given by

$$F_q^{(k)}(x_1,x_2|t) = \frac{(t[x_1]_q)^k}{k!} e^{(t[1-x_2]_{q^{-1}})} = \sum_{n=k}^{\infty} B_{k,n}(x_1,x_2|q) \frac{t^n}{n!}, \tag{14}$$

where $k \in \mathbb{N} \cup \{0\}$ (see [6,7]).

From (13), we easily get

$$B_{n-k,n}(1-x_2, 1-x_1|q^{-1}) = B_{k,n}(x_1,x_2|q), \ (0 \leq k \leq n).$$

For $1 \leq k \leq n-1$, we have the following properties (see [6,7]):

$$[1-x_2]_{q^{-1}} B_{k,n-1}(x_1,x_2|q) + [x_1]_q B_{k-1,n-1}(x_1,x_2|q) = B_{k,n}(x_1,x_2|q), \tag{15}$$

$$\frac{\partial}{\partial x_1} B_{k,n}(x_1,x_2|q) = \frac{\log q}{q-1} n \left((q-1)[x_1]_q B_{k-1,n-1}(x_1,x_2|q) + B_{k-1,n-1}(x_1,x_2|q)\right), \tag{16}$$

$$\frac{\partial}{\partial x_2} B_{k,n}(x_1,x_2|q) = \frac{\log q}{1-q} n((q-1)[x_2]_q B_{k,n-1}(x_1,x_2|q) + B_{k,n-1}(x_1,x_2|q)). \tag{17}$$

From (13) and q-Bernstein operator, we note that

$$\mathbb{B}_{n,q}(1|x_1,x_2) = \sum_{k=0}^{n} \binom{n}{k} [x_1]_q^k [1-x_2]_{q^{-1}}^{n-k} = (1+[x_1]_q - [x_2]_q)^n, \tag{18}$$

$$\mathbb{B}_{n,q}(t|x_1,x_2) = [x_1]_q (1+[x_1]_q - [x_2]_q)^{n-1},$$

$$\mathbb{B}_{n,q}(t^2|x_1,x_2) = \frac{n-1}{n} [x_1]_q^2 (1+[x_1]_q - [x_2]_q)^{n-2} + \frac{[x_1]_q}{n}(1+[x_1]_q - [x_2]_q)^{n-1},$$

and

$$\mathbb{B}_{n,q}(f|x_1,x_2) = \sum_{l=0}^{n} \binom{n}{l} [x_2]_q^l \sum_{k=0}^{l} \binom{l}{k} (-1)^{l-k} f\left(\frac{k}{n}\right) \left(\frac{[x_1]_q}{[x_2]_q}\right)^k, \tag{19}$$

where $n \in \mathbb{N}$ and f is a continuous function on $\mathbb{Z}_p$.

To see this, we first observe that

$$[1-x_2]_{q^{-1}} = 1 - [x_2]_q, \quad \binom{n}{k}\binom{n-k}{l-k} = \binom{n}{l}\binom{l}{k}.$$

Then (19) can be obtained as follows:

$$\begin{aligned}
\mathbb{B}(f|x_1,x_2) &= \sum_{k=0}^{n} f(\tfrac{k}{n}) \binom{n}{k} [x_1]_q^k (1-[x_2]_q)^{n-k} \\
&= \sum_{k=0}^{n} f(\tfrac{k}{n}) \binom{n}{k} [x_1]_q^k \sum_{l=0}^{n-k} \binom{n-k}{l}(-1)^l [x_2]_q^l \\
&= \sum_{k=0}^{n} f(\tfrac{k}{n}) \binom{n}{k} [x_1]_q^k \sum_{l=k}^{n} \binom{n-k}{l-k}(-1)^{l-k}[x_2]_q^{l-k} \\
&= \sum_{l=0}^{n} \binom{n}{l} [x_2]_q^l \sum_{k=0}^{l} \binom{l}{k}(-1)^{l-k} f(\tfrac{k}{n}) \left(\frac{[x_1]_q}{[x_2]_q}\right)^k.
\end{aligned}$$

It is easy to show that

$$\frac{1}{(1+[x_1]_q - [x_2]_q)^{n-j}} \sum_{k=j}^{n} \frac{\binom{k}{j}}{\binom{n}{j}} B_{k,n}(x_1,x_2|q) = [x_1]_q^j, \quad (20)$$

where $j \in \mathbb{N} \cup \{0\}$ and $x_1, x_2 \in \mathbb{Z}_p$.

Indeed, by making use of (18), we see that

$$\begin{aligned}
\sum_{k=j}^{n} \frac{\binom{k}{j}}{\binom{n}{j}} B_{k,n}(x_1,x_2|q) &= \sum_{k=j}^{n} \binom{n-j}{k-j} [x_1]_q^k [1-x_2]_{q^{-1}}^{n-k} \\
&= \sum_{k=0}^{n-j} \binom{n-j}{k} [x_1]_q^{k+j} [1-x_2]_{q^{-1}}^{n-j-k} \\
&= [x_1]^j (1+[x_1]_q - [x_2]_q)^{n-j}.
\end{aligned}$$

From (2), we have

$$\int_{\mathbb{Z}_p} [1-x+y]_{q^{-1}}^n d\mu_1(y) = (-1)^n q^n \int_{\mathbb{Z}_p} [x+y]_q^n d\mu_1(y), \quad (n \geq 0). \quad (21)$$

By (10) and (21), we get

$$B_{n,q^{-1}}(1-x) = (-1)^n q^n B_{n,q}(x), \quad (n \geq 0). \quad (22)$$

Again, from (11) and (12), we can derive the following equation.

$$B_{n,q}(2) = nq\frac{\log q}{q-1} + (qB_q + 1)^n = nq\frac{\log q}{q-1} + B_{n,q}, \quad (n > 1). \quad (23)$$

Thus, by (23), we obtain the following lemma.

Lemma 1. *For $n \in \mathbb{N}$ with $n > 1$, we have*

$$B_{n,q}(2) = nq\frac{\log q}{q-1} + B_{n,q}.$$

By (2), (10) and (22), we get

$$\int_{\mathbb{Z}_p} [1-x]_{q^{-1}}^n d\mu_1(x) = (-1)^n q^n \int_{\mathbb{Z}_p} [x-1]_q^n d\mu_1(x)$$
$$= (-1)^n q^n B_{n,q}(-1) \quad (24)$$
$$= B_{n,q^{-1}}(2), \ (n \geq 0).$$

For $n \in \mathbb{N}$ with $n > 1$, by (21), Lemma 1, and (24), we have

$$\int_{\mathbb{Z}_p} [1-x]_{q^{-1}}^n d\mu_1(x) = \int_{\mathbb{Z}_p} [x+2]_{q^{-1}}^n d\mu_1(x)$$
$$= (-1)^n q^n \int_{\mathbb{Z}_p} [x-1]_q^n d\mu_1(x)$$
$$= n\frac{\log q}{q-1} + \int_{\mathbb{Z}_p} [x]_{q^{-1}}^n d\mu_1(x) \quad (25)$$
$$= \frac{n \log q}{q-1} + B_{n,q^{-1}}.$$

Let us take the double p-adic integral on $\mathbb{Z}_p$ for the two variable q-Bernstein polynomials. Then we have

$$\int_{\mathbb{Z}_p} \int_{\mathbb{Z}_p} B_{k,n}(x_1, x_2|q) d\mu_1(x_1) d\mu_1(x_2)$$
$$= \binom{n}{k} \int_{\mathbb{Z}_p} \int_{\mathbb{Z}_p} [x_1]_q^k [1-x_2]_{q^{-1}}^{n-k} d\mu_1(x_1) d\mu_1(x_2)$$
$$= \binom{n}{k} B_{k,q} \int_{\mathbb{Z}_p} [1-x_2]_{q^{-1}}^{n-k} d\mu_1(x_2)$$
$$= \begin{cases} \binom{n}{k} B_{k,q}(B_{n-k,q^{-1}} + \frac{\log q}{q-1}(n-k)), & \text{if } n > k+1, \\ (k+1) B_{k,q} B_{1,q^{-1}}(2), & \text{if } n = k+1, \\ B_{k,q}, & \text{if } n = k, \\ 0, & \text{if } 0 \leq n < k. \end{cases} \quad (26)$$

Therefore, we obtain the following theorem.

Theorem 1. *For $n, k \in \mathbb{N} \cup \{0\}$, we have*

$$\int_{\mathbb{Z}_p} \int_{\mathbb{Z}_p} B_{k,n}(x_1, x_2|q) d\mu_1(x_1) d\mu_1(x_2)$$
$$= \begin{cases} \binom{n}{k} B_{k,q}(B_{n-k,q^{-1}} + \frac{\log q}{q-1}(n-k)), & \text{if } n > k+1, \\ (k+1) B_{k,q} B_{1,q^{-1}}(2), & \text{if } n = k+1, \\ B_{k,q}, & \text{if } n = k, \\ 0, & \text{if } 0 \leq n < k. \end{cases}$$

For $n, k \in \mathbb{N} \cup \{0\}$, we have

$$\int_{\mathbb{Z}_p} \int_{\mathbb{Z}_p} B_{k,n}(x_1, x_2|q) d\mu_1(x_1) d\mu_1(x_2)$$
$$= \int_{\mathbb{Z}_p} \int_{\mathbb{Z}_p} \binom{n}{k} [x_1]_q^k [1-x_2]_{q^{-1}}^{n-k} d\mu_1(x_1) d\mu_1(x_2)$$
$$= \binom{n}{k} \int_{\mathbb{Z}_p} \int_{\mathbb{Z}_p} (1 - [1-x_1]_{q^{-1}})^k [1-x_2]_{q^{-1}}^{n-k} d\mu_1(x_1) d\mu_1(x_2)$$
$$= \binom{n}{k} \sum_{l=0}^{k} \binom{k}{l} (-1)^{k-l} \int_{\mathbb{Z}_p} \int_{\mathbb{Z}_p} [1-x_1]_{q^{-1}}^{k-l} [1-x_2]_{q^{-1}}^{n-k} d\mu_1(x_1) d\mu_1(x_2) \quad (27)$$
$$= \binom{n}{k} \int_{\mathbb{Z}_p} [1-x_2]_{q^{-1}}^{n-k} d\mu_1(x_2)$$
$$\times \left(1 - k \int_{\mathbb{Z}_p} [1-x_1]_{q^{-1}} d\mu_1(x_1) + \sum_{l=0}^{k-2} \binom{k}{l} (-1)^{k-l} \left((k-l) \frac{\log q}{q-1} + B_{k-l,q^{-1}} \right) \right).$$

Thus, by (27), we get

$$\binom{n}{k}^{-1} \frac{\int_{\mathbb{Z}_p} \int_{\mathbb{Z}_p} B_{k,n}(x_1, x_2|q) d\mu_1(x_1) d\mu_1(x_2)}{\int_{\mathbb{Z}_p} [1-x_2]_{q^{-1}}^{n-k} d\mu_1(x_2)}$$
$$= 1 - k \int_{\mathbb{Z}_p} [1-x_1]_{q^{-1}} d\mu_1(x_1) + \sum_{l=0}^{k-2} \binom{k}{l} (-1)^{k-l} \left((k-l) \frac{\log q}{q-1} + B_{k-l,q^{-1}} \right) \quad (28)$$
$$= 1 - k \left(1 - \frac{\log q - q + 1}{(q-1)^2} \right)$$
$$+ k \sum_{l=0}^{k-2} \binom{k-1}{l} (-1)^{k-l} \frac{\log q}{q-1} + \sum_{l=0}^{k-2} (-1)^{k-l} \binom{k}{l} B_{k-l,q^{-1}}$$
$$= 1 - k \left(1 - \frac{\log q - q + 1}{(q-1)^2} \right) + k \frac{\log q}{q-1} + \sum_{l=0}^{k-2} (-1)^{k-l} \binom{k}{l} B_{k-l,q^{-1}} \quad (29)$$
$$= 1 - kq \left(\frac{q - \log q - 1}{(q-1)^2} \right) + \sum_{l=0}^{k-2} (-1)^{k-l} \binom{k}{l} B_{k-l,q^{-1}}.$$

Therefore, by (28), we obtain the following theorem.

Theorem 2. *For $n, k \in \mathbb{N} \cup \{0\}$ with $k > 1$, we have*

$$\frac{\int_{\mathbb{Z}_p} \int_{\mathbb{Z}_p} B_{k,n}(x_1, x_2|q) d\mu_1(x_1) d\mu_1(x_2)}{\binom{n}{k} \int_{\mathbb{Z}_p} [1-x_2]_{q^{-1}}^{n-k} d\mu_1(x_2)}$$
$$= \binom{n}{k} \left(1 - kq \left(\frac{q - \log q - 1}{(q-1)^2} \right) + \sum_{l=0}^{k-2} (-1)^{k-l} \binom{k}{l} B_{k-l,q^{-1}} \right).$$

Therefore, by Theorems 1 and 2, we obtain the following corollary.

Corollary 1. *For $k \in \mathbb{N}$ with $k > 1$, we have*

$$B_{k,q} = 1 - kq \left(\frac{q - \log q - 1}{(q-1)^2} \right) + \sum_{l=0}^{k-2} (-1)^{k-l} \binom{k}{l} B_{k-l,q^{-1}}.$$

For $m, n \in \mathbb{N} \cup \{0\}$, we have

$$\int_{\mathbb{Z}_p} \int_{\mathbb{Z}_p} B_{k,n}(x_1, x_2|q) B_{k,m}(x_1, x_2|q) d\mu_1(x_1) d\mu_1(x_2)$$
$$= \binom{n}{k}\binom{m}{k} \int_{\mathbb{Z}_p} [x_1]_q^{2k} d\mu_1(x_1) \int_{\mathbb{Z}_p} [1-x_2]_{q^{-1}}^{n+m-2k} d\mu_1(x_2). \tag{30}$$

Thus, by (29), we get

$$\frac{\int_{\mathbb{Z}_p} \int_{\mathbb{Z}_p} B_{k,n}(x_1, x_2|q) B_{k,m}(x_1, x_2|q) d\mu_1(x_1) d\mu_1(x_2)}{\int_{\mathbb{Z}_p} [1-x_2]_{q^{-1}}^{n+m-2k} d\mu_1(x_2)}$$
$$= \binom{n}{k}\binom{m}{k} B_{2k,q}.$$

Hence, we have the following proposition.

Proposition 1. *For $m, n, k \in \mathbb{N} \cup \{0\}$, we have*

$$\frac{\int_{\mathbb{Z}_p} \int_{\mathbb{Z}_p} B_{k,n}(x_1, x_2|q) B_{k,m}(x_1, x_2|q) d\mu_1(x_1) d\mu_1(x_2)}{\int_{\mathbb{Z}_p} [1-x_2]_{q^{-1}}^{n+m-2k} d\mu_1(x_2)}$$
$$= \binom{n}{k}\binom{m}{k} B_{2k,q}.$$

Let $m, n, k \in \mathbb{N} \cup \{0\}$. Then we get

$$\int_{\mathbb{Z}_p} \int_{\mathbb{Z}_p} B_{k,n}(x_1, x_2|q) B_{k,m}(x_1, x_2|q) d\mu_1(x_1) d\mu_1(x_2)$$
$$= \sum_{l=0}^{2k} \binom{n}{k}\binom{m}{k}\binom{2k}{l}(-1)^{2k-l} \tag{31}$$
$$\times \int_{\mathbb{Z}_p} \int_{\mathbb{Z}_p} [1-x_1]_{q^{-1}}^{2k-l} [1-x_2]_{q^{-1}}^{n+m-2k} d\mu_1(x_1) d\mu_1(x_2)$$

Thus, from (30), we have

$$\binom{n}{k}^{-1}\binom{m}{k}^{-1} \frac{\int_{\mathbb{Z}_p} \int_{\mathbb{Z}_p} B_{k,n}(x_1, x_2|q) B_{k,m}(x_1, x_2|q) d\mu_1(x_1) d\mu_1(x_2)}{\int_{\mathbb{Z}_p} [1-x_2]_{q^{-1}}^{n+m-2k} d\mu_1(x_2)}$$

$$= 1 - 2k \int_{\mathbb{Z}_p} [1-x_1]_{q^{-1}} d\mu_1(x_1) + \sum_{l=0}^{2k-2} \binom{2k}{l}(-1)^{2k-l} \int_{\mathbb{Z}_p} [1-x_1]_{q^{-1}}^{2k-l} d\mu_1(x_1) \tag{32}$$

$$= 1 - 2k\left(1 - \frac{\log q - q + 1}{(q-1)^2}\right) + 2k\frac{\log q}{q-1} + \sum_{l=0}^{2k-2} \binom{2k}{l}(-1)^{2k-l} B_{2k-l,q^{-1}}$$

$$= 1 - 2kq\left(\frac{q - \log q - 1}{(q-1)^2}\right) + \sum_{l=0}^{2k-2} \binom{2k}{l}(-1)^{2k-l} B_{2k-l,q^{-1}}.$$

By (31), we have the following proposition.

Proposition 2. For $m, n, k \in \mathbb{N} \cup \{0\}$, with $k \geq 1$, we have

$$\frac{\int_{\mathbb{Z}_p} \int_{\mathbb{Z}_p} B_{k,n}(x_1, x_2|q) B_{k,m}(x_1, x_2|q) d\mu_1(x_1) d\mu_1(x_2)}{\int_{\mathbb{Z}_p} [1-x_2]_{q^{-1}}^{n+m-2k} d\mu_1(x_2)}$$

$$= \binom{n}{k}\binom{m}{k}\left(1 - 2kq\left(\frac{q - \log q - 1}{(q-1)^2}\right) + \sum_{l=0}^{2k-2} \binom{2k}{l}(-1)^{2k-l} B_{2k-l,q^{-1}}\right).$$

Therefore, by Propositions 1 and 2, we obtain the following corollary.

Corollary 2. For $k \in \mathbb{N}$, we have

$$B_{2k,q} = 1 - 2k\left(\frac{q^2 - q - \log q}{(q-1)^2}\right)$$

$$+ \sum_{l=0}^{2k-2} \binom{2k}{l}(-1)^{2k-l}\left((2k-1)\frac{\log q}{q-1} + B_{2k-l,q^{-1}}\right).$$

For $m \in \mathbb{N}$, let $n_1, n_2, \cdots, n_m, k \in \mathbb{N} \cup \{0\}$. Then we note that

$$\int_{\mathbb{Z}_p} \int_{\mathbb{Z}_p} \left(\prod_{i=1}^{m} B_{k,n_i}(x_1, x_2|q)\right) d\mu_1(x_1) d\mu_1(x_2)$$

$$= \sum_{l=0}^{mk} \left(\prod_{i=1}^{m} \binom{n_i}{k}\right) \binom{mk}{l}(-1)^{mk-l} \qquad (33)$$

$$\times \int_{\mathbb{Z}_p} \int_{\mathbb{Z}_p} [1-x_1]_{q^{-1}}^{mk-l} [1-x_2]_{q^{-1}}^{n_1+n_2+\cdots+n_m-mk} d\mu_1(x_1) d\mu_1(x_2)$$

Thus, by (32), we have

$$\prod_{i=1}^{m}\binom{n_i}{k}^{-1} \frac{\int_{\mathbb{Z}_p} \int_{\mathbb{Z}_p} \left(\prod_{i=1}^{m} B_{k,n_i}(x_1, x_2|q)\right) d\mu_1(x_1) d\mu_1(x_2)}{\int_{\mathbb{Z}_p} [1-x_2]_{q^{-1}}^{n_1+n_2+\cdots+n_m-mk} d\mu_1(x_2)}$$

$$= \sum_{l=0}^{mk} \binom{mk}{l}(-1)^{mk-l} \int_{\mathbb{Z}_p} [1-x_1]_{q^{-1}}^{mk-l} d\mu_1(x_1)$$

$$= 1 - mk \int_{\mathbb{Z}_p} [1-x_1]_{q^{-1}} d\mu_1(x_1)$$

$$+ \sum_{l=0}^{mk-2} \binom{mk}{l}(-1)^{mk-l} \int_{\mathbb{Z}_p} [1-x_1]_{q^{-1}}^{mk-l} d\mu_1(x_1)$$

Therefore we obtain the following theorem.

Theorem 3. For $n_1, n_2, \cdots, n_m \in \mathbb{N} \cup \{0\}$, and $k, m \in \mathbb{N}$ with $mk > 1$, we have

$$\prod_{i=1}^{m}\binom{n_i}{k}^{-1} \frac{\int_{\mathbb{Z}_p} \int_{\mathbb{Z}_p} \left(\prod_{i=1}^{m} B_{k,n_i}(x_1, x_2|q)\right) d\mu_1(x_1) d\mu_1(x_2)}{\int_{\mathbb{Z}_p} [1-x_2]_{q^{-1}}^{n_1+n_2+\cdots+n_m-mk} d\mu_1(x_2)}$$

$$= 1 - mk\left(\frac{q^2 - q - \log q}{(q-1)^2}\right)$$

$$+ \sum_{l=0}^{mk-2} \binom{mk}{l}(-1)^{mk-l}\left((mk-1)\frac{\log q}{q-1} + B_{mk-l,q^{-1}}\right).$$

On the other hand, we easily get

$$\frac{\int_{\mathbb{Z}_p} \int_{\mathbb{Z}_p} \left(\prod_{i=1}^m B_{k,n_i}(x_1,x_2|q)\right) d\mu_1(x_1) d\mu_1(x_2)}{\int_{\mathbb{Z}_p} [1-x_2]_{q^{-1}}^{n_1+n_2+\cdots+n_m-mk} d\mu_1(x_2)} = \prod_{i=1}^m \binom{n_i}{k} B_{mk,q}. \tag{34}$$

Therefore, by Theorem 3 and (33), we obtain the following corollary.

Corollary 3. *For $m, k \in \mathbb{N}$ with $mk > 1$, we have*

$$B_{mk,q} = 1 - mk \left(\frac{q^2 - q - \log q}{(q-1)^2} \right)$$
$$+ \sum_{l=0}^{mk-2} \binom{mk}{l} (-1)^{mk-l} \left((mk-l) \frac{\log q}{q-1} + B_{mk-l,q^{-1}} \right).$$

3. Conclusions

Here we studied q-Bernoulli numbers and polynomials which are different from the classical Carlitz q-Bernoulli numbers $\beta_{n,q}$ and polynomials $\beta_{n,q}(x)$, and arise naturally from some p-adic invariant integrals on $\mathbb{Z}_p$, as was shown in (10). After investigating some of their properties, we turned our attention to two variable q-Bernstein polynomials and operators, which was introduced by Kim and generalizes the single variable q-Bernstein polynomials and operators in [6]. As a preparation, we derived several properties of these polynomials and operators.

Next, we considered the evaluation problem for the double integrals on $\mathbb{Z}_p$ of two variable q-Bernstein polynomials and showed that they can be expressed in terms of the q-Bernoulli numbers and some special values of q-Bernoulli polynomials. This was further generalized to the problem of evaluating the product of two and that of an arbitrary number of two variable q-Bernstein polynomials. It was shown again that they can be expressed in terms of the q-Bernoulli numbers. Also, some identities for q-Bernoulli numbers were found along the way.

Finally, we would like to mention that, along the same line, in [5] we studied some properties of a q-analogue of Euler numbers and polynomials arising from the p-adic fermionic integrals on $\mathbb{Z}_p$. Then we considered p-adic fermionic integrals on $\mathbb{Z}_p$ of the two variable q-Bernstein polynomials and of products of the two variable q-Bernstein polynomials, and showed that they can be expressed in terms of the q-analogues of Euler numbers.

Author Contributions: T.K. and D.S.K. conceived the framework and structured the whole paper; T.K. wrote the paper; C.S.R. and Y.Y. checked the results of the paper; D.S.K., C.S.R., Y.Y. and T.K. completed the revision of the article.

Funding: This work was supported by the National Research Foundation of Korea(NRF) grant funded by the Korea government(MEST) (No. 2017R1A2B4006092).

Acknowledgments: The authors would like to thank the referees for their valuable comments which improved the original manuscript in its present form.

Conflicts of Interest: The authors declare no conflict of interest.

References

1. Bayad, A.; Kim, T. Identities involving values of Bernstein q-Bernoulli, and q-Euler polynomials. *Russ. J. Math. Phys.* **2011**, *18*, 133–143. [CrossRef]
2. Carlitz, L. Expansions of q-Bernoulli numbers. *Duke Math. J.* **1958**, *25*, 355–364. [CrossRef]
3. Carlitz, L. q-Bernoulli and Eulerian numbers. *Trans. Am. Math. Soc.* **1954**, *76*, 332–350.
4. Kim, T. Some identities on the q-integral representation of the product of several q-Bernstein-type polynomials. *Abstr. Appl. Anal.* **2011**, *2011*, 634675. [CrossRef]
5. Jang, L.C.; Kim, T.; Kim, D.S.; Dolgy, D.V. On p-adic fermionic integrals of q-Bernstein polynomials associated with q-Euler numbers and polynomials. *Symmetry* **2018**, *10*, 311. [CrossRef]

6. Kim, T. A note on q-Bernstein polynomials. *Russ. J. Math. Phys.* **2011**, *18*, 73–82. [CrossRef]
7. Kim, T. A study on the q-Euler numbers and the fermionic q-integral of the product of several type q-Bernstein polynomials on $\mathbb{Z}_p$. *Adv. Stud. Contemp. Math.* **2013**, *23*, 5–11.
8. Kim, T. On p-adic q-Bernoulli numbers. *J. Korean Math. Soc.* **2000**, *37*, 21–30.
9. Kim, T.; Kim, H.S. Remark on the p-adic q-Bernoulli numbers. Algerbraic number theory (Hapcheon/Saga, 1996). *Adv. Stud. Contemp. Math.* **1999**, *1*, 127–136.
10. Kim, T. q-Volkenborn integration. *Russ. J. Math. Phys.* **2002**, *9*, 288–299.
11. Kurt, V. Some relation between the Bernstein polynomials and second kind Bernpolli polynomials. *Adv. Stud. Contemp. Math.* **2013**, *23*, 43–48.
12. Oruç, H.; Phillips, G.M. A generalization of Bernstein polynomials. *Proc. Edinb. Math. Soc.* **1999**, *42*, 403–413. [CrossRef]
13. Oruç, H.; Tuncer, N. On the convergence and iterates of q-Bernstein polynomials. *J. Approx. Theory* **2002**, *117*, 301–313. [CrossRef]
14. Ostrovska, S. On the q-Bernstein polynomials. *Adv. Stud. Contemp. Math.* **2015**, *11*, 193–204.
15. Ostrovska, S. On the q-Bernstein polynomials of the logarithmic function in the case $q > 1$. *Math. Slovaca* **2016**, *66*, 73–78. [CrossRef]
16. Phillips, G.M. Bernstein polynomials based on the q-integers. *Ann. Numer. Math.* **1997**, *4*, 511–518.
17. Rim, S.-H.; Joung, J.; Jin, J.-H.; Lee, S.-J. A note on the weighted Carlitz's type q-Euler numbers and q-Bernstein polynomials. *Proc. Jangjeon Math. Soc.* **2012**, *15*, 195–201.
18. Kim, D.S.; Kim, T. Some p-adic integrals on $\mathbb{Z}_p$ associated with trigonometric functions. *Russ. J. Math. Phys.* **2018**, *25*, 300–308. [CrossRef]
19. Siddiqui, M.A.; Agrawal, R.R.; Gupta, N. On a class of modified new Bernstein operators. *Adv. Stud. Contemp. Math.* **2014**, *24*, 97–107.
20. Simsek, Y. On parametrization of the q-Bernstein basis functions and their applications. *J. Inequal. Spec. Funct.* **2017**, *8*, 158–169.
21. Kim, T.; Kim, D.S. Identities for degenerate Bernoulli polynomials and Korobov polynomials of the first kind. *Sci. China Math.* **2018**. [CrossRef]
22. Taberski, R. Approximation properties of the integral Bernstein operators and their derivatives in some classes of locally integral functions. *Funct. Approx. Comment. Math.* **1992**, *21*, 85–96.

© 2018 by the authors. Licensee MDPI, Basel, Switzerland. This article is an open access article distributed under the terms and conditions of the Creative Commons Attribution (CC BY) license (http://creativecommons.org/licenses/by/4.0/).

Article

A New Sequence and Its Some Congruence Properties

Wenpeng Zhang and Xin Lin *

School of Mathematics, Northwest University, Xi'an 710127, China; wpzhang@nwu.edu.cn
* Correspondence: estelle-xin@hotmail.com

Received: 17 July 2018; Accepted: 20 August 2018; Published: date

Abstract: The aim of this paper is to study the congruence properties of a new sequence, which is closely related to Fubini polynomials and Euler numbers, using the elementary method and the properties of the second kind Stirling numbers. As results, we obtain some interesting congruences for it. This solves a problem proposed in a published paper.

Keywords: Fubini polynomials; Euler numbers; congruence; elementary method

MSC: 11B83; 11B37

1. Introduction

Let $n \geq 0$ be an integer, the famous Fubini polynomials $F_n(y)$ are defined according to the coefficients of following generating function:

$$\frac{1}{1 - y(e^t - 1)} = \sum_{n=0}^{\infty} \frac{F_n(y)}{n!} \cdot t^n, \qquad (1)$$

where $F_0(y) = 1$, $F_1(y) = y$, and so on.

These polynomials are closely related to the Stirling numbers and Euler numbers. For example, if $y = -\frac{1}{2}$, then (1) becomes

$$\frac{2}{1 + e^t} = \sum_{n=0}^{\infty} \frac{E_n}{n!} \cdot t^n, \qquad (2)$$

where E_n denotes the Euler numbers.

At the same time, the Fubini polynomials with two variables can also be defined by the following identity (see [1,2]):

$$\frac{e^{xt}}{1 - y(e^t - 1)} = \sum_{n=0}^{\infty} \frac{F_n(x, y)}{n!} \cdot t^n,$$

and $F_n(y) = F_n(0, y)$ for all integers $n \geq 0$. Many scholars have studied the properties of $F_n(x, y)$, and have obtained many important works. For example, T. Kim et al. proved a series of identities related to $F_n(x, y)$ (see [2,3]), one of which is

$$F_n(x, y) = \sum_{l=0}^{n} \binom{n}{l} x^l \cdot F_{n-l}(y), \ n \geq 0.$$

Zhao Jianhong and Chen Zhuoyu [4] studied the computational problem of the sums

$$\sum_{a_1 + a_2 + \cdots + a_k = n} \frac{F_{a_1}(y)}{(a_1)!} \cdot \frac{F_{a_2}(y)}{(a_2)!} \cdots \frac{F_{a_k}(y)}{(a_k)!},$$

where the summation in the formula above denotes all k-dimension non-negative integer coordinates $(a_1, a_2, \cdots, a_k)$ such that $a_1 + a_2 + \cdots + a_k = n$. They proved the identity

$$\sum_{a_1+a_2+\cdots+a_k=n} \frac{F_{a_1}(y)}{(a_1)!} \cdot \frac{F_{a_2}(y)}{(a_2)!} \cdots \frac{F_{a_k}(y)}{(a_k)!}$$
$$= \frac{1}{(k-1)!(y+1)^{k-1}} \cdot \frac{1}{n!} \sum_{i=0}^{k-1} C(k-1,i) F_{n+k-1-i}(y), \quad (3)$$

where the sequence $C(k,i)$ is defined for positive integer k and i with $0 \le i \le k$, $C(k,0) = 1$, $C(k,k) = k!$ and

$$C(k+1, i+1) = C(k, i+1) + (k+1)C(k, i), \text{ for all } 0 \le i < k,$$

providing $C(k,i) = 0$, if $i > k$.

For clarity, for $1 \le k \le 9$, we list values of $C(k,i)$ in the following Table 1.

Table 1. Values of $C(k,i)$.

$C(k,i)$	$i=0$	$i=1$	$i=2$	$i=3$	$i=4$	$i=5$	$i=6$	$i=7$	$i=8$	$i=9$
$k=1$	1	1								
$k=2$	1	3	2							
$k=3$	1	6	11	6						
$k=4$	1	10	35	50	24					
$k=5$	1	15	85	225	274	120				
$k=6$	1	21	175	735	1624	1764	720			
$k=7$	1	28	322	1960	6769	13,132	13,068	5040		
$k=8$	1	36	546	4536	22,449	67,284	118,124	109,584	40,320	
$k=9$	1	45	870	9450	63,273	269,325	723,680	1,172,700	1,026,576	362,880

Meanwhile, Zhao Jianhong and Chen Zhuoyu [4] proposed some conjectures related to the sequence. We believe that this sequence is meaningful because it satisfies some very interesting congruence properties, such as

$$C(p-2, i) \equiv 1 \pmod{p} \quad (4)$$

for all odd primes p and integers $0 \le i \le p-2$. The equivalent conclusion is

$$C(p-1, i) \equiv 0 \pmod{p} \quad (5)$$

for all odd primes p and positive integers $1 \le i \le p-2$. Since some related content can be found in references [5–15], we will not go through all of them here.

The aim of this paper is to prove congruence (5) by applying the elementary method and the properties of the second kind Stirling numbers. That is, we will solve the conjectures in [4], which are listed in the following.

Theorem 1. *Let p be an odd prime. For any integer $1 \le i \le p-2$, we have congruence*

$$C(p-1, i) \equiv 0 \pmod{p}.$$

From this theorem and (3), we can deduce following three corollaries:

Corollary 1. *For any positive integer n and odd prime p, we have*

$$F_{n+p-1}(y) - F_n(y) \equiv 0 \pmod{p}.$$

Corollary 2. *For any positive integer n and odd prime p, we have*

$$E_{n+p-1} - E_n \equiv 0 \pmod{p}.$$

Corollary 3. *For any odd prime p, we have the congruences*

$$2E_p \equiv -1 \pmod{p}, \quad 4E_{p+2} \equiv 1 \pmod{p}, \quad \text{and} \quad 2E_{p+4} \equiv -1 \pmod{p}.$$

Note. Since E_n is a rational number, we can denote $E_n = \dfrac{U_n}{V_n}$, where U_n and V_n are integers with $(U_n, V_n) = 1$. Based on this, in our paper, the expression $E_n \equiv 0 \pmod{p}$ means $p \mid U_n$, while $p \nmid V_n$.

2. Several Lemmas

Lemma 1. *For any positive integer k, we have the identity*

$$k! y(y+1)^{k-1} = \sum_{i=0}^{k-1} C(k-1, i) F_{k-i}(y).$$

Proof. Taking $n = 1$ in (3), and noting that $F_0(y) = 1$, $F_1(y) = y$, and the equation $a_1 + a_2 + \cdots + a_k = 1$ holds if and only if one of a_i is 1, others are 0. The number of the solutions of this equation is $\binom{k}{1} = k$. So, from (3), we have

$$\sum_{a_1+a_2+\cdots+a_k=1} \frac{F_{a_1}(y)}{(a_1)!} \cdot \frac{F_{a_2}(y)}{(a_2)!} \cdots \frac{F_{a_k}(y)}{(a_k)!} = \binom{k}{1} y = ky$$

$$= \frac{1}{(k-1)!(y+1)^{k-1}} \cdot \sum_{i=0}^{k-1} C(k-1, i) F_{k-i}(y)$$

or identity

$$k! y(y+1)^{k-1} = \sum_{i=0}^{k-1} C(k-1, i) F_{k-i}(y),$$

which proves Lemma 1. □

Lemma 2. *For any positive integer n, we have the identity*

$$F_n(y) = \sum_{k=0}^{n} S(n,k) \, k! \, y^k, \quad (n \geq 0),$$

where $S(n,k)$ are the second kind Stirling numbers, which are defined for any integer k, n with $0 \leq k \leq n$ as:

$$S(n,k) = kS(n-1, k) + S(n-1, k-1)$$

where $S(0,0) = 1$, $S(n,0) = 0$ and $S(0,k) = 0$ for $n, k > 0$.

Proof. See Reference [2]. □

Lemma 3. *For any positive integers n and k, we have*

$$S(n,k) = \frac{1}{k!} \sum_{j=0}^{k} \binom{k}{j} j^n (-1)^{k-j}.$$

Proof. See Theorem 4.3.12 of [16]. □

Lemma 4. *For any odd prime p and positive integer $2 \leq k \leq p-1$, we have the congruence*

$$k! S(p,k) \equiv 0 \pmod{p}.$$

Proof. From the definition and properties of $S(n,k)$, we have $S(n,k) = 0$, if $k > n$. For any integers $0 \leq j \leq p-1$, from the famous Fermat's little theorem, we have the congruence $j^p \equiv j \pmod{p}$. From this congruence and Lemma 3, we have

$$k! S(p,k) = \sum_{j=0}^{k} \binom{k}{j} j^p (-1)^{k-j} \equiv \sum_{j=0}^{k} \binom{k}{j} j (-1)^{k-j} \equiv k! S(1,k) \equiv 0 \pmod{p},$$

if $k \geq 2$. This completes the proof of Lemma 4. □

3. Proof of the Theorem

In this section, we will prove Theorem by mathematical induction. Taking $k = p$ in Lemma 1 and noting that $C(p-1,0) = 1$ and $C(p-1, p-1) = (p-1)!$, we have:

$$p! y (y+1)^{p-1} = \sum_{i=0}^{p-1} C(p-1,i) F_{p-i}(y)$$

$$= F_p(y) + y(p-1)! + \sum_{i=1}^{p-2} C(p-1,i) F_{p-i}(y).$$

Note that $(p-1)! + 1 \equiv 0 \pmod{p}$, which implies

$$F_p(y) - y + \sum_{i=1}^{p-2} C(p-1,i) F_{p-i}(y) \equiv 0 \pmod{p}. \tag{6}$$

From (6), we have the congruence

$$y - F_p(y) \equiv \sum_{i=1}^{p-2} C(p-1,i) F_{p-i}(y) \pmod{p}. \tag{7}$$

From Lemma 2, we have

$$F_p(y) = \sum_{k=0}^{p} S(p,k) \, k! \, y^k \tag{8}$$

and

$$F_p^{(p-1)}(0) = S(p, p-1) \, (p-1)! \cdot (p-1)!, \tag{9}$$

where $F_n^{(k)}(y)$ denotes the k-order derivative of $F_n(y)$ for variable y.

$$F_{p-1}^{(p-1)}(0) = S(p-1, p-1)\,(p-1)! \cdot (p-1)! = (p-1)! \cdot (p-1)!. \tag{10}$$

Then, applying Lemma 3 and Lemma 4 and noting that $S(1, p-1) = 0$, we have

$$\begin{aligned}(p-1)! S(p, p-1) &\equiv \sum_{j=0}^{p-1} \binom{p-1}{j} j^p (-1)^{p-1-j} \equiv \sum_{j=0}^{p-1} \binom{p-1}{j} j (-1)^{p-1-j} \\ &\equiv (p-1)! S(1, p-1) \equiv 0 \pmod{p}.\end{aligned} \tag{11}$$

Combining (7), (9), (10), and (11), we have:

$$0 \equiv -S(p, p-1)(p-1)!(p-1)! \equiv C(p-1,1)(p-1)! \cdot (p-1)! \,(\text{mod } p) \tag{12}$$

or

$$C(p-1,1) \equiv 0 \,(\text{mod } p). \tag{13}$$

That is, the theorem is true for $i = 1$.
Assume that the theorem is true for all $1 \leq i \leq s$. That is,

$$C(p-1, i) \equiv 0 \,(\text{mod } p)$$

for $1 \leq i \leq s < p-1$. It is clear that if $s = p-2$, then the theorem is true.
If $1 < s < p-2$, then from (7) we have the congruence

$$y - F_p(y) \equiv \sum_{i=s+1}^{p-2} C(p-1, i) F_{p-i}(y) \,(\text{mod } p). \tag{14}$$

In congruence (14), taking the $(p-s-1)$-order derivative with respect to t, then let $y = 0$, applying Lemma 2, we have:

$$\begin{aligned} &-S(p, p-s-1)(p-s-1)! \cdot (p-s-1)! \\ &\equiv C(p-1, s+1)(p-s-1)!(p-s-1)! \,(\text{mod } p). \end{aligned} \tag{15}$$

Note that $((p-s-1)!, p) = 1$, from Lemma 4 and (15) we have the congruence

$$C(p-1, s+1)(p-s-1)! \equiv -(p-s-1)! S(p, p-s-1) \equiv 0 \,(\text{mod } p),$$

which implies

$$C(p-1, s+1) \equiv 0 \,(\text{mod } p).$$

That is, the theorem is true for $i = s+1$. Now the proof of the theorem completes by mathematical induction.

Now, we prove Corollary 1. For any integer $n \geq 0$, taking $k = p$ in (3) and noting that

$$n! \sum_{a_1+a_2+\cdots+a_p=n} \frac{F_{a_1}(y)}{(a_1)!} \cdot \frac{F_{a_2}(y)}{(a_2)!} \cdots \frac{F_{a_p}(y)}{(a_p)!} \equiv 0 \,(\text{mod } p),$$

we have

$$\sum_{i=0}^{p-1} C(p-1, i) F_{n+p-1-i}(y) \equiv 0 \,(\text{mod } p). \tag{16}$$

From our theorem, we have

$$\sum_{i=1}^{p-2} C(p-1, i) F_{n+p-1-i}(y) \equiv 0 \,(\text{mod } p). \tag{17}$$

Note that $C(p-1,0) = 1$, $C(p-1, p-1) = (p-1)!$. Combining (16) and (17), we can deduce the congruence

$$F_{n+p-1}(y) - F_n(y) \equiv 0 \,(\mathrm{mod}\ p).$$

Now the proof of Corollary 1 completes. Since Corollarys 2 and 3 are the special situation of Corollary 1, we will not prove Corollarys 2 and 3 here.

Author Contributions: Conceptualization, W.Z.; Methodology, W.Z. and X.L.; Software, X.L.; Validation, W.Z. and X.L.; Formal Analysis, W.Z.; Investigation, X.L.; Resources, W.Z.; Data Curation, X.L.; Writing Original Draft Preparation, W.Z.; Writing Review & Editing, X.L.; Visualization, W.Z.; Supervision, W.Z.; Project Administration, X.L.; Funding Acquisition, W.Z. All authors have read and approved the final manuscript.

Funding: This research was funded by [National Natural Science Foundation of China] grant number [11771351].

Acknowledgments: The authors would like to thank the reviewers for their very detailed and helpful comments, which have significantly improved the presentation of this paper.

Conflicts of Interest: The authors declare no conflict of interest.

References

1. Kilar, N.; Simsek, Y. A new family of Fubini type numbrs and polynomials associated with Apostol-Bernoulli nujmbers and polynomials. *J. Korean Math. Soc.* **2017**, *54*, 1605–1621.
2. Kim, T.; Kim, D.S.; Jang, G.-W. A note on degenerate Fubini polynomials. *Proc. Jiangjeon Math. Soc.* **2017**, *20*, 521–531.
3. Kim, T.; Kim, D.S.; Jang, G.-W.; Kwon, J. Symmetric identities for Fubini polynomials. *Symmetry* **2018**, *10*, 219.
4. Zhao, J.-H.; Chen, Z.-Y. Some symmetric identities involving Fubini polynomials and Euler numbers. *Symmetry* **2018**, *10*, 303.
5. Chen, L.; Zhang, W.-P. Chebyshev polynomials and their some interesting applications. *Adv. Differ. Equ.* **2017**, *2017*, 303.
6. Clemente, C. Identities and generating functions on Chebyshev polynomials. *Georgian Math. J.* **2012**, *19*, 427–440.
7. He, Y. Symmetric identities for Calitz's q-Bernoulli numbers and polynomials. *Adv. Differ. Equ.* **2013**, *2013*, 246.
8. Kim, T. Symmetry of power sum polynomials and multivariate fermionic p-adic invariant integral on Z_p. *Russ. J. Math. Phys.* **2009**, *16*, 93–96.
9. Kim, T.; Kim, D.S. An identity of symmetry for the degernerate Frobenius-Euler polynomials. *Math. Slovaca* **2018**, *68*, 239–243.
10. Li, X.-X. Some identities involving Chebyshev polynomials. *Math. Probl. Eng.* **2015**, *2015*, 950695.
11. Rim, S.-H.; Jeong, J.-H.; Lee, S.-J.; Moon, E.-J.; Jin, J.-H. On the symmetric properties for the generalized twisted Genocchi polynomials. *ARS Comb.* **2012**, *105*, 267–272.
12. Wang, T.-T.; Zhang, W.-P. Some identities involving Fibonacci, Lucas polynomials and their applications. *Bull. Math. Soc. Sci. Math. Roum.* **2012**, *55*, 95–103.
13. Yi, Y.; Zhang, W.P. Some identities involving the Fibonacci polynomials. *Fibonacci Q.* **2002**, *40*, 314–318.
14. Zhang, W.-P. Some identities involving the Euler and the central factorial numbers. *Fibonacci Q.* **1998**, *36*, 154–157.
15. Kim, D.S.; Park, K.H. Identities of symmetry for Bernoulli polynomials arising from quotients of Volkenborn integrals invariant under S_3. *Appl. Math. Comput.* **2013**, *219*, 5096–5104.
16. Feng R.-Q.; Song C.-W. *Combinatorial Mathematics*; Beijing University Press: Beijing, China, 2015.

© 2018 by the authors. Licensee MDPI, Basel, Switzerland. This article is an open access article distributed under the terms and conditions of the Creative Commons Attribution (CC BY) license (http://creativecommons.org/licenses/by/4.0/).

Article

On p-Adic Fermionic Integrals of q-Bernstein Polynomials Associated with q-Euler Numbers and Polynomials [†]

Lee-Chae Jang [1], Taekyun Kim [2,*], Dae San Kim [3] and Dmitry Victorovich Dolgy [4,5]

1. Graduate School of Education, Konkuk University, Seoul 143-701, Korea; Lcjang@konkuk.ac.kr
2. Department of Mathematics, Kwangwoon University, Seoul 139-701, Korea
3. Department of Mathematics, Sogang University, Seoul 121-742, Korea; dskim@sogang.ac.kr
4. Kwangwoon Institute for Advanced Studies, Kwangwoon University, Seoul 139-701, Korea; d_dol@mail.ru
5. Institute of National Sciences, Far Eastern Federal University, 690950 Vladivostok, Russia
* Correspondence: tkkim@kw.ac.kr
† 2010 *Mathematics Subject Classication.* 11B83; 11S80.

Received: 6 July 2018; Accepted: 23 July 2018; Published: 1 August 2018

Abstract: We study a q-analogue of Euler numbers and polynomials naturally arising from the p-adic fermionic integrals on $\mathbb{Z}_p$ and investigate some properties for these numbers and polynomials. Then we will consider p-adic fermionic integrals on $\mathbb{Z}_p$ of the two variable q-Bernstein polynomials, recently introduced by Kim, and demonstrate that they can be written in terms of the q-analogues of Euler numbers. Further, from such p-adic integrals we will derive some identities for the q-analogues of Euler numbers.

Keywords: two variable q-Berstein polynomial; two variable q-Berstein operator; q-Euler number; q-Euler polynomial

1. Introduction

As is well known, the classical Bernstein polynomial of order n for $f \in C[0,1]$ is defined by (see [1–3]),

$$\mathbb{B}_n(f|x) = \sum_{k=0}^{n} f\left(\frac{k}{n}\right) B_{k,n}(x), \quad 0 \leq x \leq 1, \tag{1}$$

where $\mathbb{B}_n$ is called the Bernstein operater of order n, and (see [4–30]),

$$B_{k,n}(x) = \binom{n}{k} x^k (1-x)^{n-k}, \quad n,k \geq 0, \tag{2}$$

are called the Bernstein basis polynomials (or Bernstein polynomials of degree n).

The Weierstrass approximation theorem states that every continuous function defined on $[0,1]$ can be uniformly approximated as closely as desired by a polynomial function. In 1912, S. N. Bernstein explicitly constructed a sequence of polynomials that uniformly approximates any given continuous function f on $[0,1]$. Namely, he showed that $\mathbb{B}_n(f|x)$ tends uniformly to $f(x)$ as $n \to \infty$ on $[0,1]$ (see [3]). For $q \in \mathbb{C}$, with $0 < |q| < 1$, and $n,k \in \mathbb{Z}_{\geq 0}$, with $n \geq k$, the q-Bernstein polynomials of degree n are defined by Kim as (see [8])

$$B_{k,n}(x,q) = \binom{n}{k} [x]_q^k [1-x]_{\frac{1}{q}}^{n-k}, \tag{3}$$

where $[x]_q = \frac{1-q^x}{1-q}$. For any $f \in C[0,1]$, the q-Bernstein operator of order n is defined as

$$\mathbb{B}_{n,q}(f|x) = \sum_{k=0}^{n} f\left(\frac{k}{n}\right) B_{k,n}(x,q) = \sum_{k=0}^{n} f\left(\frac{k}{n}\right) \binom{n}{k} [x]_q^k [1-x]_{\frac{1}{q}}^{n-k}, \quad (4)$$

where $0 \leq x \leq 1$, and $n \in \mathbb{Z}_{\geq 0}$, (see [8,13]).

Here we note that a different version of q-Bernstein polynomials from Kim's was introduced earlier in 1997 by Phillips (see [22]). His q-Bernstein polynomial of order n for f is defined by

$$\mathbb{B}_n(f,q;x) = \sum_{k=0}^{n} f\left(\frac{[k]_q}{[n]_q}\right) \begin{bmatrix} n \\ k \end{bmatrix}_q x^k \prod_{s=0}^{n-1-k} (1-q^s x),$$

where f is a function defined on $[0,1]$, q is any positive real number, and

$$\begin{bmatrix} n \\ k \end{bmatrix}_q = \frac{[n]_q!}{[k]_q![n-k]_q!}, \quad [n]_q! = [1]_q[2]_q \cdots [n]_q, (n \geq 1), \quad [0]_q! = 1.$$

The properties of Phillips' q-Bernstein polynomilas for $q \in (0,1)$ were treated for example in [6,15,16,22–24], while those for $q > 1$ were developed for instance in [17–20].

A Bernoulli trial is an experiment where only two outcomes, whether a particular event A occurs or not, are possible. Flipping of coin is an example of Bernoulli trial, where only two outcomes, namely head and tail, are possible. Conventionally, it is said that the outcome of Bernoulli trial is a "success" if A occurs and a "failure" otherwise. Let $P_n(k)$ denote the probability of k successes in n independent Bernoulli trials with the probability of success r. Then it is given by the binomial probability law

$$P_n(k) = \binom{n}{k} r^k (1-r)^{n-k}, \text{ for } k = 0, 1, 2, \cdots, n. \quad (5)$$

We remark here that the Bernstein basis is the probability mass function of the binomial distribution from the definition of Bernstein polynomials. Let p be a fixed odd prime number. Throughout this paper, we will use the notations $\mathbb{Z}_p, \mathbb{Q}_p$, and $\mathbb{C}_p$ to denote respectively the ring of p-adic integers, the field of p-adic rational numbers and the completion of the algebraic closure of $\mathbb{Q}_p$. The p-adic norm in $\mathbb{C}_p$ is normalized in such a way that $|p|_p = \frac{1}{p}$. It is known that in terms of the recurrence relation the Euler numbers are given as follows (see [10,11]):

$$E_0 = 1, \quad (E+1)^n + E_n = 2\delta_{0,n}, \quad (6)$$

where $\delta_{n,k}$ is the Kronecker's symbol. Then the Euler polynomials can be given as (see [10])

$$E_n(x) = \sum_{l=0}^{n} \binom{n}{l} E_l x^{n-l}, \quad (7)$$

The q-Euler polynomials, considered by L. Carlitz, are given by

$$\mathcal{E}_{0,q} = 1, \quad q(q\mathcal{E}_q + 1)^n + \mathcal{E}_{n,q} = \begin{cases} [2]_q, & \text{if } n = 0, \\ 0, & \text{if } n > 0, \end{cases} \quad (8)$$

with the understanding that $\mathcal{E}_q^n$ is to be replaced by $\mathcal{E}_{n,q}$ (see [5]). Note that $\lim_{q \to 1} \mathcal{E}_{n,q} = E_n$, $(n \geq 0)$.

Let $f(x)$ be a continuous function on $\mathbb{Z}_p$. Then the p-adic fermionic integral on $\mathbb{Z}_p$ is defined by Kim as (see [12])

$$I_{-1}(f) = \int_{\mathbb{Z}_p} f(x) d\mu_{-1}(x) = \lim_{N \to \infty} \sum_{x=0}^{p^N-1} f(x)(-1)^x, \qquad (9)$$

where we notice that $\mu_{-1}(x + p^N \mathbb{Z}_p) = (-1)^x$ is a measure.
From (9), we note that (see [12])

$$I_{-1}(f_1) + I_{-1}(f) = 2f(0), \qquad (10)$$

where $f_1(x) = f(x+1)$. By (10), we easily get (see [25])

$$\int_{\mathbb{Z}_p} (x+y)^n d\mu_{-1}(y) = E_n(x), \ (n \geq 0), \qquad (11)$$

When $x = 0$, we note that $\int_{\mathbb{Z}_p} x^n d\mu_{-1}(x) = E_n$, $(n \geq 0)$. Let q be an indeterminate in $\mathbb{C}_p$ with $|1-q|_p < p^{-\frac{1}{p-1}}$. Taking (11) into consideration, we may investigate a q-analogue of Euler polynomials which are given by (see [12,26])

$$\int_{\mathbb{Z}_p} [x+y]_q^n d\mu_{-1}(y) = E_{n,q}(x), \ (n \geq 0), \qquad (12)$$

When $x = 0$, $E_{n,q} = E_{n,q}(0)$, $(n \geq 0)$ are said to be the q-Euler numbers. Using (9), we can easily see that

$$\int_{\mathbb{Z}_p} [x]_q^n d\mu_{-1}(x) = \frac{2}{(1-q)^n} \sum_{l=0}^n \binom{n}{l} (-1)^l \frac{1}{1+q^l}$$
$$= 2 \sum_{m=0}^\infty (-1)^m [m]_q^n, \ (n \geq 0). \qquad (13)$$

Thus, by (13), we get

$$E_{n,q} = 2 \sum_{m=0}^\infty (-1)^m [m]_q^n = \frac{2}{(1-q)^n} \sum_{l=0}^n \binom{n}{l} (-1)^l \frac{1}{1+q^l}. \qquad (14)$$

For $n, k \geq 0$, with $n \geq k$, and $q \in \mathbb{C}_p$, with $|1-q|_p < p^{-\frac{1}{p-1}}$, we define the p-adic q-Bernstein polynomials as follows:

$$B_{k,n}(x,q) = \binom{n}{k} [x]_q^k [1-x]_{\frac{1}{q}}^{n-k}. \qquad (15)$$

Then we consider the p-adic q-Bernstein operator defined for continuous functions f on $\mathbb{Z}_p$ and given by

$$\mathbb{B}_{n,q}(f|x) = \sum_{k=0}^n f\left(\frac{k}{n}\right) B_{k,n}(x,q), \ (x \in \mathbb{Z}_p). \qquad (16)$$

We study a q-analogue of Euler numbers and polynomials naturally arising from the p-adic fermionic integrals on $\mathbb{Z}_p$ and investigate some properties for these numbers and polynomials. Then we will consider p-adic fermionic integrals on $\mathbb{Z}_p$ of the two variable q-Bernstein polynomials, recently introduced by Kim in [8], and demonstrate that they can be written in terms of the q-analogues of Euler numbers. Further, from such p-adic integrals we will derive some identities for the q-analogues of Euler numbers.

2. q-Bernstein Polynomials Associated with q-Euler Numbers and Polynomials

We assume that $q \in \mathbb{C}_p$, with $|1-q|_p < p^{-\frac{1}{p-1}}$, throughout this section. From (12), we notice that

$$\sum_{n=0}^{\infty} E_{n,q}(x) \frac{t^n}{n!} = \sum_{m=0}^{\infty} (-1)^m e^{[m+x]_q t}. \tag{17}$$

By (10), we get

$$\int_{\mathbb{Z}_p} [x+1]_q^n d\mu_{-1}(x) + \int_{\mathbb{Z}_p} [x]_q^n d\mu_{-1}(x) = 2\delta_{0,n}, \quad (n \geq 0). \tag{18}$$

Thus, from (12), we have

$$E_{n,q}(1) + E_{n,q} = \begin{cases} 2, & \text{if } n=0, \\ 0, & \text{if } n>0. \end{cases} \tag{19}$$

On the other hand,

$$\begin{aligned} E_{n,q}(x) &= \int_{\mathbb{Z}_p} [x+y]_q^n d\mu_{-1}(y) \\ &= \sum_{l=0}^{n} \binom{n}{l} [x]_q^{n-l} q^{lx} \int_{\mathbb{Z}_p} [y]_q^l d\mu_{-1}(y) \\ &= \sum_{l=0}^{n} \binom{n}{l} q^{lx} E_{l,q} [x]_q^{n-l} = (q^x E_q + [x]_q)^n, \end{aligned} \tag{20}$$

with the understanding that E_q^n is to be replaced by $E_{n,q}$. From (19) and (20), we note that

$$E_{0,q} = 1, \ (qE_q + 1)^n + E_{n,q} = \begin{cases} 2, & \text{if } n=0, \\ 0, & \text{if } n>0. \end{cases} \tag{21}$$

Now, we observe that

$$\begin{aligned} E_{n,q}(2) &= (q^2 E_q + 1 + q)^n = (q(qE_q+1)+1)^n \\ &= \sum_{l=0}^{n} q^l (qE_q+1)^l \binom{n}{l} = 2 - E_{0,q} - \sum_{l=1}^{n} q^l E_{l,q} \binom{n}{l} \\ &= 2 - \sum_{l=0}^{n} q^l E_{l,q} \binom{n}{l} = 2 - (qE_q+1)^n. \end{aligned} \tag{22}$$

Now, by combining (21) with (22), we have the following theorem.

Theorem 1. For any $n \geq 0$, we have

$$E_{n,q}(2) = 2 + E_{n,q}, \ (n>0), \ E_{0,q}(2) = 1. \tag{23}$$

Invoking (9), we can derive the following equation

$$\int_{\mathbb{Z}_p} [1-x+y]_{q^{-1}}^n d\mu_{-1}(y) = (-1)^n q^n \int_{\mathbb{Z}_p} [x+y]_q^n d\mu_{-1}(y), \tag{24}$$

where n is a nonnegative integer. By (12) and (24), we get

$$E_{n,q^{-1}}(1-x) = (-1)^n q^n E_{n,q}(x), \ (n>0). \tag{25}$$

On the other hand, we have

$$\int_{\mathbb{Z}_p} [1-x]_{q^{-1}}^n d\mu_{-1}(x) = (-1)^n q^n \int_{\mathbb{Z}_p} [x-1]_q^n d\mu_{-1}(x) \qquad (26)$$
$$= (-1)^n q^n E_{n,q}(-1),$$

as $[-x]_{q^{-1}} = -q[x]_q$. By (25) and (26), we get

$$\int_{\mathbb{Z}_p} [1-x]_{q^{-1}}^n d\mu_{-1}(x) = (-1)^n q^n E_{n,q}(-1) = E_{n,q^{-1}}(2). \qquad (27)$$

Therefore, by (23) and (27), we have

Theorem 2. *For any $n > 0$, we have*

$$\int_{\mathbb{Z}_p} [1-x]_{q^{-1}}^n d\mu_{-1}(x) = 2 + E_{n,q^{-1}}. \qquad (28)$$

For $q \in \mathbb{C}_p$, with $|1-q|_p < p^{-\frac{1}{p-1}}$, and $x_1, x_1 \in \mathbb{Z}_p$, the two variable q-Bernstein polynomials are defined by

$$B_{k,n}(x_1, x_2|q) = \begin{cases} \binom{n}{k} [x_1]_q^k [1-x_2]_{q^{-1}}^{n-k}, & \text{if } n \geq k, \\ 0, & \text{if } n < k, \end{cases} \qquad (29)$$

where $n, k \geq 0$. From (29), we note that

$$B_{n-k,n}(1-x_2, 1-x_1|q^{-1}) = B_{k,n}(x_1, x_2|q), \quad B_{k,n}(x, x|q) = B_{k,n}(x, q), \qquad (30)$$

where $n, k \geq 0$ and $x_1, x_2 \in \mathbb{Z}_p$. For continuous functions f on $\mathbb{Z}_p$, the two variable q-Bernstein operator of order n is defined by

$$\mathbb{B}_{n,q}(f|x_1, x_2) = \sum_{k=0}^n f\left(\frac{k}{n}\right) \binom{n}{k} [x_1]_q^k [1-x_2]_{q^{-1}}^{n-k}$$
$$= \sum_{k=0}^n f\left(\frac{k}{n}\right) B_{k,n}(x_1, x_2|q), \qquad (31)$$

where $n, k \in \mathbb{Z}_{\geq 0}$, and $x_1, x_2 \in \mathbb{Z}_p$. In particular, if $f = 1$, then we have

$$\mathbb{B}_{n,q}(1|x_1, x_2) = \sum_{k=0}^n \binom{n}{k} [x_1]_q^k [1-x_2]_{q^{-1}}^{n-k}$$
$$= (1 + [x_1]_q - [x_2]_q)^n, \qquad (32)$$

where we used the fact

$$[1-x]_{q^{-1}} = 1 - [x]_q. \qquad (33)$$

Taking the double p-adic fermionic integral on $\mathbb{Z}_p$ as in the following, we have

$$\int_{\mathbb{Z}_p}\int_{\mathbb{Z}_p} B_{k,n}(x_1,x_2|q)d\mu_{-1}(x_1)d\mu_{-1}(x_2)$$
$$= \binom{n}{k}\int_{\mathbb{Z}_p}[x_1]_q^k d\mu_{-1}(x_1)\int_{\mathbb{Z}_p}[1-x_2]_{q^{-1}}^{n-k}d\mu_{-1}(x_2) \quad (34)$$
$$= \begin{cases} \binom{n}{k}E_{k,q}(2+E_{n-k,q^{-1}}), & \text{if } n>k, \\ E_{k,q}, & \text{if } n=k. \end{cases}$$

Therefore, from (34) we obtain the next theorem.

Theorem 3. *For any $n,k \in \mathbb{Z}_{\geq 0}$, with $n \geq k$, we have*
$$\int_{\mathbb{Z}_p}\int_{\mathbb{Z}_p} B_{k,n}(x_1,x_2|q)d\mu_{-1}(x_1)d\mu_{-1}(x_2)$$
$$= \begin{cases} \binom{n}{k}E_{n,q}(2+E_{n,q^{-1}}), & \text{if } n>k, \\ E_{k,q}, & \text{if } n=k. \end{cases} \quad (35)$$

Making the use of the definition of the two variable q-Bernstein polynomials and from (33), we notice that

$$\int_{\mathbb{Z}_p}\int_{\mathbb{Z}_p} B_{k,n}(x_1,x_2|q)d\mu_{-1}(x_1)d\mu_{-1}(x_2)$$
$$= \sum_{l=0}^{k}\binom{n}{n-k}\binom{k}{l}(-1)^{k+l}\int_{\mathbb{Z}_p}\int_{\mathbb{Z}_p}[1-x_1]_{q^{-1}}^{k-l}[1-x_2]_{q^{-1}}^{n-k}d\mu_{-1}(x_1)d\mu_{-1}(x_2)$$
$$= \binom{n}{k}\int_{\mathbb{Z}_p}[1-x_2]_{q^{-1}}^{n-k}d\mu_{-1}(x_2)\sum_{l=0}^{k}\binom{k}{l}(-1)^{k-l}\int_{\mathbb{Z}_p}[1-x_1]_{q^{-1}}^{k-l}d\mu_{-1}(x_1) \quad (36)$$
$$= \binom{n}{k}\int_{\mathbb{Z}_p}[1-x_2]_{q^{-1}}^{n-k}d\mu_{-1}(x_2)\left\{1+\sum_{l=0}^{k-1}\binom{k}{l}(2+E_{k-l,q^{-1}})\right\}.$$

Therefore, from (34) and (36) we deduce the following theorem.

Theorem 4. *For any $k \geq 0$, we have*
$$E_{k,q} = 2(2^k-1) + \sum_{l=0}^{k}\binom{k}{l}E_{k-l,q^{-1}}. \quad (37)$$

For $m,n,k \in \mathbb{Z}_{\geq 0}$, we observe that
$$\int_{\mathbb{Z}_p}\int_{\mathbb{Z}_p} B_{k,n}(x_1,x_2|q)B_{k,m}(x_1,x_2|q)d\mu_{-1}(x_1)d\mu_{-1}(x_2)$$
$$= \binom{n}{k}\binom{m}{k}\int_{\mathbb{Z}_p}[x_1]_q^{2k}d\mu_{-1}(x_1)\int_{\mathbb{Z}_p}[1-x_2]_{q^{-1}}^{n+m-2k}d\mu_{-1}(x_2) \quad (38)$$
$$= \binom{n}{k}\binom{m}{k}E_{2k,q}\int_{\mathbb{Z}_p}[1-x_2]_{q^{-1}}^{n+m-2k}d\mu_{-1}(x_2).$$

On the other hand,

$$\int_{\mathbb{Z}_p} \int_{\mathbb{Z}_p} B_{k,n}(x_1, x_2|q) B_{k,m}(x_1, x_2|q) d\mu_{-1}(x_1) d\mu_{-1}(x_2)$$

$$= \sum_{l=0}^{2k} \binom{n}{k}\binom{m}{k}\binom{2k}{l}(-1)^{2k-l}$$

$$\times \int_{\mathbb{Z}_p} \int_{\mathbb{Z}_p} [1-x_1]_{q^{-1}}^{2k-l} [1-x_2]_{q^{-1}}^{n+m-2k} d\mu_{-1}(x_1) d\mu_{-1}(x_2) \tag{39}$$

$$= \binom{n}{k}\binom{m}{k} \int_{\mathbb{Z}_p} [1-x_2]_{q^{-1}}^{n+m-2k} d\mu_{-1}(x_2)$$

$$\times \left\{ 1 + \sum_{l=0}^{2k-1} \binom{2k}{l}(-1)^{2k-l} \int_{\mathbb{Z}_p} [1-x_1]_{q^{-1}}^{2k-l} d\mu_{-1}(x_1) \right\}.$$

Hence, by (28), (38) and (39), we arrive at the following theorem.

Theorem 5. *For any $k \in \mathbb{N}$, we have*

$$E_{2k,q} = -2 + \sum_{l=0}^{2k} \binom{2k}{l}(-1)^{2k-l} E_{2k-l,q^{-1}}. \tag{40}$$

Let $n_1, n_2, \ldots, n_s, k \in \mathbb{Z}_{\geq 0}$, with $s \in \mathbb{N}$. Then we clearly have

$$\int_{\mathbb{Z}_p} \int_{\mathbb{Z}_p} \prod_{i=1}^{s} B_{k,n_i}(x_1, x_2|q) d\mu_{-1}(x_1) d\mu_{-1}(x_2)$$

$$= \prod_{i=1}^{s} \binom{n_i}{k} \int_{\mathbb{Z}_p} \int_{\mathbb{Z}_p} [x_1]_q^{sk} [1-x_2]_{q^{-1}}^{n_1+\cdots+n_s-sk}(x_1) d\mu_{-1}(x_2) \tag{41}$$

$$= \prod_{i=1}^{s} \binom{n_i}{k} E_{sk,q} \int_{\mathbb{Z}_p} [1-x_2]_{q^{-1}}^{n_1+\cdots+n_s-sk} d\mu_{-1}(x_2).$$

On the other hand,

$$\int_{\mathbb{Z}_p} \int_{\mathbb{Z}_p} \prod_{i=1}^{s} B_{k,n_i}(x_1, x_2|q) d\mu_{-1}(x_1) d\mu_{-1}(x_2)$$

$$= \sum_{l=0}^{sk} \prod_{i=1}^{s} \binom{n_i}{k}\binom{sk}{l}(-1)^{sk-l} \tag{42}$$

$$\times \int_{\mathbb{Z}_p} \int_{\mathbb{Z}_p} [1-x_1]_{q^{-1}}^{sk-l} [1-x_2]_{q^{-1}}^{n_1+\cdots+n_s-sk} d\mu_{-1}(x_1) d\mu_{-1}(x_2).$$

By (41) and (42), we get

$$E_{sk,q} = \sum_{l=0}^{sk} \binom{sk}{l}(-1)^{sk-l} \int_{\mathbb{Z}_p} \int_{\mathbb{Z}_p} [1-x_1]_{q^{-1}}^{sk-l} d\mu_{-1}(x_1)$$

$$= 1 + \sum_{l=0}^{sk-1} \binom{sk}{l}(-1)^{sk-l} \int_{\mathbb{Z}_p} \int_{\mathbb{Z}_p} [1-x_1]_{q^{-1}}^{sk-l} d\mu_{-1}(x_1). \tag{43}$$

Hence (28) and (43) together yield the next theorem.

Theorem 6. *For any $s \in \mathbb{N}$, we have*

$$E_{sk,q} = -2 + \sum_{l=0}^{sk} \binom{sk}{l}(-1)^{sk-l} E_{sk-l,q^{-1}}. \tag{44}$$

3. Conclusions

In the previous paper [8], the q-Bernstein polynomials were introduced as a generalization of the classical Bernstein polynomials. Here we studied some properties of a q-analogue of Euler numbers and polynomials arising from the p-adic fermionic integrals on $\mathbb{Z}_p$. Then we considered p-adic fermionic integrals on $\mathbb{Z}_p$ of the two variable q-Bernstein polynomials, recently introduced by Kim, and show that they can be expressed in terms of the q-analogues of Euler numbers. Along the same line, we can introduce a new q-Bernoulli numbers and polynomials, different from the classical Carlitz q-Bernoulli numbers $\beta_{n,q}$ and polynomials $\beta_{n,q}(x)$, by considering the Volkenborn integrals in lieu of the p-adic fermionic integrals on $\mathbb{Z}_p$. Then we may investigate Volkenborn integrals on $\mathbb{Z}_p$ of the q-Bernstein polynomials and unveil their connections with those new q-Bernoulli numbers which is our ongoing project.

Author Contributions: T.K. and D.S.K. conceived the framework and structured the whole paper; T.K. wrote the paper; L.C.J. and D.V.D. checked the results of the paper; D.S.K. and L.-C.J. completed the revision of the article.

Funding: This paper was supported by Konkuk University in 2017.

Acknowledgments: The authors would like to express their sincere gratitude to the referees for their valuable comments which improved the original manuscript in its present form.

Conflicts of Interest: The authors declare no conflict of interest.

References

1. Açikgöz, M.; Erdal, D.; Araci, S. A new approach to q-Bernoulli numbers and q-Bernoulli polynomials related to q-Bernstein polynomials. *Adv. Differ. Equ.* **2010**, *2015*, 272. [CrossRef]
2. Araci, S.; Acikgoz, M. A note on the Frobenius-Euler numbers and polynomials associated with Bernstein polynomials. *Adv. Stud. Contemp. Math. (Kyungshang)* **2012**, *22*, 399–406.
3. Bernstein, S. Démonstration du théorème de Weierstrass fondée sur le calcul des probabilités. *Commun. Soc. Math. Kharkov* **1912**, *2*, 1–2.
4. Bayad, A.; Kim, T. Identities involving values of Bernstein, q-Bernoulli, and q-Euler polynomials. *Russ. J. Math. Phys.* **2011**, *18*, 133–143. [CrossRef]
5. Carlitz, L. Degenerate Stirling, Bernoulli and Eulerian numbers. *Utilitas Math.* **1979**, *15*, 51–88.
6. Goodman, T.N.T.; Oruç, H.; Phillips, G.M. Convexity and generalized Bernstein polynomials. *Proc. Edinb. Math. Soc.* **1999**, *42*, 179–190. [CrossRef]
7. Kim, T.; Choi, J.; Kim, Y.-H. On the k-dimensional generalization of q-Bernstein polynomials. *Proc. Jangjeon Math. Soc.* **2011**, *14*, 199–207.
8. Kim, T. A note on q-Bernstein polynomials. *Russ. J. Math. Phys.* **2011**, *18*, 73–82. [CrossRef]
9. Kim, T.; Choi, J.; Kim, Y.-H. Some identities on the q-Bernstein polynomials, q-Stirling numbers and q-Bernoulli numbers. *Adv. Stud. Contemp. Math. (Kyungshang)* **2010**, *20*, 335–341.
10. Kim, T.; Lee, B.; Choi, J.; Kim, Y.-H.; Rim, S.H. On the q-Euler numbers and weighted q-Bernstein polynomials. *Adv. Stud. Contemp. Math. (Kyungshang)* **2011**, *21*, 13–18.
11. Kim, T. Identities on the weighted q-Euler numbers and q-Bernstein polynomials. *Adv. Stud. Contemp. Math. (Kyungshang)* **2012**, *22*, 7–12. [CrossRef]
12. Kim, T. A study on the q-Euler numbers and the fermionic q-integral of the product of several type q-Bernstein polynomials on $\mathbb{Z}_p$. *Adv. Stud. Contemp. Math. (Kyungshang)* **2013**, *23*, 5–11.
13. Kim, T.; Ryoo, C. S.; Yi, H. A note on q-Bernoulli numbers and q-Bernstein polynomials. *Ars Comb.* **2012**, *104*, 437–447.
14. Kurt, V. Some relation between the Bernstein polynomials and second kind Bernoulli polynomials. *Adv. Stud. Contemp. Math. (Kyungshang)* **2013**, *23*, 43–48.
15. Oruç, H.; Phillips, G.M. A generalization of Bernstein polynomials. *Proc. Edinb. Math. Soc.* **1999**, *42*, 403–413. [CrossRef]
16. Oruç, H.; Tuncer, N. On the convergence and iterates of q-Bernstein polynomials. *J. Approx. Theory* **2002**, *117*, 301–313. [CrossRef]
17. Ostrovska, S. q-Bernstein polynomials and their iterates *J. Approx. Theory* **2003** *123*, 232–255. [CrossRef]
18. Ostrovska, S. On the q-Bernstein polynomials. *Adv. Stud. Contemp. Math. (Kyungshang)* **2005**, *11*, 193–204.

19. Ostrovska, S. On the *q*-Bernstein polynomials of the logarithmic function in the case $q > 1$. *Math. Slovaca* **2016**, *66*, 73–78. [CrossRef]
20. Ostrovska, S. On the approximation of analytic functions by the *q*-Bernstein polynomials in the case $q > 1$. *Electron. Trans. Numer. Anal.* **2010**, *37*, 105–112.
21. Park, J.-W.; Pak, H.K.; Rim, S.-H.; Kim, T.; Lee, S.-H. A note on the *q*-Bernoulli numbers and *q*-Bernstein polynomials. *J. Comput. Anal. Appl.* **2013**, *15*, 722–729.
22. Phillips, G.M. Bernstein polynomials based on the *q*-integers. *Ann. Numer. Math.* **1997**, *4*, 511–518.
23. Phillips, G.M. A de Casteljau algorithm for generalized Bernstein polynomials. *BIT* **1997**, *37*, 232–236. [CrossRef]
24. Phillips, G.M. A generalization of the Bernstein polynomials based on the *q*-integers. *ANZIAM J.* **2000**, *42*, 79–86. [CrossRef]
25. Rim, S.-H.; Joung, J.; Jin, J.-H.; Lee, S.-J. A note on the weighted Carlitz's type *q*-Euler numbers and *q*-Bernstein polynomials. *Proc. Jangjeon Math. Soc.* **2012**, *15*, 195–201.
26. Ryoo, C.S. Some identities of the twisted *q*-Euler numbers and polynomials associated with *q*-Bernstein polynomials. *Proc. Jangjeon Math. Soc.* **2011**, *14*, 239–248.
27. Simsek, Y.; Gunay, M. On Bernstein type polynomials and their applications. *Adv. Differ. Equ.* **2015**, *2015*, 79. [CrossRef]
28. Siddiqui, M.A.; Agrawal, R.R.; Gupta, N. On a class of modified new Bernstein operators. *Adv. Stud. Contemp. Math. (Kyungshang)* **2014**, *24*, 97–107.
29. Srivastava, H.M.; Mursaleen, M.; Alotaibi, A.M.; Nasiruzzaman, M.; Al-Abied, A.A.H. Some approximation results involving the *q*-Szász-Mirakyan-Kantorovich type operators via Dunkl's generalization. *Math. Methods Appl. Sci.* **2017**, *40*, 5437–5452. [CrossRef]
30. Tunc, T.; Simsek, E. Some approximation properties of Szász-Mirakjan-Bernstein operators. *Eur. J. Pure Appl. Math.* **2014**, *7*, 419–428.

© 2018 by the authors. Licensee MDPI, Basel, Switzerland. This article is an open access article distributed under the terms and conditions of the Creative Commons Attribution (CC BY) license (http://creativecommons.org/licenses/by/4.0/).

Article

Some Symmetric Identities Involving Fubini Polynomials and Euler Numbers

Zhao Jianhong [1] and Chen Zhuoyu [2,*]

1. Department of Teachers Education, Lijiang Teachers College, Lijiang 674199, China; zjh3004@163.com
2. School of Mathematics, Northwest University, Xi'an 710127, China
* Correspondence: chenzymath@163.com

Received: 30 June 2018; Accepted: 24 July 2018; Published: 1 August 2018

Abstract: The aim of this paper is to use elementary methods and the recursive properties of a special sequence to study the computational problem of one kind symmetric sums involving Fubini polynomials and Euler numbers, and give an interesting computational formula for it. At the same time, we also give a recursive calculation method for the general case.

Keywords: Fubini polynomials; Euler numbers; symmetric identities; elementary method; computational formula

MSC: 11B83; 11B37

1. Introduction

For any integer $n \geq 0$, the Fubini polynomials $\{F_n(y)\}$ are defined by the coefficients of the generating function

$$\frac{1}{1 - y(e^t - 1)} = \sum_{n=0}^{\infty} \frac{F_n(y)}{n!} \cdot t^n, \tag{1}$$

where $F_0(y) = 1$, $F_1(y) = y$, and so on. $F_n(1) = F_n$ are called Fubini numbers. These polynomials and numbers are closely connected with the Stirling numbers. Some contents and properties of Stirling numbers can be found in reference [1]. T. Kim et al. [2] proved the identity

$$F_n(y) = \sum_{k=0}^{n} S_2(n,k)\, k!\, y^k,\ (n \geq 0),$$

where $S_2(n,k)$ are the Stirling numbers of the second kind. It not only associated Fubini polynomials with Stirling numbers, but also stressed the importance of researching Fubini polynomials.

Please note that the identity (see [3,4])

$$\frac{2e^{tx}}{1 + e^t} = \sum_{n=0}^{\infty} \frac{E_n(x)}{n!} \cdot t^n, \tag{2}$$

where $E_n(x)$ signifies the Euler polynomials.

It is distinct that if taking $y = -\frac{1}{2}$ in (1) and $x = 0$ in (2), then from (1) and (2) we can get the identity

$$E_n(0) = F_n\left(-\frac{1}{2}\right),\ n \in N^*0 \tag{3}$$

where $E_n(0) = E_n$ is the Euler number (see [5] for related contents).

On the other hand, two variable Fubini polynomials are defined by means of the following (see [2,6])

$$\frac{e^{xt}}{1 - y(e^t - 1)} = \sum_{n=0}^{\infty} \frac{F_n(x,y)}{n!} \cdot t^n,$$

and $F_n(y) = F_n(0,y)$ for all integers $n \geq 0$. About the properties of $F_n(x,y)$, several scholars have also researched it, especially T. Kim and others have done a large amount of vital works. For instance, they proved a series of identities linked to $F_n(x,y)$ (see [2,7]), one of which is

$$F_n(x,y) = \sum_{l=0}^{n} \binom{n}{l} x^l \cdot F_{n-l}(y), n \in N^*0.$$

These polynomials occupy indispensable positions in the theory and application of mathematics. In particular, they are widely used in combinatorial mathematics. Therefore, several scholars have researched their various properties, and acquired a series of vital results. Some involved contents can be found in references [5,7–17].

The goal of this paper is to use elementary methods and recursive properties of a special sequenc to research the computational problem of the sums

$$\sum_{a_1+a_2+\cdots+a_k=n} \frac{F_{a_1}(y)}{(a_1)!} \cdot \frac{F_{a_2}(y)}{(a_2)!} \cdots \frac{F_{a_k}(y)}{(a_k)!}, \tag{4}$$

where the summation is over all k-tuples with non-negative integer coordinates $(a_1, a_2, \cdots, a_k)$ such that $a_1 + a_2 + \cdots + a_k = n$.

About this content, it seems there is no valid method to solve the computational problem of (4). However, this problem is significant, it can reveal the structure of Fubini polynomials itself and its internal relations, at least it can reflect the combination properties of Fubini polynomials.

In this paper, we will take elementary methods and the properties of $F_n(y)$ to obtain a fascinating computational formula for (4). Simultaneously, we can also acquire a recursive calculation method for the general case. That is, we are going to prove the following major result:

Theorem 1. *For any positive integers n and k, we have the identity*

$$\sum_{a_1+a_2+\cdots+a_k=n} \frac{F_{a_1}(y)}{(a_1)!} \cdot \frac{F_{a_2}(y)}{(a_2)!} \cdots \frac{F_{a_k}(y)}{(a_k)!}$$

$$= \frac{1}{(k-1)!(y+1)^{k-1}} \cdot \frac{1}{n!} \sum_{i=0}^{k-1} C(k-1,i) F_{n+k-1-i}(y),$$

where the sequence $\{C(k,i)\}$ is defined as follows: For any positive integer k and integers $0 \leq i \leq k$, we define $C(k,0) = 1$, $C(k,k) = k!$ and

$$C(k+1, i+1) = C(k, i+1) + (k+1)C(k,i), \text{ for all } 0 \leq i < k,$$

providing $C(k,i) = 0$, if $i > k$.

The characteristic of this theorem is to represent a complex sum of Fubini polynomials as a linear combination of a single Fubini polynomial. Of course, our method can also be further generalized, provided a corresponding results for $F_n(x,y)$. It is just that its form is not so pretty, so we are not listing it here. If taking $k = 3, 4$ and 5, then from our theorem we may instantly deduce the following several corollaries:

Corollary 1. *For any positive integer n, we have the identity*

$$\sum_{a+b+c=n} \frac{F_a(y)}{a!} \cdot \frac{F_b(y)}{b!} \cdot \frac{F_c(y)}{c!} = \frac{1}{2 \cdot n! \cdot (y+1)^2} \left(F_{n+2}(y) + 3F_{n+1}(y) + 2F_n(y) \right).$$

Corollary 2. *For any positive integer n, we have the identity*

$$\sum_{a+b+c+d=n} \frac{F_a(y)}{a!} \cdot \frac{F_b(y)}{b!} \cdot \frac{F_c(y)}{c!} \cdot \frac{F_d(y)}{d!}$$
$$= \frac{1}{6 \cdot n! \cdot (y+1)^3} \left(F_{n+3}(y) + 6F_{n+2}(y) + 11F_{n+1}(y) + 6F_n(y) \right).$$

Corollary 3. *For any positive integer n, we have the identity*

$$\sum_{a+b+c+d+e=n} \frac{F_a(y)}{a!} \cdot \frac{F_b(y)}{b!} \cdot \frac{F_c(y)}{c!} \cdot \frac{F_d(y)}{d!} \cdot \frac{F_e(y)}{e!}$$
$$= \frac{1}{24 \cdot n! \cdot (y+1)^4} \left(F_{n+4}(y) + 10F_{n+3}(y) + 35F_{n+2}(y) + 50F_{n+1}(y) + 24F_n(y) \right).$$

If taking $y = -\frac{1}{2}$ in our theorem, then from (3) we can also infer the following:

Corollary 4. *For any positive integers n and $k \geq 2$, we have the identity*

$$\sum_{a_1+a_2+\cdots+a_k=n} \frac{E_{a_1}}{(a_1)!} \cdot \frac{E_{a_2}}{(a_2)!} \cdots \frac{E_{a_k}}{(a_k)!} = \frac{2^{k-1}}{(k-1)!} \cdot \frac{1}{n!} \sum_{i=0}^{k-1} C(k-1,i) E_{n+k-1-i}.$$

If $n = p$ is an odd prime, then taking $y = 1$ in Corollarys 1 and 2, we also have the following congruences.

Corollary 5. *For any odd prime p, we have the congruence*

$$22F_p \equiv F_{p+2} + 3F_{p+1} \pmod{p}.$$

Corollary 6. *For any odd prime p, we have the congruence*

$$186F_p \equiv F_{p+3} + 6F_{p+2} + 11F_{p+1} \pmod{p}.$$

2. A Simple Lemma

For purpose of proving our theorem, we need a uncomplicated lemma. As a matter of convenience, we first present a new sequence $\{C(k,i)\}$ as follows. For any positive integer k and integers $0 \leq i \leq k$, we define $C(k,0) = 1$, $C(k,k) = k!$ and
$C(k+1, i+1) = C(k, i+1) + (k+1)C(k, i)$, $1 \leq i \leq k$, $C(k, i) = 0$, if $i > k$.
For clarity, for $1 \leq k \leq 9$, we list values of $C(k, i)$ in the Table 1.

Table 1. Values of $C(k, i)$.

$C(k,i)$	$i=0$	$i=1$	$i=2$	$i=3$	$i=4$	$i=5$	$i=6$	$i=7$	$i=8$	$i=9$
$k=1$	1	1								
$k=2$	1	3	2							
$k=3$	1	6	11	6						
$k=4$	1	10	35	50	24					
$k=5$	1	15	85	225	274	120				
$k=6$	1	21	175	735	1624	1764	720			
$k=7$	1	28	322	1960	6769	13,132	13,068	5040		
$k=8$	1	36	546	4536	22,449	67,284	118,124	109,584	40,320	
$k=9$	1	45	870	9450	63,273	269,325	723,680	1,172,700	1,026,576	362,880

Obviously, the values of $C(k, i)$ can be easily calculated by using a computer program. Hence, for any positive integer k, the computational problem of (4) can be solved fully.

In this table of numerical values, we also find that for prime $p = 3, 5$ and 7, we have the congruence

$$C(p-1, i) \equiv 0 \pmod{p} \text{ for all } 1 \leq i \leq p-2.$$

For all prime $p > 7$ is true? This is an enjoyable open problem.

If this congruence is true, then we can also deduce that for any positive integer n and odd prime p, one has the congruence

$$F_{n+p-1}(y) + F_n(y) \equiv 0 \pmod{p}.$$

Now let function $f(t) = \frac{1}{1-y(e^t-1)}$. Then we have the following

Lemma 1. *For any positive integer k, we have the identity*

$$\sum_{i=0}^{k} C(k,i) f^{(k-i)}(t) = k!(y+1)^k f^{k+1}(t),$$

where $f^{(0)}(t) = f(t)$, $f^{(r)}(t)$ denotes the r-order derivative of $f(t)$ for variable t.

Proof. Now we prove this lemma by induction. From the definition of the derivative we acquire

$$f'(t) = \frac{ye^t}{(1-y(e^t-1))^2} = -f(t) + (y+1)f^2(t) \quad (5)$$

or

$$f'(t) + f(t) = (y+1)f^2(t). \quad (6)$$

Please note that $C(1, 0) = 1$ and $C(1, 1) = 1$, so the lemma is true for $k = 1$.
Suppose that the lemma is true for all integer $k \geq 1$. That is,

$$\sum_{i=0}^{k} C(k,i) f^{(k-i)}(t) = k!(y+1)^k f^{k+1}(t). \quad (7)$$

Then take the derivative for t in (7) and applying (5) and (7) we obtain

$$\begin{aligned}
\sum_{i=0}^{k} C(k,i) f^{(k+1-i)}(t) &= (k+1)!(y+1)^k f^k(t) \cdot f'(t) \\
&= (k+1)!(y+1)^k f^k(t) \cdot (-f(t) + (y+1)f^2(t)) \\
&= (k+1)!(y+1)^{k+1} f^{k+2}(t) - (k+1)!(y+1)^k f^{k+1}(t) \\
&= (k+1)!(y+1)^{k+1} f^{k+2}(t) - (k+1) \left(\sum_{i=0}^{k} C(k,i) f^{(k-i)}(t) \right).
\end{aligned} \quad (8)$$

It is evident that (8) implies

$$
\begin{aligned}
(k+1)!&(y+1)^{k+1}f^{k+2}(t) \\
&= C(k,0)f^{(k+1)}(t) + \sum_{i=0}^{k-1}\left(C(k,i+1)+(k+1)C(k,i)\right)f^{(k-i)}(t) + (k+1)!f(t) \\
&= C(k,0)f^{(k+1)}(t) + \sum_{i=0}^{k-1} C(k+1,i+1)f^{(k-i)}(t) + (k+1)!f(t) \\
&= \sum_{i=0}^{k+1} C(k+1,i)f^{(k+1-i)}(t),
\end{aligned} \quad (9)
$$

where we have used the identities $C(k,0) = 1$ and $C(k,k) = k!$. Now the lemma follows from (9) and mathematical induction. □

3. Proof of the Theorem

In this section, the proof of our theorem will be completed. Firstly, for any positive integer k, from the definition of $f(t)$ and the properties of the power series we obtain

$$f^{(k)}(t) = \sum_{n=0}^{\infty} \frac{F_{n+k}(y)}{n!} \cdot t^n \qquad (10)$$

and

$$
\begin{aligned}
f^k(t) &= \left(\sum_{a_1=0}^{\infty} \frac{F_{a_1}(y)}{a_1!} \cdot t^{a_1}\right)\left(\sum_{a_2=0}^{\infty} \frac{F_{a_2}(y)}{a_2!} \cdot t^{a_2}\right)\cdots\left(\sum_{a_k=0}^{\infty} \frac{F_{a_k}(y)}{a_k!} \cdot t^{a_k}\right) \\
&= \left(\sum_{a_1=0}^{\infty}\sum_{a_2=0}^{\infty}\cdots\sum_{a_k=0}^{\infty} \frac{F_{a_1}(y)}{(a_1)!}\cdot\frac{F_{a_2}(y)}{(a_2)!}\cdots\frac{F_{a_k}(y)}{(a_k)!} \cdot t^{a_1+a_2+\cdots+a_k}\right) \\
&= \sum_{n=0}^{\infty}\left(\sum_{a_1+a_2+\cdots+a_k=n} \frac{F_{a_1}(y)}{(a_1)!}\cdot\frac{F_{a_2}(y)}{(a_2)!}\cdots\frac{F_{a_k}(y)}{(a_k)!}\right)\cdot t^n.
\end{aligned} \quad (11)
$$

From (10), (11) and Lemma we acquire

$$
\begin{aligned}
&\frac{1}{(k-1)!(y+1)^{k-1}}\cdot\sum_{i=0}^{k-1} C(k-1,i)\sum_{n=0}^{\infty}\frac{F_{n+k-1-i}(y)}{n!}\cdot t^n \\
&= \sum_{n=0}^{\infty}\left(\sum_{a_1+a_2+\cdots+a_k=n} \frac{F_{a_1}(y)}{(a_1)!}\cdot\frac{F_{a_2}(y)}{(a_2)!}\cdots\frac{F_{a_k}(y)}{(a_k)!}\right)\cdot t^n.
\end{aligned} \quad (12)
$$

Comparing the coefficients of t^n in (12) we have the identity

$$
\sum_{a_1+a_2+\cdots+a_k=n} \frac{F_{a_1}(y)}{(a_1)!}\cdot\frac{F_{a_2}(y)}{(a_2)!}\cdots\frac{F_{a_k}(y)}{(a_k)!}
= \frac{1}{(k-1)!(y+1)^{k-1}}\cdot\frac{1}{n!}\sum_{i=0}^{k-1} C(k-1,i)F_{n+k-1-i}(y).
$$

This completes the proof of our Theorem.

Author Contributions: Writing-original draft: J.Z.; Writing-review and editing: Z.C.

Funding: This research was funded by the N. S. F. (11771351) of China.

Acknowledgments: The author would like to thank the referees for their very helpful and detailed comments, which have significantly improved the presentation of this paper.

Conflicts of Interest: The authors declare no conflict of interest.

References

1. Feng, R.-Q.; Song, C.-W. *Combinatorial Mathematics*; Beijing University Press: Beijing, China, 2015.
2. Kim, T.; Kim, D.S.; Jang, G.-W. A note on degenerate Fubini polynomials. *Proc. Jangjeon Math. Soc.* **2017**, *20*, 521–531.
3. Kim, T. Symmetry of power sum polynomials and multivariate fermionic p-adic invariant integral on Z_p. *Russ. J. Math. Phys.* **2009**, *16*, 93–96. [CrossRef]
4. Kim, D.S.; Park, K.H. Identities of symmetry for Bernoulli polynomials arising from quotients of Volkenborn integrals invariant under S_3. *Appl. Math. Comput.* **2013**, *219*, 5096–5104. [CrossRef]
5. Zhang, W. Some identities involving the Euler and the central factorial numbers. *Fibonacci Q.* **1998**, *36*, 154–157.
6. Kilar, N.; Simesk, Y. A new family of Fubini type numbrs and polynomials associated with Apostol-Bernoulli nujmbers and polynomials. *J. Korean Math. Soc.* **2017**, *54*, 1605–1621.
7. Kim, T.; Kim, D. S.; Jang, G.-W.; Kwon, J. Symmetric identities for Fubini polynomials. *Symmetry* **2018**, *10*, 219. [CrossRef]
8. Kim, T.; Kim, D.S. An identity of symmetry for the degernerate Frobenius-Euler polynomials. *Math. Slovaca* **2018**, *68*, 239–243. [CrossRef]
9. He, Y. Symmetric identities for Calitz's q-Bernoulli numbers and polynomials. *Adv. Differ. Equ.* **2013**, *2013*, 246. [CrossRef]
10. Rim, S.-H.; Jeong, J.-H.; Lee, S.-J.; Moon, E.-J.; Jin, J.-H. On the symmetric properties for the generalized twisted Genocchi polynomials. *ARS Comb.* **2012**, *105*, 267–272.
11. Yi, Y.; Zhang, W. Some identities involving the Fibonacci polynomials. *Fibonacci Q.* **2002**, *40*, 314–318.
12. Ma, R.; Zhang, W. Several identities involving the Fibonacci numbers and Lucas numbers. *Fibonacci Q.* **2007**, *45*, 164–170.
13. Wang, T.; Zhang, W. Some identities involving Fibonacci, Lucas polynomials and their applications. *Bull. Math. Soc. Sci. Math. Roum.* **2012**, *55*, 95–103.
14. Chen, L.; Zhang, W. Chebyshev polynomials and their some interesting applications. *Adv. Differ. Equ.* **2017**, *2017*, 303.
15. Li, X.X. Some identities involving Chebyshev polynomials. *Math. Probl. Eng.* **2015**, *2015*, 950695. [CrossRef]
16. Ma, Y.; Lv, X.-X. Several identities involving the reciprocal sums of Chebyshev polynomials. *Math. Probl. Eng.* **2017**, *2017*, 4194579. [CrossRef]
17. Clemente, C. Identities and generating functions on Chebyshev polynomials. *Georgian Math. J.* **2012**, *19*, 427–440.

© 2018 by the authors. Licensee MDPI, Basel, Switzerland. This article is an open access article distributed under the terms and conditions of the Creative Commons Attribution (CC BY) license (http://creativecommons.org/licenses/by/4.0/).

Article

Representing Sums of Finite Products of Chebyshev Polynomials of Third and Fourth Kinds by Chebyshev Polynomials

Taekyun Kim [1], Dae San Kim [2], Dmitry V. Dolgy [3] and Cheon Seoung Ryoo [4,*]

[1] Department of Mathematics, Kwangwoon University, Seoul 139-701, Korea; tkkim@kw.ac.kr
[2] Department of Mathematics, Sogang University, Seoul 121-742, Korea; dskim@sogang.ac.kr
[3] Institute of Natural Sciences, Far Eastern Federal University, 690950 Vladivostok, Russia; d_dol@mail.ru
[4] Department of Mathematics, Hannam University, Daejeon 306-791, Korea
* Correspondence: ryoocs@hnu.kr

Received: 05 June 2018; Accepted: 30 June 2018; Published: 3 July 2018

Abstract: Here, we consider the sums of finite products of Chebyshev polynomials of the third and fourth kinds. Then, we represent each of those sums of finite products as linear combinations of the four kinds of Chebyshev polynomials, which involve the hypergeometric function $_3F_2$.

Keywords: Chebyshev polynomials; sums of finite products; hypergeometric function

MSC: 11B68; 33C45

---

1. Introduction and Preliminaries

We first recall here that, for any nonnegative integer n, the falling factorial polynomials $(x)_n$ and the rising factorial polynomials $<x>_n$ are respectively given by:

$$(x)_n = x(x-1)\cdots(x-n+1), \quad (n \geq 1), \quad (x)_0 = 1, \tag{1}$$

$$<x>_n = x(x+1)\cdots(x+n-1), \quad (n \geq 1), \quad <x>_0 = 1. \tag{2}$$

The two factorial polynomials are related by:

$$(x)_n = (-1)^n <-x>_n, \quad <x>_n = (-1)^n (-x)_n. \tag{3}$$

We will make use of the following.

$$\frac{(2n-2s)!}{(n-s)!} = \frac{2^{2n-2s}(-1)^s <\frac{1}{2}>_n}{<\frac{1}{2}-n>_s}, \tag{4}$$

for any integers n, s with $n \geq s \geq 0$.

$$B(x,y) = \int_0^1 t^{x-1}(1-t)^{y-1}dt = \frac{\Gamma(x)\Gamma(y)}{\Gamma(x+y)}, \quad (\mathrm{Re}(x), \mathrm{Re}(y) > 0), \tag{5}$$

$$\Gamma\left(n+\frac{1}{2}\right) = \frac{(2n)!\Gamma(\frac{1}{2})}{2^{2n}n!}, \quad (n \geq 0). \tag{6}$$

Here, $B(x,y)$ and $\Gamma(x)$ are respectively the Beta and Gamma functions.

The hypergeometric function ${}_pF_q\left(\begin{matrix} a_1, \cdots, a_p \\ b_1, \cdots, b_q \end{matrix}; x\right)$ is defined by (see [1]):

$${}_pF_q\left(\begin{matrix} a_1, \cdots, a_p \\ b_1, \cdots, b_q \end{matrix}; x\right) = \sum_{n=0}^{\infty} \frac{<a_1>_n \cdots <a_p>_n}{<b_1>_n \cdots <b_q>_n} \frac{x^n}{n!} \tag{7}$$

$$(p \leq q+1, \quad |x| < 1).$$

In this paper, we will need only some basic knowledge about Chebyshev polynomials, which we recall here in below. The interested reader may want to refer to [1–3] for full accounts of this fascinating area of orthogonal polynomials.

The Chebyshev polynomials of the first, second, third and fourth kinds are respectively defined by the following generating functions.

$$\frac{1-xt}{1-2xt+t^2} = \sum_{n=0}^{\infty} T_n(x) t^n, \tag{8}$$

$$\frac{1}{1-2xt+t^2} = \sum_{n=0}^{\infty} U_n(x) t^n, \tag{9}$$

$$F(t, x) = \frac{1-t}{1-2xt+t^2} = \sum_{n=0}^{\infty} V_n(x) t^n, \tag{10}$$

$$G(t, x) = \frac{1+t}{1-2xt+t^2} = \sum_{n=0}^{\infty} W_n(x) t^n. \tag{11}$$

One way of deriving their generating functions is from their trigonometric formulas. For example, those formulas for $V_n(x)$ and $W_n(x)$ are given by:

$$V_n(\cos\theta) = \frac{\cos(n+\tfrac{1}{2})\theta}{\cos\tfrac{\theta}{2}},$$

$$W_n(\cos\theta) = \frac{\sin(n+\tfrac{1}{2})\theta}{\sin\tfrac{\theta}{2}}.$$

They are explicitly expressed as in the following.

$$T_n(x) = {}_2F_1\left(-n, n; \frac{1}{2}; \frac{1-x}{2}\right)$$
$$= \frac{n}{2}\sum_{l=0}^{[\frac{n}{2}]} (-1)^l \frac{1}{n-l} \binom{n-l}{l} (2x)^{n-2l}, \quad (n \geq 1), \tag{12}$$

$$U_n(x) = (n+1)\, {}_2F_1\left(-n, n+2; \frac{3}{2}; \frac{1-x}{2}\right)$$
$$= \sum_{l=0}^{[\frac{n}{2}]} (-1)^l \binom{n-l}{l} (2x)^{n-2l}, \quad (n \geq 0), \tag{13}$$

$$V_n(x) = {}_2F_1\left(-n, n+1; \frac{1}{2}; \frac{1-x}{2}\right)$$
$$= \sum_{l=0}^{n} \binom{2n-l}{l} 2^{n-l}(x-1)^{n-l}, \quad (n \geq 0), \tag{14}$$

$$W_n(x) = (2n+1) \, _2F_1\left(-n, n+1; \frac{3}{2}; \frac{1-x}{2}\right)$$

$$= (2n+1) \sum_{l=0}^{n} \frac{2^{n-l}}{2n-2l+1} \binom{2n-l}{l} (x-1)^{n-l}, \quad (n \geq 0). \tag{15}$$

The Chebyshev polynomials of the first, second, third and fourth kinds are also given by Rodrigues' formulas.

$$T_n(x) = \frac{(-1)^n 2^n n!}{(2n)!} (1-x^2)^{\frac{1}{2}} \frac{d^n}{dx^n} (1-x^2)^{n-\frac{1}{2}}, \tag{16}$$

$$U_n(x) = \frac{(-1)^n 2^n (n+1)!}{(2n+1)!} (1-x^2)^{-\frac{1}{2}} \frac{d^n}{dx^n} (1-x^2)^{n+\frac{1}{2}}, \tag{17}$$

$$(1-x)^{-\frac{1}{2}}(1+x)^{\frac{1}{2}} V_n(x) = \frac{(-1)^n 2^n n!}{(2n)!} \frac{d^n}{dx^n} (1-x)^{n-\frac{1}{2}} (1+x)^{n+\frac{1}{2}}, \tag{18}$$

$$(1-x)^{\frac{1}{2}}(1+x)^{-\frac{1}{2}} W_n(x) = \frac{(-1)^n 2^n n!}{(2n)!} \frac{d^n}{dx^n} (1-x)^{n+\frac{1}{2}} (1+x)^{n-\frac{1}{2}}. \tag{19}$$

They have the following orthogonalities with respect to various weight functions.

$$\int_{-1}^{1} (1-x^2)^{-\frac{1}{2}} T_n(x) T_m(x) dx = \frac{\pi}{\epsilon_n} \delta_{n,m}, \tag{20}$$

$$\int_{-1}^{1} (1-x^2)^{\frac{1}{2}} U_n(x) U_m(x) dx = \frac{\pi}{2} \delta_{n,m}, \tag{21}$$

$$\int_{-1}^{1} \left(\frac{1+x}{1-x}\right)^{\frac{1}{2}} V_n(x) V_m(x) dx = \pi \delta_{n,m}, \tag{22}$$

$$\int_{-1}^{1} \left(\frac{1-x}{1+x}\right)^{\frac{1}{2}} W_n(x) W_m(x) dx = \pi \delta_{n,m}, \tag{23}$$

where:

$$\epsilon_n = \begin{cases} 1, & \text{if } n = 0, \\ 2, & \text{if } n \geq 1, \end{cases} \quad \delta_n = \begin{cases} 0, & \text{if } n \neq m, \\ 1, & \text{if } n = m. \end{cases} \tag{24}$$

To proceed further, we let:

$$\alpha_{n,r}(x) = \sum_{l=0}^{n} \sum_{i_1+i_2+\cdots+i_{r+1}=l} \binom{r-1+n-l}{r-1} V_{i_1}(x) V_{i_2}(x) \cdots V_{i_{r+1}}(x), \tag{25}$$

$$(n \geq 0, r \geq 1),$$

$$\beta_{n,r}(x) = \sum_{l=0}^{n} \sum_{i_1+i_2+\cdots+i_{r+1}=l} (-1)^{n-l} \binom{r-1+n-l}{r-1} W_{i_1}(x) W_{i_2}(x) \cdots W_{i_{r+1}}(x), \tag{26}$$

$$(n \geq 0, r \geq 1).$$

We note here that both $\alpha_{n,r}(x)$ and $\beta_{n,r}(x)$ are polynomials of degree n.

In the following, we assume that the polynomials with subscript n, like $p_n(x), q_n(x)$ and $r_n(x)$, have degree n.

The linearization problem in general consists of determining the coefficients $c_{nm}(k)$ in the expansion of the product of two polynomials $q_n(x)$ and $r_m(x)$ in terms of an arbitrary polynomial sequence $\{p_k(x)\}_{k \geq 0}$:

$$q_n(x) r_m(x) = \sum_{k=0}^{n+m} c_{nm}(k) p_k(x).$$

A special problem of this is the case when $p_n(x) = q_n(x) = r_n(x)$, which is called either the standard linearization or the Clebsch–Gordan-type problem.

Another particular case is when $r_m(x) = 1$, which is the so-called connection problem. If further $q_n(x) = x^n$, it is called the inversion problem for the sequence $\{p_k(x)\}_{k \geq 0}$.

In this paper, we will consider the sums of finite products of Chebyshev polynomials of the third and fourth kinds in (25) and (26). Then, we are going to express each of them as linear combinations of the four kinds of Chebyshev polynomials $T_n(x)$, $U_n(x)$, $V_n(x)$ and $W_n(x)$. Thus, our problem may be regarded as a generalization of the linearization problem. We obtain them by explicit computations and using Propositions 1 and Lemma 1. The general formulas in Proposition 1 can be derived by using orthogonalities and Rodrigues' formulas for Chebyshev polynomials and integration by parts.

Finally, we note that many problems in physics and engineering can be solved with the help of special functions; for instance, we let the reader refer to the excellent papers [4–6] in this direction.

The next two theorems are our main results in which the terminating hypergeometric functions $_3F_2\left(\begin{smallmatrix}-n,\ a,\ b\\ d,\ e\end{smallmatrix};1\right)$ appear.

Theorem 1. *Let n, r be integers with $n \geq 0$, $r \geq 1$. Then we have following.*

$$\sum_{l=0}^{n} \sum_{i_1+i_2+\cdots+i_{r+1}=l} \binom{r-1+n-l}{r-1} V_{i_1}(x) V_{i_2}(x) \cdots V_{i_{r+1}}(x)$$

$$= \frac{(-1)^n (2n+2r)!}{r! 2^{2r} (n+r-\tfrac{1}{2})_r} \tag{27}$$

$$\times \sum_{k=0}^{n} \frac{(-1)^k \epsilon_k}{(n-k)!(n+k)!} {}_3F_2\left(\begin{matrix}k-n,\ -k-n,\ \tfrac{1}{2}-n-r\\ \tfrac{1}{2}-n,\ -2n-2r\end{matrix};1\right) T_k(x)$$

$$= \frac{(-1)^n (2n+2r)!}{r! 2^{2r-2} (n+r-\tfrac{1}{2})_{r-1}}$$

$$\times \sum_{k=0}^{n} \frac{(-1)^k (k+1)}{(n-k)!(n+k+2)!} {}_3F_2\left(\begin{matrix}k-n,\ -k-n-2,\ \tfrac{1}{2}-n-r\\ -\tfrac{1}{2}-n,\ -2n-2r\end{matrix};1\right) U_k(x) \tag{28}$$

$$= \frac{(-1)^n (2n+2r)!}{r! 2^{2r} (n+r-\tfrac{1}{2})_r}$$

$$\times \sum_{k=0}^{n} \frac{(-1)^k (2k+1)}{(n-k)!(n+k+1)!} {}_3F_2\left(\begin{matrix}k-n,\ -k-n-1,\ \tfrac{1}{2}-n-r\\ \tfrac{1}{2}-n,\ -2n-2r\end{matrix};1\right) V_k(x) \tag{29}$$

$$= \frac{(-1)^n (2n+2r)!}{r! 2^{2r-1} (n+r-\tfrac{1}{2})_{r-1}}$$

$$\times \sum_{k=0}^{n} \frac{(-1)^k}{(n-k)!(n+k+1)!} {}_3F_2\left(\begin{matrix}k-n,\ -k-n-1,\ \tfrac{1}{2}-n-r\\ -\tfrac{1}{2}-n,\ -2n-2r\end{matrix};1\right) W_k(x). \tag{30}$$

Theorem 2. *Let n, r be integers with $n \geq 0$, $r \geq 1$. Then we have following.*

$$\sum_{l=0}^{n} \sum_{i_1+i_2+\cdots+i_{r+1}=l} (-1)^{n-l} \binom{r-1+n-l}{r-1} W_{i_1}(x) W_{i_2}(x) \cdots W_{i_{r+1}}(x)$$

$$= \frac{(-1)^n (2n+2r)!}{r! 2^{2r} (n+r+\tfrac{1}{2})_r} \tag{31}$$

$$\times \sum_{k=0}^{n} \frac{(-1)^k \epsilon_k}{(n-k)!(n+k)!} {}_3F_2\left(\begin{matrix}k-n,\ -k-n,\ -\tfrac{1}{2}-n-r\\ \tfrac{1}{2}-n,\ -2n-2r\end{matrix};1\right) T_k(x)$$

$$= \frac{(-1)^n(2n+1)(2n+2r)!}{r!2^{2r-1}(n+r+\frac{1}{2})_r}$$

$$\times \sum_{k=0}^{n} \frac{(-1)^k(k+1)}{(n-k)!(n+k+2)!}{}_3F_2\left(\begin{array}{c} k-n,\ -k-n-2,\ -\frac{1}{2}-n-r \\ -\frac{1}{2}-n,\ -2n-2r \end{array};1\right)U_k(x) \quad (32)$$

$$= \frac{(-1)^n(2n+2r)!}{r!2^{2r}(n+r+\frac{1}{2})_r}$$

$$\times \sum_{k=0}^{n} \frac{(-1)^k(2k+1)}{(n-k)!(n+k+1)!}{}_3F_2\left(\begin{array}{c} k-n,\ -k-n-1,\ -\frac{1}{2}-n-r \\ \frac{1}{2}-n,\ -2n-2r \end{array};1\right)V_k(x) \quad (33)$$

$$= \frac{(-1)^n(2n+1)(2n+2r)!}{r!2^{2r}(n+r+\frac{1}{2})_r}$$

$$\times \sum_{k=0}^{n} \frac{(-1)^k}{(n-k)!(n+k+1)!}{}_3F_2\left(\begin{array}{c} k-n,\ -k-n-1,\ -\frac{1}{2}-n-r \\ -\frac{1}{2}-n,\ -2n-2r \end{array};1\right)W_k(x). \quad (34)$$

As we know, the Bernoulli polynomials are not orthogonal polynomials, but Appell polynomials. In [7], the sums of finite products of Chebyshev polynomials in (25) and (26) were expressed as linear combinations of Bernoulli polynomials. Furthermore, the same has been done for the sums of finite products of Bernoulli, Euler and Genocchi polynomials in [8–10]. All of these were found by deriving Fourier series expansions for the functions closely connected with those various sums of finite products. For some other applications of Chebyshev polynomials, we let the reader refer to [11–13].

2. Proof of Theorem 1

Here, we will prove Theorem 1. For this purpose, we first state Proposition 1 and Lemma 1 that will be used in Sections 2 and 3.

The results in Proposition 1 can be derived by using the orthogonalities in (20)–(23) and the Rodrigues formulas in (16)–(19). The statements (a) and (b) in Proposition 1 are respectively from the Equations (23) and (35) of [14], while (c) and (d) are respectively from the Equations (22) and (37) of [15].

Proposition 1. *Let $q(x) \in \mathbb{R}[x]$ be a polynomial of degree n. Then, we have the following.*

(a) $q(x) = \sum_{k=0}^{n} c_{k,1} T_k(x)$,

where $c_{k,1} = \frac{(-1)^k 2^k k! \epsilon_k}{(2k)! \pi} \int_{-1}^{1} q(x) \frac{d^k}{dx^k}(1-x^2)^{k-\frac{1}{2}} dx$,

(b) $q(x) = \sum_{k=0}^{n} c_{k,2} U_k(x)$,

where $c_{k,2} = \frac{(-1)^k 2^{k+1}(k+1)!}{(2k+1)! \pi} \int_{-1}^{1} q(x) \frac{d^k}{dx^k}(1-x^2)^{k+\frac{1}{2}} dx$,

(c) $q(x) = \sum_{k=0}^{n} c_{k,3} V_k(x)$,

where $c_{k,3} = \frac{(-1)^k 2^k k!}{(2k)! \pi} \int_{-1}^{1} q(x) \frac{d^k}{dx^k}(1-x)^{k-\frac{1}{2}}(1+x)^{k+\frac{1}{2}} dx$,

(d) $q(x) = \sum_{k=0}^{n} c_{k,A} W_k(x),$

where $c_{k,A} = \dfrac{(-1)^k 2^k k!}{(2k)! \pi} \displaystyle\int_{-1}^{1} q(x) \dfrac{d^k}{dx^k} (1-x)^{k+\frac{1}{2}} (1+x)^{k-\frac{1}{2}} dx.$

Lemma 1. *Let l, m be nonnegative integers. Then, we have the following.*

$$\int_{-1}^{1} (1-x)^{m-\frac{1}{2}} (1+x)^{l-\frac{1}{2}} dx$$
$$= \dfrac{2^{l+m}}{(l+m)!} \Gamma(l+\tfrac{1}{2}) \Gamma(m+\tfrac{1}{2}) \qquad (35)$$
$$= \dfrac{(2l)! (2m)! \pi}{2^{l+m} (l+m)! \, l! \, m!}.$$

Proof. By changing the variables $1+x = 2y$, the integral in (35) becomes:

$$2^{l+m} \int_0^1 y^{l+\frac{1}{2}-1} (1-y)^{m+\frac{1}{2}-1} dy = 2^{l+m} \dfrac{\Gamma(l+\tfrac{1}{2})\Gamma(m+\tfrac{1}{2})}{\Gamma(l+m+1)}$$
$$= \dfrac{2^{l+m} (2l)! \, \Gamma(\tfrac{1}{2}) (2m)! \, \Gamma(\tfrac{1}{2})}{(l+m)! \, 2^{2l} \, l! \, 2^{2m} \, m!},$$

where we used (5) and (6). □

As was shown in [7], the following lemma can be obtained by differentiating Equation (10). It expresses the sums of finite products in (25) very neatly, which plays an important role in the following discussion.

Lemma 2. *Let n, r be integers with $n \geq 0, r \geq 1$. Then, we have the identity.*

$$\sum_{l=0}^{n} \sum_{i_1+i_2+\cdots+i_{r+1}=l} \binom{r-1+n-l}{r-1} V_{i_1}(x) \cdots V_{i_{r+1}}(x) = \dfrac{1}{2^r \, r!} V_{n+r}^{(r)}(x), \qquad (36)$$

where the inner sum runs over all nonnegative integers $i_1, i_2, \cdots, i_{r+1}$, with $i_1 + i_2 + \cdots + i_{r+1} = l$.

From (14), the r-th derivative of $V_n(x)$ is given by:

$$V_n^{(r)}(x) = \sum_{l=0}^{n-r} \binom{2n-l}{l} 2^{n-l} (n-l)_r (x-1)^{n-l-r}. \qquad (37)$$

In particular, we have:

$$V_{n+r}^{(r+k)}(x) = \sum_{l=0}^{n-k} \binom{2n+2r-l}{l} 2^{n+r-l} (n+r-l)_{r+k} (x-1)^{n-k-l}. \qquad (38)$$

$$V_{n+r}^{(r+k)}(x) = \sum_{l=0}^{n-k} \binom{2n+2r-l}{l} 2^{n+r-l} (n+r-l)_{r+k} (x-1)^{n-k-l}. \qquad (38)$$

Here, we will show only (28) of Theorem 1, since (27), (29) and (30) can be proved similarly to (28). With $\alpha_{n,r}(x)$ as in (25), we let:

$$\alpha_{n,r}(x) = \sum_{k=0}^{n} c_{k,2} U_k(x). \qquad (39)$$

Then, from (b) of Proposition 1, (36), (38) and integration by parts k times, we have:

$$\begin{aligned}
c_{k,2} &= \frac{(-1)^k 2^{k+1}(k+1)!}{(2k+1)!\pi} \int_{-1}^{1} \alpha_{n,r}(x) \frac{d^k}{dx^k}(1-x^2)^{k+\frac{1}{2}} dx \\
&= \frac{(-1)^k 2^{k+1}(k+1)!}{(2k+1)!\pi 2^r r!} \int_{-1}^{1} V_{n+r}^{(r)}(x) \frac{d^k}{dx^k}(1-x^2)^{k+\frac{1}{2}} dx \\
&= \frac{2^{k+1}(k+1)!}{(2k+1)!\pi 2^r r!} \int_{-1}^{1} V_{n+r}^{(r+k)}(x)(1-x^2)^{k+\frac{1}{2}} dx \qquad (40) \\
&= \frac{2^{k+1}(k+1)!}{(2k+1)!\pi 2^r r!} \sum_{l=0}^{n-k} (-1)^{n-k-l}\binom{2n+2r-l}{l} 2^{n+r-l} \\
&\quad \times (n+r-l)_{r+k}\int_{-1}^{1}(1-x)^{n-l+1-\frac{1}{2}}(1+x)^{k+1-\frac{1}{2}} dx.
\end{aligned}$$

From (40), (35), we get:

$$\begin{aligned}
c_{k,2} &= \frac{2^{k+1}(k+1)!}{(2k+1)!\pi 2^r r!} \\
&\quad \times \sum_{l=0}^{n-k} \frac{(-1)^{n-k-l}(2n+2r-l)! 2^{n+r-l}(n+r-l)!(2k+2)!(2n-2l+2)!\pi}{l!(2n+2r-2l)!(n-k-l)! 2^{n-l+k+2}(n-l+k+2)!(n-l+1)!(k+1)!} \\
&= \frac{(-1)^{n-k}(k+1)}{r!} \qquad (41) \\
&\quad \times \sum_{l=0}^{n-k} \frac{(-1)^l (2n+2r-l)!(n+r-l)!(2n+2-2l)!}{l!(n-k-l)!(n+k-l+2)!(2n+2r-2l)!(n+1-l)!}.
\end{aligned}$$

Using (3) and (4), (41) is equal to:

$$\begin{aligned}
c_{k,2} &= \frac{(-1)^{n-k}(k+1)(2n+2r)!}{r!(n-k)!(n+k+2)!} \\
&\quad \times \sum_{l=0}^{n-k} \frac{(-1)^l (n-k)_l (n+k+2)_l <\frac{1}{2}-n-r>_l 2^{2n-2l+2}(-1)^l <\frac{1}{2}>_{n+1}}{l!(2n+2r)_l 2^{2n+2r-2l}(-1)^l <\frac{1}{2}>_{n+r}<\frac{1}{2}-n-1>_l} \\
&= \frac{(-1)^n (2n+2r)!}{r! 2^{2r-2}(n+r-\frac{1}{2})_{r-1}} \\
&\quad \times \frac{(-1)^k(k+1)}{(n-k)!(n+k+2)!} \sum_{l=0}^{n-k} \frac{<k-n>_l <-k-n-2>_l <\frac{1}{2}-n-r>_l}{<-\frac{1}{2}-n>_l <-2n-2r>_l l!} \qquad (42) \\
&= \frac{(-1)^n(2n+2r)!}{r! 2^{2r-2}(n+r-\frac{1}{2})_{r-1}} \\
&\quad \times \frac{(-1)^k(k+1)}{(n-k)!(n+k+2)!} {}_3F_2\!\left(\begin{matrix} k-n, \ -k-n-2, \ \frac{1}{2}-n-r \\ -\frac{1}{2}-n, \ -2n-2r \end{matrix}; 1\right).
\end{aligned}$$

Now, the Equation (28) in Theorem 1 follows from (39) and (42).

3. Proof of Theorem 2

In this section, we will show (31) of Theorem 2, as (32)–(34) can be treated analogously to (31). The following lemma can be obtained by differentiating (11) and is stated as Lemma 3 in [7].

Lemma 3. Let n, r be integers with $n \geq 0, r \geq 1$. Then, we have the following identity.

$$\sum_{l=0}^{n} \sum_{i_1+i_2+\cdots+i_{r+1}=l} (-1)^{n-l} \binom{r-1+n-l}{r-1} W_{i_1}(x) W_{i_2}(x) \cdots W_{i_{r+1}}(x) \qquad (43)$$
$$= \frac{1}{2^r r!} W_{n+r}^{(r)}(x),$$

where the inner sum runs over all nonnegative integers $i_1, i_2, \cdots, i_{r+1}$, with $i_1 + i_2 + \cdots + i_{r+1} = l$.

From (15), the r-th derivative of $W_n(x)$ is given by:

$$W_n^{(r)}(x) = (2n+1) \sum_{l=0}^{n-r} \frac{2^{n-l}}{2n+1-2l} \binom{2n-l}{l}(n-l)_r (x-1)^{n-l-r}. \qquad (44)$$

In particular,

$$W_{n+r}^{(r+k)}(x)$$
$$= (2n+1) \sum_{l=0}^{n-k} \frac{2^{n+r-l}}{2n+2r+1-2l} \binom{2n+2r-l}{l}(n+r-l)_{r+k}(x-1)^{n-k-l}. \qquad (45)$$

Here, we will show only (31) of Theorem 2, since (32)–(34) can be proven analogously to (31). With $\beta_{n,r}(x)$ as in (26), we put:

$$\beta_{n,r}(x) = \sum_{k=0}^{n} c_{k,1} T_k(x). \qquad (46)$$

Then, from (a) of Proposition 1, (43), (45) and integration by parts k times, we have:

$$c_{k,1} = \frac{(-1)^k 2^k k! \epsilon_k}{(2k)! \pi} \int_{-1}^{1} \beta_{n,r}(x) \frac{d^k}{dx^k}(1-x^2)^{k-\frac{1}{2}} dx$$
$$= \frac{(-1)^k 2^k k! \epsilon_k}{(2k)! \pi 2^r r!} \int_{-1}^{1} W_{n+r}^{(r)}(x) \frac{d^k}{dx^k}(1-x^2)^{k-\frac{1}{2}} dx$$
$$= \frac{2^k k! \epsilon_k}{(2k)! \pi 2^r r!} \int_{-1}^{1} W_{n+r}^{(r+k)}(x)(1-x^2)^{k-\frac{1}{2}} dx \qquad (47)$$
$$= \frac{(2n+1) 2^k k! \epsilon_k}{(2k)! \pi 2^r r!} \sum_{l=0}^{n-k} \frac{(-1)^{n-k-l} 2^{n+r-l}}{2n+2r+1-2l} \binom{2n+2r-l}{l}$$
$$\times (n+r-l)_{r+k} \int_{-1}^{1}(1-x)^{n-l-\frac{1}{2}}(1+x)^{k-\frac{1}{2}} dx.$$

From (47), (35) and after some simplifications, we get:

$$c_{k,2} = \frac{(2n+1)\epsilon_k (-1)^{n-k}}{r!}$$
$$\times \sum_{l=0}^{n-k} \frac{(-1)^l (2n+2r-1)!(n+r-l)!}{l!(n-k-l)!(n+k-l)!(2n+2r-2l+1)!(n-l)!} \qquad (48)$$
$$= \frac{2(2n+1)\epsilon_k (-1)^{n-k}}{r!}$$
$$\times \sum_{l=0}^{n-k} \frac{(-1)^l (2n+2r-1)!(n+r-l+1)!(2n-2l)!}{l!(n-k-l)!(n+k-l)!(2n+2r-2l+2)!(n-l)!}.$$

216

Using (3) and (4), (48) is equal to:

$$c_{k,1} = \frac{2(2n+1)(2n+2r)!\epsilon_k(-1)^{n-k}}{r!(n-k)!(n+k)!}$$

$$\times \sum_{l=0}^{n-k} \frac{(-1)^l(n-k)_l(n+k)_l <\frac{1}{2}-n-r-1>_l\, 2^{2n-2l}(-1)^l <\frac{1}{2}>_n}{l!(2n+2r)_l 2^{2n+2r+2-2l}(-1)^l <\frac{1}{2}>_{n+r+1}<\frac{1}{2}-n>_l}$$

$$= \frac{(2n+1)(-1)^n(2n+2r)!}{r! 2^{2r+1}(n+r+\frac{1}{2})_{r+1}}$$

$$\times \frac{(-1)^k \epsilon_k}{(n-k)!(n+k)!} \sum_{l=0}^{k} \frac{<k-n>_l<-k-n>_l<-\frac{1}{2}-n-r>_l}{<\frac{1}{2}-n>_l<-2n-2r>_l}$$

$$= \frac{(-1)^n(2n+2r)!}{r! 2^{2r}(n+r+\frac{1}{2})_r}$$

$$\times \frac{(-1)^k \epsilon_k}{(n-k)!(n+k)!} {}_3F_2\left(\begin{matrix} k-n,\ -k-n,\ -\frac{1}{2}-n-r \\ \frac{1}{2}-n,\ -2n-2r \end{matrix}; 1\right). \tag{49}$$

Now, Equation (31) in Theorem 2 follows from (46) and (49).

Remark 1. *As we noted earlier, Lemmas 2 and 3 play crucial roles and express sums of finite products in (25) and (26) very neatly as higher-order derivatives of $V_n(x)$ and $W_n(x)$. These could be derived by noting that Chebyshev polynomials are special cases of Jacobi polynomials and using the general formula for the derivative of Jacobi polynomials. Indeed, their Jacobi polynomial expressions and the derivatives of the Jacobi polynomials are as follows:*

$$V_n(x) = P_n^{(-1/2,1/2)}(x)/P_n^{(-1/2,1/2)}(1),$$
$$W_n(x) = P_n^{(1/2,-1/2)}(x)/P_n^{(1/2,-1/2)}(1),$$
$$\frac{d}{dx}P_n^{(a,b)}(x) = \frac{1}{2}(n+a+b+1)P_{n-1}^{(a+1,b+1)}(x).$$

4. Conclusions

The linearization problem in general consists of determining the coefficients $c_{nm}(k)$ in the expansion of the product of two polynomials $q_n(x)$ and $r_m(x)$ in terms of an arbitrary polynomial sequence $\{p_k(x)\}_{k\geq 0}$:

$$q_n(x)r_m(x) = \sum_{k=0}^{n+m} c_{nm}(k) p_k(x).$$

Along this line and as a generalization of this, we considered sums of finite products of Chebyshev polynomials of the third and fourth kinds and represented each of those sums of finite products as linear combinations of the four kinds of Chebyshev polynomials, which involve the hypergeometric function ${}_3F_2$. It is certainly possible to represent such sums of finite products by other orthogonal polynomials, which is our ongoing project.

Author Contributions: T.K. and D.S.K. conceived the framework and structured the whole paper; T.K. wrote the paper; C.S.R. and D.D.V. checked the results of the paper; D.S.K. and C.S.R. completed the revision of the article.

Funding: This work was supported by the National Research Foundation of Korea (NRF) grant funded by the Korean government (MEST) (No. 2017R1A2B4006092).

Acknowledgments: The authors would like to thank the referees for their valuable comments, which improved the original manuscript in its present form.

Conflicts of Interest: The authors declare no conflict of interest.

References

1. Andrews, G.E.; Askey, R.; Roy, R. *Special Functions*; Encyclopedia of Mathematics and Its Applications 71; Cambridge University Press: Cambridge, UK, 1999.
2. Beals, R.; Wong, R. *Special Functions and Orthogonal Polynomials*; Cambridge Studies in Advanced Mathematics 153; Cambridge University Press: Cambridge, UK, 2016.
3. Wang, Z.X.; Guo, D.R. *Special Functions*; Guo, D.R., Xia, X.J., Trans.; World Scientific Publishing Co., Inc.: Teaneck, NJ, USA, 1989.
4. Marin, M. A temporally evolutionary equation in elasticity of micropolar bodies with voids. *Politehn. Univ. Buchar. Sci. Bull. Ser. A Appl. Math. Phys.* **1998**, *60*, 3–12.
5. Marin, M.; Stan, G. Weak solutions in elasticity of dipolar bodies with stretch. *Carpath. J. Math.* **2013**, *29*, 33–40.
6. Marin, D.; Baleanu, D. On vibrations in thermoelasticity without energy dissipation for micropolar bodies. *Bound. Value Probl.* **2016**, *2016*, 111. [CrossRef]
7. Kim, T.; Kim, D.S.; Dolgy, D.V.; Kwon, J. Representing Sums of finite products of Chebyshev polynomials of the third and fourth kinds by Chebyshev Polynomials. *Preprints* **2018**, 2018060079. [CrossRef]
8. Agarwal, R.P.; Kim, D.S.; Kim, T.; Kwon, J. Sums of finite products of Bernoulli functions. *Adv. Differ. Equ.* **2017**, *2017*, 15. [CrossRef]
9. Kim, T.; Kim, D.S.; Jang, G.-W.; Kwon, J. Sums of finite products of Euler functions. In *Advances in Real and Complex Analysis with Applications, Trends in Mathematics*; Springer: New York, NY, USA, 2017; pp. 243–260.
10. Kim, T.; Kim, D.S.; Jang, L.C.; Jang, G.-W. Sums of finite products of Genocchi functions. *Adv. Differ. Equ.* **2017**, *2017*, 17. [CrossRef]
11. Doha, E.H.; Abd-Elhameed, W.M.; Alsuyuti, M.M. On using third and fourth kinds Chebyshev polynomials for solving the integrated forms of high odd-order linear boundary value problems. *J. Egyptian Math. Soc.* **2015**, *23*, 397–405. [CrossRef]
12. Kruchinin, D.V.; Kruchinin, V.V. Application of a composition of generating functions for obtaining explicit formulas of polynomials. *J. Math. Anal. Appl.* **2013**, *404*, 161–171. [CrossRef]
13. Mason, J.C. Chebyshev polynomials of the second, third and fourth kinds in approximation, indefinite integration, and integral transforms. *J. Comput. Appl. Math.* **1993**, *49*, 169–178. [CrossRef]
14. Kim, D.S.; Kim, T.; Lee, S.-H. Some identities for Berounlli polynomials involving Chebyshev polynomials. *J. Comput. Anal. Appl.* **2014**, *16*, 172–180.
15. Kim, D.S.; Dolgy, D.V.; Kim, T.; Rim, S.-H. Identities involving Bernoulli and Euler polynomials arising from Chebyshev polynomials. *Proc. Jangjeon Math. Soc.* **2012**, *15*, 361–370.

© 2018 by the authors. Licensee MDPI, Basel, Switzerland. This article is an open access article distributed under the terms and conditions of the Creative Commons Attribution (CC BY) license (http://creativecommons.org/licenses/by/4.0/).

Article
Symmetric Identities for Fubini Polynomials

Taekyun Kim [1,2], Dae San Kim [3], Gwan-Woo Jang [2] and Jongkyum Kwon [4,*]

1. Department of Mathematics, College of Science, Tianjin Polytechnic University, Tianjin 300160, China; tkkim@kw.ac.kr or kwangwoonmath@hanmail.net
2. Department of Mathematics, Kwangwoon University, Seoul 139-701, Korea; gwjang@kw.ac.kr
3. Department of Mathematics, Sogang University, Seoul 121-742, Korea; dskim@sogang.ac.kr
4. Department of Mathematics Education and ERI, Gyeongsang National University, Jinju, Gyeongsangnamdo 52828, Korea
* Correspondence: mathkjk26@gnu.ac.kr

Received: 20 April 2018; Accepted: 13 June 2018; Published: 14 June 2018

Abstract: We represent the generating function of w-torsion Fubini polynomials by means of a fermionic p-adic integral on $\mathbb{Z}_p$. Then we investigate a quotient of such p-adic integrals on $\mathbb{Z}_p$, representing generating functions of three w-torsion Fubini polynomials and derive some new symmetric identities for the w-torsion Fubini and two variable w-torsion Fubini polynomials.

Keywords: Fubini polynomials; w-torsion Fubini polynomials; fermionic p-adic integrals; symmetric identities

1. Introduction and Preliminaries

In recent years, various p-adic integrals on $\mathbb{Z}_p$ have been used in order to find many interesting symmetric identities related to some special polynomials and numbers. The relevant p-adic integrals are the Volkenborn, fermionic, q-Volkenborn, and q-fermionic integrals of which the last three were discovered by the first author T. Kim (see [1–3]). They have been used by a good number of researchers in various contexts and especially in unfolding new interesting symmetric identities. This verifies the usefulness of such p-adic integrals. Moreover, we can expect that people will find some further applications of these p-adic integrals in the years to come. The present paper is an effort in this direction. Assume that p is any fixed odd prime number. Throughout our discussion, we will use the standard notations $\mathbb{Z}_p$, $\mathbb{Q}_p$, and $\mathbb{C}_p$ to denote the ring of p-adic integers, the field of p-adic rational numbers and the completion of the algebraic closure of $\mathbb{Q}_p$, respectively. The p-adic norm $|\cdot|_p$ is normalized as $|p|_p = \frac{1}{p}$. Assume that $f(x)$ is a continuous function on $\mathbb{Z}_p$. Then the fermionic p-adic integral of $f(x)$ on $\mathbb{Z}_p$ was introduced by Kim (see [2]) as

$$\int_{\mathbb{Z}_p} f(x) d\mu_{-1}(x) = \lim_{N \to \infty} \sum_{x=0}^{p^N - 1} f(x)(-1)^x, \qquad (1)$$

where $\mu_{-1}(x + p^N \mathbb{Z}_p) = (-1)^x$.

We can easily deduce from (1) that (see [2,3])

$$\int_{\mathbb{Z}_p} f(x+1) d\mu_{-1}(x) + \int_{\mathbb{Z}_p} f(x) d\mu_{-1}(x) = 2f(0). \qquad (2)$$

By invoking (2), we easily get (see [2,4])

$$\int_{\mathbb{Z}_p} e^{(x+y)t} d\mu_{-1}(y) = \frac{2}{e^t + 1} e^{xt} = \sum_{n=0}^{\infty} E_n(x) \frac{t^n}{n!}, \qquad (3)$$

where $E_n(x)$ are the usual Euler polynomials.

As is known, the two variable Fubini polynomials are defined by means of the following (see [5,6])

$$\sum_{n=0}^{\infty} F_n(x,y)\frac{t^n}{n!} = \frac{1}{1-y(e^t-1)}e^{xt}. \tag{4}$$

When $x = 0$, $F_n(y) = F_n(0,y)$, $(n \geq 0)$, are called Fubini polynomials. Further, if $y = 1$, then $Ob_n = F_n(0,1)$ are the ordered Bell numbers (also called Frobenius numbers). They first appeared in Cayley's work on a combinatorial counting problem in 1859 and have many different combinatorial interpretations. For example, the ordered Bell numbers count the possible outcomes of a multi-candidate election. From (3) and (4), we note that $F_n(x,-1/2) = E_n(x)$, $(n \geq 0)$. By (4), we easily get (see [6]),

$$F_n(y) = \sum_{k=0}^{n} S_2(n,k)k!y^k, \ (n \geq 0), \tag{5}$$

where $S_2(n,k)$ are the Stirling numbers of the second kind.

For $w \in \mathbb{N}$, we define the two variable w-torsion Fubini polynomials given by

$$\frac{1}{1-y^w(e^t-1)^w}e^{xt} = \sum_{n=0}^{\infty} F_{n,w}(x,y)\frac{t^n}{n!}. \tag{6}$$

In particular, for $x = 0$, $F_{n,w}(y) = F_{n,w}(0,y)$ are called the w-torsion Fubini polynomials. It is obvious that $F_{n,1}(x,y) = F_n(x,y)$.

We represent the generating function of w-torsion Fubini polynomials by means of a fermionic p-adic integral on $\mathbb{Z}_p$. Then we investigate a quotient of such p-adic integrals on $\mathbb{Z}_p$, representing generating functions of three w-torsion Fubini polynomials and derive some new symmetric identities for the w-torsion Fubini and two variable w-torsion Fubini polynomials. Recently, a number of researchers have studied symmetric identities for some special polynomials. The reader may refer to [7–11] as an introduction to this active area of research. Some symmetric identities for q-special polynomials and numbers were treated in [12–15], including q-Bernoulli, q-Euler, and q-Genocchi numbers and polynomials. While some identities of symmetry for degenerate special polynomials were discussed in the more recent papers [6,16,17]. Finally, interested readers may want to have a glance at [18,19] as general references on polynomials.

2. Symmetric Identities for w-torsion Fubini and Two Variable w-torsion Fubini Polynomials

From (2), we note that

$$\int_{\mathbb{Z}_p} (-1)^x (y(e^t-1))^x d\mu_{-1}(x) = \frac{2}{1-y(e^t-1)} = 2\sum_{n=0}^{\infty} F_n(y)\frac{t^n}{n!}, \tag{7}$$

and

$$e^{xt}\int_{\mathbb{Z}_p} (-1)^z (y(e^t-1))^z d\mu_{-1}(z) = \frac{2}{1-y(e^t-1)}e^{xt} = 2\sum_{n=0}^{\infty} F_n(x,y)\frac{t^n}{n!}. \tag{8}$$

From (7) and (8), we note that

$$\left(\sum_{l=0}^{\infty} x^l \frac{t^l}{l!}\right)\left(\sum_{m=0}^{\infty} 2F_m(y)\frac{t^m}{m!}\right) = e^{xt}\int_{\mathbb{Z}_p}(-1)^z(y(e^t-1))^z d\mu_{-1}(z)$$

$$= \sum_{n=0}^{\infty} 2F_n(x,y)\frac{t^n}{n!}. \tag{9}$$

Thus, by (9), we easily get

$$\sum_{l=0}^{n}\binom{n}{l}x^l F_{n-l}(y) = F_n(x,y), \quad (n \geq 0). \tag{10}$$

Now, we observe that

$$\frac{1-y^k(e^t-1)^k}{1-y(e^t-1)} = \sum_{i=0}^{k-1} y^i(e^t-1)^i = \sum_{i=0}^{k-1}\sum_{l=0}^{i}\binom{i}{l}(-1)^{i-l}y^i e^{lt}$$

$$= \sum_{n=0}^{\infty}\left(\sum_{i=0}^{k-1}\sum_{l=0}^{i}\binom{i}{l}(-1)^{i-l}y^i l^n\right)\frac{t^n}{n!} \tag{11}$$

$$= \sum_{n=0}^{\infty}\left(\sum_{i=0}^{k-1} y^i \Delta^i 0^n\right)\frac{t^n}{n!},$$

where $\Delta f(x) = f(x+1) - f(x)$.

For $w \in \mathbb{N}$, the w-torsion Fubini polynomials are represented by means of the following fermionic p-adic integral on $\mathbb{Z}_p$:

$$\int_{\mathbb{Z}_p}(-y^w(e^t-1)^w)^x d\mu_{-1}(x) = \frac{2}{1-y^w(e^t-1)^w} = \sum_{n=0}^{\infty} 2F_{n,w}(y)\frac{t^n}{n!}, \tag{12}$$

From (7) and (12), we have

$$\frac{\int_{\mathbb{Z}_p}(-y(e^t-1))^x d\mu_{-1}(x)}{\int_{\mathbb{Z}_p}(-y^{w_1}(e^t-1)^{w_1})^x d\mu_{-1}(x)} = \frac{1-y^{w_1}(e^t-1)^{w_1}}{1-y(e^t-1)} = \sum_{i=0}^{w_1-1} y^i(e^t-1)^i$$

$$= \sum_{n=0}^{\infty}\left(\sum_{i=0}^{w_1-1} y^i \Delta^i 0^n\right)\frac{t^n}{n!}, \quad (w_1 \in \mathbb{N}). \tag{13}$$

For $w_1, w_2 \in \mathbb{N}$, we let

$$I = \frac{\int_{\mathbb{Z}_p}(-y^{w_1}(e^t-1)^{w_1})^{x_1} d\mu_{-1}(x_1)\int_{\mathbb{Z}_p}(-y^{w_2}(e^t-1)^{w_2})^{x_2} d\mu_{-1}(x_2)}{\int_{\mathbb{Z}_p}(-y^{w_1 w_2}(e^t-1)^{w_1 w_2})^x d\mu_{-1}(x)}. \tag{14}$$

Here it is important to observe that (14) has the built-in symmetry. Namely, it is invariant under the interchange of w_1 and w_2.

Then, by (14), we get

$$I = \left(\int_{\mathbb{Z}_p}(-y^{w_1}(e^t-1)^{w_1})^x d\mu_{-1}(x)\right) \times \left(\frac{\int_{\mathbb{Z}_p}(-y^{w_2}(e^t-1)^{w_2})^x d\mu_{-1}(x)}{\int_{\mathbb{Z}_p}(-y^{w_1 w_2}(e^t-1)^{w_1 w_2})^x d\mu_{-1}(x)}\right). \tag{15}$$

First, we observe that

$$\frac{\int_{\mathbb{Z}_p}(-y^{w_2}(e^t-1)^{w_2})^x d\mu_{-1}(x)}{\int_{\mathbb{Z}_p}(-y^{w_1w_2}(e^t-1)^{w_1w_2})^x d\mu_{-1}(x)} = \frac{1-y^{w_1w_2}(e^t-1)^{w_1w_2}}{1-y^{w_2}(e^t-1)^{w_2}} = \sum_{i=0}^{w_1-1} y^{w_2 i}(e^t-1)^{w_2 i}$$

$$= \sum_{i=0}^{w_1-1} y^{w_2 i} \sum_{l=0}^{w_2 i} \binom{w_2 i}{l}(-1)^{w_2 i - l} e^{lt} \qquad (16)$$

$$= \sum_{n=0}^{\infty}\left(\sum_{i=0}^{w_1-1} y^{w_2 i} \Delta^{w_2 i} 0^n\right)\frac{t^n}{n!}.$$

From (15) and (16), we can derive the following equation.

$$I = \left(\int_{\mathbb{Z}_p}(-y^{w_1}(e^t-1)^{w_1})^x d\mu_{-1}(x)\right) \times \left(\frac{\int_{\mathbb{Z}_p}(-y^{w_2}(e^t-1)^{w_2})^x d\mu_{-1}(x)}{\int_{\mathbb{Z}_p}(-y^{w_1w_2}(e^t-1)^{w_1w_2})^x d\mu_{-1}(x)}\right)$$

$$= \left(\sum_{m=0}^{\infty} 2F_{m,w_1}(y)\frac{t^m}{m!}\right) \times \left(\sum_{k=0}^{\infty}(\sum_{i=0}^{w_1-1} y^{w_2 i}\Delta^{w_2 i}0^k)\frac{t^k}{k!}\right) \qquad (17)$$

$$= \sum_{n=0}^{\infty}\left(2\sum_{k=0}^{n}\sum_{i=0}^{w_1-1} y^{w_2 i}\Delta^{w_2 i}0^k F_{n-k,w_1}(y)\binom{n}{k}\right)\frac{t^n}{n!}.$$

Interchanging the roles of w_1 and w_2, by (14), we get

$$I = \left(\int_{\mathbb{Z}_p}(-y^{w_2}(e^t-1)^{w_2})^x d\mu_{-1}(x)\right) \times \left(\frac{\int_{\mathbb{Z}_p}(-y^{w_1}(e^t-1)^{w_1})^x d\mu_{-1}(x)}{\int_{\mathbb{Z}_p}(-y^{w_1w_2}(e^t-1)^{w_1w_2})^x d\mu_{-1}(x)}\right). \qquad (18)$$

We note that

$$\frac{\int_{\mathbb{Z}_p}(-y^{w_1}(e^t-1)^{w_1})^x d\mu_{-1}(x)}{\int_{\mathbb{Z}_p}(-y^{w_1w_2}(e^t-1)^{w_1w_2})^x d\mu_{-1}(x)} = \frac{1-y^{w_1w_2}(e^t-1)^{w_1w_2}}{1-y^{w_1}(e^t-1)^{w_1}} = \sum_{i=0}^{w_2-1} y^{w_1 i}(e^t-1)^{w_1 i}$$

$$= \sum_{n=0}^{\infty}\left(\sum_{i=0}^{w_2-1} y^{w_1 i} \Delta^{w_1 i} 0^n\right)\frac{t^n}{n!}. \qquad (19)$$

Thus, by (18) and (19), we get

$$I = \left(\int_{\mathbb{Z}_p}(-y^{w_2}(e^t-1)^{w_2})^x d\mu_{-1}(x)\right) \times \left(\frac{\int_{\mathbb{Z}_p}(-y^{w_1}(e^t-1)^{w_1})^x d\mu_{-1}(x)}{\int_{\mathbb{Z}_p}(-y^{w_1w_2}(e^t-1)^{w_1w_2})^x d\mu_{-1}(x)}\right)$$

$$= \left(\sum_{m=0}^{\infty} 2F_{m,w_2}(y)\frac{t^m}{m!}\right) \times \left(\sum_{k=0}^{\infty}(\sum_{i=0}^{w_2-1} y^{w_1 i}\Delta^{w_1 i}0^k)\frac{t^k}{k!}\right) \qquad (20)$$

$$= \sum_{n=0}^{\infty}\left(2\sum_{k=0}^{n}\sum_{i=0}^{w_2-1} y^{w_1 i}\Delta^{w_1 i}0^k F_{n-k,w_2}(y)\binom{n}{k}\right)\frac{t^n}{n!}.$$

The following theorem is now obtained by Equations (17) and (20).

Theorem 1. *For $w_1, w_2 \in \mathbb{N}$ with $w_1 \equiv 1 \pmod 2$, $w_2 \equiv 1 \pmod 2$, $n \geq 0$, we have*

$$\sum_{k=0}^{n}\sum_{i=0}^{w_1-1}\binom{n}{k}F_{n-k,w_1}(y)y^{w_2 i}\Delta^{w_2 i}0^k = \sum_{k=0}^{n}\sum_{i=0}^{w_2-1}\binom{n}{k}F_{n-k,w_2}(y)y^{w_1 i}\Delta^{w_1 i}0^k. \qquad (21)$$

Remark 1. *In particular, for $w_1 = 1$, we have*

$$F_n(y) = \sum_{k=0}^{n} \sum_{i=0}^{w_2-1} \binom{n}{k} F_{n-k,w_2}(y) y^i \Delta^i 0^k. \tag{22}$$

By expressing I in a different way, we have

$$I = \left(\int_{\mathbb{Z}_p} (-y^{w_1}(e^t - 1)^{w_1})^x d\mu_{-1}(x) \right) \times \left(\frac{\int_{\mathbb{Z}_p} (-y^{w_2}(e^t - 1)^{w_2})^x d\mu_{-1}(x)}{\int_{\mathbb{Z}_p} (-y^{w_1 w_2}(e^t - 1)^{w_1 w_2})^x d\mu_{-1}(x)} \right)$$

$$= \left(\int_{\mathbb{Z}_p} (-y^{w_1}(e^t - 1)^{w_1})^x d\mu_{-1}(x) \right) \times \left(\frac{1 - y^{w_1 w_2}(e^t - 1)^{w_1 w_2}}{1 - y^{w_2}(e^t - 1)^{w_2}} \right)$$

$$= \left(\sum_{i=0}^{w_1-1} y^{w_2 i}(e^t - 1)^{w_2 i} \right) \times \left(\frac{2}{1 - y^{w_1}(e^t - 1)^{w_1}} \right) \tag{23}$$

$$= \sum_{i=0}^{w_1-1} \sum_{l=0}^{w_2 i} \binom{w_2 i}{l} y^{w_2 i}(-1)^l \frac{2}{1 - y^{w_1}(e^t - 1)^{w_1}} e^{(w_2 i - l)t}$$

$$= 2 \sum_{n=0}^{\infty} \left(\sum_{i=0}^{w_1-1} \sum_{l=0}^{w_2 i} \binom{w_2 i}{l} y^{w_2 i}(-1)^l F_{n,w_1}(w_2 i - l, y) \right) \frac{t^n}{n!}.$$

Interchanging the roles of w_1 and w_2, by (14), we get

$$I = \left(\int_{\mathbb{Z}_p} (-y^{w_2}(e^t - 1)^{w_2})^x d\mu_{-1}(x) \right) \times \left(\frac{\int_{\mathbb{Z}_p} (-y^{w_1}(e^t - 1)^{w_1})^x d\mu_{-1}(x)}{\int_{\mathbb{Z}_p} (-y^{w_1 w_2}(e^t - 1)^{w_1 w_2})^x d\mu_{-1}(x)} \right)$$

$$= \left(\int_{\mathbb{Z}_p} (-y^{w_2}(e^t - 1)^{w_2})^x d\mu_{-1}(x) \right) \times \left(\frac{1 - y^{w_1 w_2}(e^t - 1)^{w_1 w_2}}{1 - y^{w_1}(e^t - 1)^{w_1}} \right)$$

$$= \left(\sum_{i=0}^{w_2-1} y^{w_1 i}(e^t - 1)^{w_1 i} \right) \times \left(\frac{2}{1 - y^{w_2}(e^t - 1)^{w_2}} \right) \tag{24}$$

$$= \sum_{i=0}^{w_2-1} \sum_{l=0}^{w_1 i} y^{w_1 i} \binom{w_1 i}{l} (-1)^l \frac{2}{1 - y^{w_2}(e^t - 1)^{w_2}} e^{(w_1 i - l)t}$$

$$= 2 \sum_{n=0}^{\infty} \left(\sum_{i=0}^{w_2-1} \sum_{l=0}^{w_1 i} y^{w_1 i} \binom{w_1 i}{l} (-1)^l F_{n,w_2}(w_1 i - l, y) \right) \frac{t^n}{n!}.$$

Hence, by Equations (23) and (24), we obtain the following theorem.

Theorem 2. *For $w_1, w_2 \in \mathbb{N}$ with $w_1 \equiv 1 \pmod{2}$, $w_2 \equiv 1 \pmod{2}$, $n \geq 0$, we have*

$$\sum_{i=0}^{w_1-1} \sum_{l=0}^{w_2 i} y^{w_2 i} \binom{w_2 i}{l} (-1)^l F_{n,w_1}(w_2 i - l, y) = \sum_{i=0}^{w_2-1} \sum_{l=0}^{w_1 i} y^{w_1 i} \binom{w_1 i}{l} (-1)^l F_{n,w_2}(w_1 i - l, y). \tag{25}$$

Remark 2. *Especially, if we take $w_1 = 1$, then by Theorem 2, we get*

$$F_n(y) = \sum_{i=0}^{w_2-1} \sum_{l=0}^{i} \binom{i}{l} y^i (-1)^l F_{n,w_2}(i - l, y). \tag{26}$$

3. Conclusions

In this paper, we introduced w-torsion Fubini polynomials as a generalization of Fubini polynomials and expressed the generating function of w-torsion Fubini polynomials by means of a fermionic p-adic integral on $\mathbb{Z}_p$. Then we derived some new symmetric identities for the w-torsion Fubini and two variable w-torsion Fubini polynomials by investigating a quotient of such p-adic integrals on $\mathbb{Z}_p$, representing generating functions of three w-torsion Fubini polynomials. It seems that they are the first double symmetric identities on Fubini polynomials. As was done, for example in [4,20,21], we expect that this result can be extended to the case of triple symmetric identities. That is one of our next projects.

Author Contributions: T.K. and D.S.K. conceived the framework and structured the whole paper; T.K. wrote the paper; G.-W.J. and J.K. checked the results of the paper; D.S.K. and J.K. completed the revision of the article.

Conflicts of Interest: The authors declare no conflict of interest.

References

1. Kim, T. q-Volkenborn integration. *Russ. J. Math. Phys.* **2002**, *9*, 288–299.
2. Kim, T. Symmetry of power sum polynomials and multivariate fermionic p-adic invariant integral on $\mathbb{Z}_p$. *Russ. J. Math. Phys.* **2009**, *16*, 93–96. [CrossRef]
3. Kim, T. A study on the q-Euler numbers and the fermionic q-integral of the product of several type q-Bernstein polynomials on $\mathbb{Z}_p$. *Adv. Stud. Contemp. Math.* **2013**, *23*, 5–11.
4. Kim, D.S.; Park, K.H. Identities of symmetry for Bernoulli polynomials arising from quotients of Volkenborn integrals invariant under S_3. *Appl. Math. Comput.* **2013**, *219*, 5096–5104. [CrossRef]
5. Kilar, N.; Simsek, Y. A new family of Fubini type numbers and polynomials associated with Apostol-Bernoulli numbers and polynomials. *J. Korean Math. Soc.* **2017**, *54*, 1605–1621.
6. Kim, T.; Kim, D.S.; Jang, G.-W. A note on degenerate Fubini polynomials. *Proc. Jangjeon Math. Soc.* **2017**, *20*, 521–531.
7. Kim, Y.-H.; Hwang, K.-H. Symmery of power sum and twisted Bernoulli polynomials. *Adv. Stud. Contemp. Math. (Kyungshang)* **2009**, *18*, 127–133.
8. Lee, J.G.; Kwon, J.; Jang, G.-W.; Jang, L.-C. Some identities of λ-Daehee polynomials. *J. Nonlinear Sci. Appl.* **2017**, *10*, 4137–4142. [CrossRef]
9. Rim, S.-H.; Jeong, J.-H.; Lee, S.-J.; Moon, E.-J.; Jin, J.-H. On the symmetric properties for the generalized twisted Genocchi polynomials. *ARS Comb.* **2012**, *105*, 267–272.
10. Rim, S.-H.; Moon, E.-J.; Jin, J.-H.; Lee, S.-J. On the symmetric properties for the generalized Genocchi polynomials. *J. Comput. Anal. Appl.* **2011**, *13*, 1240–1245.
11. Seo, J.J.; Kim, T. Some identities of symmetry for Daehee polynomials arising from p-adic invariant integral on $\mathbb{Z}_p$. *Proc. Jangjeon Math. Soc.* **2016**, *19*, 285–292.
12. Ağyüz, E.; Acikgoz, M.; Araci, S. A symmetric identity on the q-Genocchi polynomials of higher-order under third dihedral group D_3. *Proc. Jangjeon Math. Soc.* **2015**, *18*, 177–187.
13. He, Y. Symmetric identities for Calitz's q-Bernoulli numbers and polynomials. *Adv. Differ. Equ.* **2013**, *2013*, 246. [CrossRef]
14. Moon, E.-J.; Rim, S.-H.; Jin, J.-H.; Lee, S.-J. On the symmetric properties of higher-order twisted q-Euler numbers and polynoamials. *Adv. Differ. Equ.* **2010**, *2010*, 765259. [CrossRef]
15. Ryoo, C.S. An identity of the symmetry for the second kind q-Euler polynomials. *J. Comput. Anal. Appl.* **2013**, *15*, 294–299.
16. Kim, T.; Kim, D.S. An identity of symmetry for the degenerate Frobenius-Euler polynomials. *Math. Slovaca* **2018**, *68*, 239–243. [CrossRef]
17. Kim, T.; Kim, D.S. Identities of symmetry for degenerate Euler polynomials and alternating generalized falling factorial sums. *Iran J. Sci. Technol. Trans. A Sci.* **2017**, *41*, 939–949. [CrossRef]
18. Carlitz, L. Eulerian numbers and polynomials. *Math. Mag.* **1959**, *32*, 247–260. [CrossRef]
19. Milovanović, G.V.; Mitrinović, D.S.; Rassias, T.M. *Topics in Polynomials: Extremal Problems, Inequalities, Zeros*; World Scientific Publishing Co., Inc.: River Edge, NJ, USA, 1994.

20. Kim, D.S.; Kim, T. Triple symmetric identities for w-Catalan polynomials. *J. Korean Math. Soc.* **2017**, *54*, 1243–1264.
21. Kim, D.S.; Lee, N.; Na, J.; Park, K.H. Identities of symmetry for higher-order Euler polynomials in the three varibles (I). *Adv. Stud. Contemp. Math. (Kyungshang)* **2012**, *22*, 51–74.

© 2018 by the authors. Licensee MDPI, Basel, Switzerland. This article is an open access article distributed under the terms and conditions of the Creative Commons Attribution (CC BY) license (http://creativecommons.org/licenses/by/4.0/).

MDPI
St. Alban-Anlage 66
4052 Basel
Switzerland
Tel. +41 61 683 77 34
Fax +41 61 302 89 18
www.mdpi.com

Symmetry Editorial Office
E-mail: symmetry@mdpi.com
www.mdpi.com/journal/symmetry